W0246288

Analytical Methods for Dynamic Modelers

Analytical Methods for Dynamic Modelers

edited by Hazhir Rahmandad, Rogelio Oliva,
and Nathaniel D. Osgood

foreword by George Richardson

The MIT Press
Cambridge, Massachusetts
London, England

© 2015 Massachusetts Institute of Technology

All rights reserved. No part of this book may be reproduced in any form by any electronic or mechanical means (including photocopying, recording, or information storage and retrieval) without permission in writing from the publisher.

This book was set in Palatino by Toppan Best-set Premedia Limited. Printed and bound in the United States of America.

Library of Congress Cataloging-in-Publication Data

Analytical methods for dynamic modelers / edited by Hazhir Rahmandad, Rogelio Oliva, and Nathaniel D. Osgood.
    p.   cm
Includes bibliographical references and index.
ISBN 978-0-262-02949-0 (hardcover : alk. paper)
1. Simulation methods. 2. System analysis. I. Rahmandad, Hazhir. II. Oliva, Rogelio. III. Osgood, Nathaniel D., 1968–
T57.62.A466   2015
511'.8—dc23
2015009374

10   9   8   7   6   5   4   3   2   1

To Jay Forrester and John Sterman, who taught us rigor and relevance in modeling.

Contents

Overview of Online Appendices

Ch1-MLE-Supplement
    R code files and instructions to perform maximum likelihood and bootstrapping analysis, as well as all the files (R code files, data, instructions, etc.) to replicate the examples and challenge in chapter 1.

Ch2-MSM-Supplement
    Vensim and MATLAB files required for replicating the example in the chapter and conducting analysis using the method of simulated moments as discussed in chapter 2, in addition to solution of the exercise.

Ch3-SEM-Supplement
    R code example for the structural equation modeling package and results as well as the Vensim models for examples provided in the chapter.

Ch4-FLT-Supplement
    Models and other files necessary to replicate the results in the chapter on filtering and state resetting. The models contained here are all Vensim models and were developed using Vensim DSS, but can also be opened in the Vensim Model Reader.

Ch5-MCMC-Supplement
    A library bridging between R and Vensim, R code to ease the application of MCMC with Vensim models, definition of the example model in that framework, and R code for the specific steps undertaken in the chapter and the exercises.

Ch6-PRG-Supplement
    Files and software necessary to conduct pattern-based model calibration, testing, and behavior analysis as discussed in the chapter, as well as the instructions to set up and use these. In addition, the test model that is used in the chapter for demonstrative purposes can be found in this electronic supplement.

Ch7-EEA-Supplement   Mathematica noteboooks and utilities to perform
                            eigenvalue elasticity analysis, as well as all the files
                            (models, runs, etc.) and instructions to replicate the
                            examples and proposed exercises.

Ch8-OPT-Supplement   Powersim Studio and the compiled SOPS files for the
                            optimization models in the chapter. The archive also
                            contains the user manual for SOPS and all Powersim
                            Studio models and compiled SOPS files used in the
                            user manual.

Ch9-DA-Supplement   Includes hybrid decision analysis/Vensim executable
                            jar file, example decision tree, example simulation
                            model interfacing with decision tree, and software user
                            manual.

Ch10-ROP-Supplement   System dynamics model and sensitivity files (in
                            Vensim), decision tree files (in DPL), macro-enabled
                            Excel template files with embedded VBA programs,
                            and other Excel files (some with embedded @Risk
                            functions) to calculate the middle steps of the
                            algorithms in chapter.

Ch11-OCT-Supplement   Mathematica notebook for solving the example in the
                            text using Ricatti equations.

Ch12-DGM-Supplement   MATLAB and Vensim files for replicating the example
                            differential game discussed in the chapter as well as for
                            solving the exercise.

Foreword

Many years ago, as a doctoral student concentrating in system dynamics and statistics, I facilitated a session of our technical seminar, asking the assembled group of current and soon-to-be experts what technical knowledge a professional modeler needs to know. We filled many sheets of flip chart paper with our thoughts, from differential equations to LaGrange multipliers, along with eigenvalues (of course), sophisticated statistical methods for analysis and calibration, Kalman filtering, and so on. It was a daunting collection—so daunting, in fact, that the founder of our field, Jay Forrester, felt moved to assure us, somewhat tongue in cheek, that he didn't know any of those things.

In the more than 40 years since, we have made great strides in our studies of complex dynamic systems and the policy puzzles they pose. In the process, some scholars have pushed forward the use of richly applicable technical tools. But the widespread use of analytic methods in dynamic modeling has been held back by the lack of texts presenting these tools in a language and style specifically suiting the needs of dynamic modelers.

This marvelous collection fills that need. Compiled by three talented dynamic modelers, it focuses squarely on technical tools that will enhance building, analyzing, and establishing confidence in formal models of complex dynamic systems. The project was superbly collaborative: the editors solicited suggestions for content from a wide range of modeling experts, wrote some of the chapters themselves, and engaged gifted colleagues for the rest. The result is a statement of the state of the art of technical tools for dynamic modelers.

The first five chapters address a powerful array of system-identification and parameter-estimation methods. The next two cover tools for model analysis, identified in the system dynamics literature more than 20 years ago as the first of eight "problems for the future" and now finally appearing in a text. Topics here include analytic methods for pattern recognition as well as eigenvalue elasticity analysis. The remaining five chapters cover optimization (including optimization among policy options as well as fitting a model to data by optimization), dynamic decision analysis, real options analysis, optimal control, and the use of differential games to aid modeling and analyzing competitive dynamics.

We have been waiting for this text a long time—so long, in fact, that a number of the tools and techniques treated here were not even invented yet when that technical seminar addressed our needs more than 40 years ago. I look forward to the contributions dynamic modelers will make with the insights these tools will help to reveal.

George P. Richardson

Preface

The impetus toward this volume was our unsuccessful quest for a single text that covered the most common analytical methods of which a dynamic modeler should be aware. We clearly felt the need for such a volume in training our students as well as informing our own scholarly research. From various model estimation methods to model analysis and decision support tools, many important methodological pieces we knew about were spread across different bodies of literature and were often not presented in a format easily accessible to dynamic modelers new to that specific literature. Our experience in editing and reviewing for various modeling journals and conferences further underlined the need for bridging this gap, as we frequently came across missed opportunities to leverage existing analytical methods in the context of dynamic modeling applications. Further hearing about this need from several colleagues and over a few discussions during the 2011 system dynamics conference, we decided to put together an edited volume that targeted this gap—a hands-on introduction to the key analytical methods that dynamic modelers can regularly benefit from in their work. The current text is the result of this effort.

Putting this volume together would not have been possible without the extensive contributions of many colleagues and organizations. We are sincerely thankful to all of them. John Sterman offered advice and encouragement instrumental to the inception of the project. Several colleagues, including Ed Anderson, Gokhan Dogan, Jim Duggan, Bob Eberlein, David Ford, Shayne Gary, Paulo Gonçalves, Burak Güneralp, Jim Hines, Jack Homer, Peter Hovmand, Nitin Joglekar, Christian Kampmann, David Lane, Len Malczynski, Ignacio Martinez, Mark Paich, David Peterson, Erik Pruyt, George Richardson, Scott Rockart, Mohammed Saleh, John Sterman, Jeroen Struben, Burcu Tan, and Imrana Umar provided input into the selection of topics to cover. John

Covell at MIT Press supported the project from the beginning and put up with delays and challenges in finalizing a multiauthor product. Anonymous reviewers encouraged and enabled the publication of this book by MIT Press. Most importantly, we are extremely grateful to the authors of the chapters, who accepted a significant commitment, applied themselves across multiple iterations, and delivered high-quality work, which we assemble here.

Over the years, the editors have found much camaraderie, intellectual stimulation, and many research collaborations that would not have been possible without the events and contributions of the International System Dynamics Society. We hope this book is a small step in giving back to this community, and as a token of our appreciation we donate any proceeds from this volume to the society.

Hazhir Rahmandad
Rogelio Oliva
Nathaniel Osgood

Introduction

Simulation modeling is increasingly integrated into research and policy analysis of complex sociotechnical systems. Model-based analysis and policy design is now informing a wide range of applications, from assessing national economic policy to designing market entry strategies, investment planning, evaluating community health interventions, developing better project-control tools, and preventing and controlling epidemic outbreaks, to name just a few. This rise in application is partly due to the increasingly complex nature of the common problems faced in corporations and public policy debates, where human intuition fails and simulation models are indispensable for both making decisions and informing the "mental models" of decision makers.

Advances in engineering and the social sciences have informed the application of realistic simulation models to understanding and managing sociotechnical systems. However, the supply of modelers that master this trade has been constrained by at least two factors. First, modeling sociotechnical systems is inherently more complex than modeling physical systems since it relies on interdisciplinary teams that include diverse stakeholders and depends on much tacit knowledge and approximations that depend on modelers' skills.

The second constraint on the supply of proficient modelers is the diversity of analytical methods that can facilitate dynamic modeling and analysis for social systems. These methods—needed for model formulation, estimation, analysis, and decision support—span a wide range of tools and techniques that have roots in diverse literatures, control theory, computational statistics, operations research, and econometrics, to name just a few. Keeping abreast of these developments across various fields is a challenge for any modeler focusing on a specific application domain. Furthermore, when pursued through traditional pedagogical avenues, learning all the relevant analytical tools will require taking several graduate-level courses in different departments, few of which directly focus on dynamics of social systems.

As a result of these two challenges, the time investment for becoming a proficient modeler of sociotechnical systems is very high. The required training is rarely fully completed within formal education. Nevertheless, because of the rapid evolution of contributing fields and the shifting balance of their project portfolios, many modelers need to expand their methodological toolbox on an ongoing basis throughout their career.

In short, the need for high-quality models of sociotechnical systems is high and growing, yet the supply of such models is significantly constrained by the complexity of learning all the required components for becoming a skillful modeler. In this edited volume, we assemble a hands-on introduction to the key analytical tools that should be familiar to a sociotechnical systems modeler. By bringing together these synergistic tools—typically covered in different courses and texts—this volume contributes to removing a major barrier to training effective modelers of dynamic social phenomena. By making these tools available to modelers, we hope to not only improve their modeling practice by allowing them to leverage more advanced tools for the task, but also to improve the impact of their contributions, since the methods will enhance the credibility of their work and facilitate the communication of modeling insights across disciplines.

What Distinguishes This Book?

We have identified, through a survey of skilled modelers in the field, the most important tools of which a modeler should be aware for data-driven model formulation, model calibration and parameter estimation, model analysis, and optimization and decision support. For each of these methods, we solicited a chapter from a group of authors with experience in the application of the method to models of complex sociotechnical systems. We asked authors to follow a similar logic in the presentation of their ideas, and we edited for consistency of length and complexity across chapters.

Each chapter is written so that readers with a solid college-level calculus and statistics background can follow its logic. As a result, the book does not include entirely new theoretical concepts, or in-depth mathematical expositions of these tools. Rather, we want to enable the reader to appreciate the basic intuition behind each method, and learn through tutorial-style examples how to apply the method to typical problems. We particularly focused on methods that lack a canonical existing description, and present them here in a uniform format that highlights the synergies and connections across different tools and provides an intuitive and simple introduction, along with concrete examples relevant to modeling sociotechnical systems. Each chapter also presents a literature review—covering content such as the history of the development of the method and references to more detailed presentations of their analytical intricacies—so that an interested reader can use the chapter as a starting point.

Finally, all chapters include hands-on tutorials as well as proposed exercises to further develop intuition regarding the methods. In many instances, the execution of these analyses required the use of custom-made software developed by the authors. In all cases, we asked the authors to make the code accessible to the reader (either actual visibility of the documented code, pseudo-code, or by reference to existing features of readily available software) so that readers can adopt it to their specific application. As such, the book includes a substantial online appendix, including computer code, models, solutions to exercises, and other complementary resources.

Audience and Use Cases

This volume targets two audiences. First, it is intended as a textbook for teaching analytical methods in simulation modeling courses. In this role, the text provides students with an accessible introduction to several different analytical tools as they relate to building and applying simulation models of social systems. These students are typically taking graduate-level courses on simulation of social systems (under topics such as simulation, system dynamics, agent-based modeling, and systems science methods). These courses are often taught in business and economics schools as well as departments of industrial and systems engineering and computer science. A few schools of humanities and social sciences also include such courses in their sociology, psychology, public policy, and social work departments, as do a growing number of schools of public health. Moreover, there is an increasing number of multidisciplinary centers in different universities associated with programs offering such courses (centers focused on simulation, complex systems, social dynamics, etc.). The contributing authors to this volume are largely affiliated with the system dynamics community of researchers, although some circulate widely in other communities as well. Therefore, most chapters may build on differential equations modeling architecture, with the notations of stocks and flows (equivalent to state variables and rates in an ordinary differential equation context) figuring more prominently. Nevertheless, the chapters are written with the general audience above in mind, providing expositions that are responsive to the needs of agent-based and discrete event modeling projects as well.

In using the book as a textbook, specific chapters could be assigned as reading. Exercises from each chapter can be used as assignments, and detailed examples can enrich classroom discussion. Depending on the type of the course one may use a few of the chapters, or organize a full-semester course around analytical methods for dynamic modelers. In fact, the authors have conducted PhD seminars at Virginia Tech and the Massachusetts Institute of Technology, where several sessions have been organized around topics covered here, using the chapters as core reading material. Our experience suggests that teaching this material at the depth required for actual application calls for at least a three-hour instruction and discussion session per chapter.

The second key target group includes independent modelers who want to learn about and start applying tools outside of their current skillset but relevant for the modeling challenges they face. This audience is spread across universities, national and research labs, and consulting practices. They are expected to use the text as a reference, focusing on a subset of chapters most relevant to their current project. Examples and tutorial-like explanations will allow readers to be quickly reminded of key concepts and practical issues in using a method. Some in this audience may also plan a more rigorous independent learning experience for themselves, covering all the chapters of the text in a systematic manner. In this case, replication of examples in the text and completing chapter exercises could be beneficial.

Overview of Material

The focus of this volume is on quantitative techniques that can facilitate the work of dynamic modelers during model construction and analysis. Dynamic modeling projects typically separate the development of the structure of the model (causal relationships captured and the functional forms used to reflect those relationships) from the process of estimating those relationships, understanding the structural sources of different behavior patterns observed, and developing policy recommendations. The first step, model development, largely builds on qualitative data, domain knowledge and modelers' abstractions, and often relies more on the skills and expertise of the modelers than on any analytical tool. The output of this step is usually a model that is fully specified, even if some parameter values are uncertain or unknown. This model is then used in the next steps of model estimation, understanding, and policy analysis, allowing for the application of more rigorous analytical methods across these three activities. We therefore have organized the methods that we cover within three sections, Calibration and Estimation; Model Analysis; and Decision Support and Optimization. Naturally, most dynamic modeling projects include multiple iterations in which the modeler uses insights from estimation and analysis to revisit the model structure, and thus these tools indeed inform the model structure as well.

Dynamic simulation models of sociotechnical systems can be formulated using different computational architectures, treating time and state variables as continuous or discrete, and putting different levels of emphasis on stochastic components. Canonical modeling approaches such as differential equations, agent-based modeling, and discrete event simulation emerge as a result of these architectures. Several of the tools covered in this volume can be implemented uniformly across these architectures. A few chapters require assumptions regarding the computational architecture, in which case we focus on implementing the tool in conjunction with differential equations models, the dominant architecture used by system dynamics modelers. Even in such cases, extensions to agent-based and discrete event simulations are often (but not always) feasible.

Within each section, we have included chapters that introduce the advanced quantitative methods useful for the corresponding step in the dynamic modeling process. Next we provide an overview of these three sections and the included chapters, offering the reader a broad outline of how these methods can benefit dynamic modeling projects.

Calibration and Estimation

Once the modeler identifies the causal relationships relevant to the problem at hand and their functional forms, model parameters need to be specified to create a fully operational model. Many academic communities consider this the most important step in the research process. In fact, typical statistical research includes models with simple, explicit structures, while the focus of study is on collecting the data and devising the estimation strategy that best tease out the strength of a handful of causal links (i.e., estimate a few model parameters). In dynamic modeling, parameter values can come from many different sources, including prior literature and statistical estimates, data at lower levels of aggregation (i.e., the physics of the problem), and expert judgments, or could be statistically estimated from quantitative data. Statistical estimation becomes increasingly important when the strength of some causal links cannot be found through other methods and those strengths matter for the problem at hand. Well-executed statistical estimation may indeed be a requirement for publishing dynamic modeling results in many quantitatively inclined journals. Estimation could be conducted at the level of partial model components or the full model.

Estimation of model parameters using quantitative data is the focus in this section. Specifically, we include five chapters that address different methods and considerations in estimation of model parameters of dynamic models. The first two methods we introduce—maximum likelihood estimation (MLE) and method of simulated moments—are the workhorses of statistics and econometrics. They provide concrete methods that can be applied to many estimation problems dynamic modelers regularly face, and provide the conceptual background useful for understanding the more advanced methods introduced in chapters on structural equation modeling, Markov chain Monte Carlo (MCMC), and Kalman filtering and state resetting.

The basic idea behind statistical estimation is to compare the outcomes of model simulations against observed data, and change the model parameters to minimize the discrepancy between the two. It is commonly assumed that variations in observed data are due to a structural component, which is captured in the model, and a random component that is not included in the model. In the absence of the random component, one could expect the model to perfectly match the data. In its presence, the difference between the model and data is attributed to this noise term. The MLE method imposes a distribution on this noise component (e.g., normal), and calculates as a data–model fit measure the likelihood of observing the data given a model parameterization. If this likelihood function can be calculated with defensible assumptions, the resulting estimates have many desirable features: they are asymptotically unbiased and efficient, confidence intervals are easy to calculate, and statistics for comparing alternative models are readily available. MLE can be applied to a wide range of problems, such as when time-series data are available for elements modeled at both individual and aggregate levels. In the chapter by Struben, Sterman, and Keith, MLE is introduced along with different methods for identifying the confidence intervals of estimated parameters and concrete examples in dynamic modeling domains.

In the next chapter, Jalali, Rahmandad, and Ghoddusi introduce another commonly used estimation approach, the method of simulated moments (MSM). This method relies on matching the model-generated statistics (moments), such as mean, standard deviation, and covariance, for different variables against their empirical counterparts. Confidence intervals and goodness-of-fit measures are easy to calculate for most applications. MSM does not require the calculation of the likelihood function and therefore is more general. However, finding appropriate moments is potentially ad hoc and limits the utility of the method to modeling situations in which such moments can be identified. Moments are typically calculated for a single element over time, with the assumption that the system is in equilibrium (not a common use case for dynamic modelers), or over a population of elements at the same point in time, allowing dynamic modelers to use cross-sectional data at the population level for estimating dynamic models. This is a potentially important extension of the toolbox most familiar to dynamic modelers, and may enable many novel applications of dynamic modeling to organizational and individual phenomena in which data is largely cross-sectional.

The next three chapters introduce different extensions and nuances that could be relevant to many common estimation problems. Structural equation modeling is increasingly used in social and behavioral sciences to simultaneously estimate networks of causal relationships (rather than having a single dependent variable). Hovmand and Chalise provide an overview of structural equation modeling with multiple examples of increasing levels of complexity. This method identifies the linear model that best matches the observed covariance (or correlation) matrix among the variables in the causal network. The focus of the method on causal networks, with potential for incorporating unobserved variables, makes it a good fit for many dynamic modeling applications. Moreover, the data requirements are modest, only requiring the covariance matrices. On the other hand, the method is not yet fully developed for estimating feedback loops and nonlinear relationships, limiting its application to partial model calibration of subsets of model variables.

In many real-world applications, noisy measurements combine with randomness in the underlying processes and lead to significant divergence of deterministic models from empirical observations. Kalman filtering is one of the first and most commonly used predictor-corrector methods developed for inferring the evolution of the underlying system based on noisy observations and correcting for such evolution in subsequent predictions. For example, one can use Kalman filtering to get a recurrent series of more precise estimate for the location of a car on a road based on an inevitably flawed model and a successive series of noisy GPS measures of that location. Originally designed for linear systems, it has been extended to nonlinear systems and incorporated in some dynamic modeling software packages, including Vensim, to facilitate its application. Extended Kalman filtering can be combined with estimation to find parameter estimates for dynamic models. This is especially important if the system's dynamics include noise drift; that is, the simulation's trajectory changes significantly if different streams of random numbers (from the same distribution) are used for process noises that drive the model, or if measurement errors are significant. The chapter by Eberlein provides an applied introduction to the use of Kalman filtering and estimation, and the idea of state-resetting, which motivates the use of Kalman filtering.

One of the challenges in using the statistically efficient MLE methods for more complex models is the fact that, for nonlinear models, it is typically not possible to reliably transform MLE estimates of parameters (including confidence intervals) into measures of likelihood of different model outcomes or scenarios. Underlying this challenge is the fact that outputs from nonlinear models frequently exhibit multimodal distributions, among which the MLE estimate highlights just a single plausible estimate of many. Therefore, we need a way to assess the plausibility of alternative parameters, outcomes, or even models, given a set of empirical observations. Given empirical data, a probabilistic model specifying a likelihood function relating model parameters and outputs (on the one hand) to empirical data (on the other) and—in their Bayesian form—a prior distribution over parameters, MCMC methods provide a general and straightforward means of sampling from the implied (*posterior*) distributions. MCMC can sample such distributions over parameters, over model outputs, over cross-scenario differences in model output, and even over models themselves. In the final chapter in this section, Osgood and Liu discuss the theory and tools for practical application of MCMC to dynamic models, and illustrate such application using a series of examples.

Model Analysis

Understanding the link between model structure and its behavior is central to the applications of dynamic models. Two chapters provide distinct tools to facilitate this step of modeling process. First, in a chapter on pattern recognition by Yücel and Barlas, a set of tools are introduced that allow modelers to automatically identify key modes of behavior, such as oscillation, S-shaped growth, and exponential growth. Given the central role of these patterns in the modeling process (e.g., in identifying the reference modes a model is expected to replicate), the automated pattern recognition can facilitate both model building and model analysis significantly. For example, it can allow for automatic identification of regions of parameter space that lead to one type of reference modes versus others, a task otherwise requiring manual inspection of thousands of sensitivity runs. The authors provide tools for conducting this analysis as well as concrete examples.

The second chapter in this section discusses the use of eigenvalues and eigenvectors for connecting model behavior to the responsible model structures. Building on the mathematics of linear systems, any behavior that emerges from a simulation can be associated with eigenvalues of the model's state transition matrix. Therefore, the sensitivity of those eigenvalues and corresponding eigenvectors to changes in strengths of model links and feedback loops offers a direct connection between model behavior and its structure. Nonlinear models can be linearized at different points in time to utilize this method. While the basic idea behind this method has been around for many years, it is only more recently that research has specified the methods clearly and the tools for conducting the analysis have become more viable for general users. In the chapter by Oliva, the background and definition of two of these methods, loop eigenvalue elasticity analysis and dynamic decomposition weight analysis, are provided along with examples and software code that can facilitate implementation.

Decision Support and Optimization

Once a model is estimated and understood, modelers are often interested in designing policies that can improve actual system performance in the real world. This step usually requires some version of optimization, that is, changing some model parameters—including parameters that could activate additional structures in a model—to secure better outputs based on prespecified goals. The section on decision support and optimization includes five chapters that discuss the different tools and concepts useful for this step of the modeling process.

In the chapter on stochastic optimization, Moxnes discusses the extension of deterministic optimization frameworks to cases in which model processes and inputs can include stochastic components. For these types of problems, he discusses the policy-optimization framework where a policy structure is specified to regulate how decisions should be made, and the parameters of this policy structure are optimized using a Monte Carlo ensemble of simulations. This method is appealing for many practical problems due to its conceptual and practical simplicity. Moreover, if the policy structure is designed carefully, the results would be satisfactory for many practical applications. However, this method has some downsides as well: the quality of its solutions are only as good as the policy structure imposed, and the computational costs for covering the uncertainty space, especially when rare events are critical in the problem at hand, could be huge. Decision trees provide a different avenue for solving the optimization problem, and are particularly attractive in those practical situations in which exogenous events unfolding over time have a major bearing on tradeoffs between policies, and where adaptive, incremental decision making is sought to adjust policy choices over time in light of observed events. The model can then be simulated for each choice to find the best course of action under different exogenous shocks. These choices can then be backtracked from the end of the problem time horizon to the beginning of time, establishing the best response policy for different contingencies the decision maker may face. Osgood et al. provide details on how the formalism of decision trees can be applied to dynamic models, and offer tools for implementation as well as concrete examples.

Another way by which stochastic outcomes may influence decisions relates to the investments decision makers can choose to make over time. Some investments—say, in building a new factory—will offer the decision maker the option to benefit in the future, for example, from sales of products made in the factory. Yet the investment may be wasted if the realized customer demand is low. Evaluation of such options could follow a decision tree method and net present value calculations, or, with some assumptions regarding future outcomes, can benefit from more compact solutions borrowed from the finance literature. The evaluation of the current value of such investment options is the topic of the chapter on real options by Anderson and Tan.

The general case of optimally managing a dynamic system is discussed in the chapter on optimal control. This chapter discusses the basic problem definition and intuition for solving the optimal control problem for deterministic systems. Detailed solution methods are presented for linear systems with a quadratic payoff function, and approximation methods for the nonlinear cases based on machine learning and dynamic programming are discussed. Finally, the chapter by Rahmandad and Spiteri builds on the ideas discussed in the optimal control and stochastic optimization chapters to introduce differential games. Whereas the previous chapters on optimization focus on a single decision maker, many real-world problems include competing decision makers who interact with each other over time. Dynamic games provide a general framework to analyze this problem, assuming rational decision makers. Differential games are set up where dynamics of the system are described using differential equations, the common setting for many dynamic modelers. While this topic is in general analytically involved, the chapter provides the intuition and simple approximate solution ideas that can be used in many problems with satisfactory results.

Closing Remarks

We hope this volume can provide the diverse communities of dynamic modelers with a user-friendly introduction to some of the most applicable analytical tools for model building, estimation, understanding, and policy analysis. We expect the benefits of this text to be threefold. First, knowledge and application of these methods will expand the range of problems dynamic modelers can tackle. From novel estimation methods to more in-depth model analysis tools and policy design approaches, the techniques and ideas discussed here can enable more rigorous analysis when applied to common dynamic modeling problems, and open up new opportunities for application of dynamic modeling to novel problem settings. Moreover, the use of formal analytical methods in different stages of dynamic modeling projects will provide a common language that is understood and appreciated by researchers and practitioners with expertise in statistics, econometrics, operations research, and other analytic disciplines. Building such bridges will be crucial for the healthy growth of dynamic modeling disciplines and offers many synergistic opportunities for academic and applied collaborative research. Finally, we hope this primer provides aspiring researchers with the stepping stones for building new techniques that further expand the analytical toolbox of dynamic modelers.

Besides the online appendix that accompanies the current volume and is hosted by MIT Press, we have made arrangements with the System Dynamics Society to host diverse analytical software that facilitates the application of methods reported here, as well as extensions that will be developed by the modeling community. We hope these additional resources will contribute to further methodological dynamism of the dynamic modeling community.

I   Estimation of Model Parameters

1 Parameter Estimation Through Maximum Likelihood and Bootstrapping Methods

Jeroen Struben, John Sterman, and David Keith

Estimating model parameters, and the uncertainty in these estimates, has long been central to good dynamic modeling practice (Oliva 2003). Modelers use a range of techniques to do so, ranging from formal statistical parameter estimation to expert judgment (Barlas 1989, 1996; Croson et al. 2014; Dogan 2007; Forrester and Senge 1978; Sterman 1989, 2000). The goal is to bring the widest range of relevant information to bear on the difficult problem of parameter estimation. With the advent of "big data," the data required for formal statistical estimation of model parameters are increasingly available, making formal estimation both more feasible and more important. Robust parameter and confidence interval estimation in dynamic models is critical, first, in the use of models and data to discriminate among competing hypotheses, and second, to provide decision makers with models that are rigorously grounded in data and for which important uncertainties are quantified.

Models must be grounded in data if modelers are to provide reliable advice to policymakers. The ability of a model to replicate the history of the relevant system provides one important test of a model's fidelity and utility.[1] Ideally, as discussed by Homer (2012), Oliva (2003) and Graham (1980), one should estimate model parameters using data below the level of aggregation of the model; that is, data that are independent of the model behavior. For example, a fishery model should use data derived from biological studies for maturation times and mortality rates; a model of workforce skill development should use independent data on how long it takes a new hire to master a skill. Often, however, direct estimation from independent data is not possible. In practice, modelers must frequently estimate at least some model parameters using the historical data itself, finding the set of parameters that minimize the difference between the historical and simulated time series. The ability of a model to fit historical data when those same data were used to find the parameters that maximize the fit is a weaker, but still vital, test of model adequacy, and should be carried out by rigorous and appropriate formal methods wherever possible.

Many approaches are available to estimate parameters and confidence intervals in dynamic models, including, for estimation, the generalized methods of moments and maximum likelihood, and, for confidence intervals, likelihood-based methods and bootstrapping. Although popular simulation software packages for dynamic modeling (e.g., Vensim, Powersim) include tools for parameter estimation that can deal appropriately with nonlinear feedback systems, methods to address violations of standard maintained hypotheses such as independence and normality of errors, homoscedasticity, and so on are needed. Still less attention has been paid to the problems of statistical inference, hypothesis testing, and finding confidence intervals around the estimated parameters (Dogan 2007; Oliva 2003).

Violations of standard assumptions are common in nonlinear, dynamic models, including nonnormal, autocorrelated errors and (conditional) heteroscedasticity, rendering the problem of estimating the confidence intervals around best-fit parameters and hypothesis testing difficult. Although bootstrapping to generate confidence intervals has gained some popularity in the world of dynamic modeling (Dogan 2007; Keith 2012), bootstrapping is computation-intensive. Although computational power keeps increasing, models for decision-making and "big data" sets—enabled by gains in information processing and data storage technology—are also growing in complexity (McAfee and Brynjolfsson 2012; Simini et al. 2012). As modeling pioneer Mihajlo Mesarovic presciently noted, "No matter how many resources one has, one can envision a complex enough model to render resources insufficient to the task" (Meadows et al. 1982: p. 197). Thus, even as computation becomes faster and cheaper, data availability and model complexity grow as well, requiring methods for formal parameter estimation that are both appropriate given the attributes of the models and data and feasible in terms of computation time and cost.

Maximum likelihood estimation has become increasingly important for nonlinear models. Modelers need to know how the various methods work, from a theoretical and computational perspective, what their advantages and limitations are, how to code specific models, and how to take existing code and adapt it to represent new situations (Train

2003). Here we highlight the basic properties of and compare two important approaches, bootstrapping and likelihood-based methods. The generalized method of moments is discussed elsewhere in this volume (see chapter 2). We consider the advantages and disadvantages of the methods through applications from actual modeling projects.

Specifically, we draw upon field data from a study of service quality in the banking industry (Dogan 2007; Oliva and Sterman 2001). Further, we provide a challenge that uses experimental data from the beer distribution game (Croson et al. 2014; Dogan 2007; Sterman 1989). We first describe basic maximum likelihood estimation and bootstrapping with step-by-step guidance. Next, we discuss the practical issues that arise in applying these methods in different nonlinear dynamic modeling contexts—including feedbacks, accumulations, nonlinear formulations including nonnegativity constraints and saturation nonlinearities, limits on plausible/allowable values of estimated parameters, autocorrelation, heteroscedasticity, and so on. We comment on differences between methods and results, and their respective advantages and disadvantages in different contexts. We provide a step-by-step process for assessment of the uncertainty around parameter estimates in dynamic models. We close with recommendations for modelers seeking to develop robust estimates of confidence intervals and call for the automation of these methods when using dynamic modeling software.

The full code, written using the open-source statistical software R (R Core Team 2012), data, and information required to replicate all the analysis in this chapter, including solutions to the challenge, are provided in the electronic supplement. We include customized R-scripts and functions developed by the authors for the purpose of flexible maximum likelihood estimation and bootstrap analysis.[2] Appendix A1.0 in the online supplement provides an overview of the documents, data, and codes.

Parameter and Confidence Interval Estimation Using Maximum Likelihood and Bootstrapping

Maximum Likelihood Estimation Overview

Maximum likelihood estimation (MLE) offers a unified approach to estimation (Wooldrige 2002). Simply put, given a set of data and a hypothesized model, the method of maximum likelihood selects values of the model parameters that make the observed data most likely (i.e., parameters that maximize the likelihood function). MLE is conceptually simple and can be applied to a broad set of problems, providing a method to find not only the best parameter estimates but also to assess the uncertainty around them (Stigler 2007). MLE development took off after Fisher (1920, 1922) demonstrated its important characteristic of "sufficiency": that is, complete information about the parameter of interest is contained in its maximum likelihood estimate (Aldrich 1997). While attempts at proofs guaranteeing an optimum under general conditions have failed, its usefulness across many applications is unchallenged, and its power lies in the applicability of the technique in complex situations (Stigler 2007).

MLE has been applied in many fields, from sociology and economics to biology and physics. MLE is effective for uses from parameter estimation to curve fitting, hypothesis testing, and confidence interval estimation, with canonical applications in settings where closed-form solutions for the likelihood function are known, such as certain probabilistic choice models, including logit choice. However, increasing computational power and algorithm development relax requirements for closed-form solutions, enabling much wider application. For canonical applications and examples see, for example, Bates and Watts (1988), Meeker and Escobar (1995), and Millar (2011). Millar (2011) offers applications with examples in R and SAS software (SAS Institute Inc. 2011). For examples with discrete choice models and simulated methods see Train (2003). We now highlight the basic process for parameter and confidence interval estimation.

MLE Definition

MLE finds the parameters of a model that maximize the likelihood of obtaining the observed data given that model. The likelihood function L is the joint probability density function of observable random variables, viewed as a function of the parameters given the realized random variables. Formally (Adda and Cooper 2003; Casella and Berger 2001), if $X_i \ldots X_n$ are an i.i.d. sample from a population with probability density function or frequency function $f(x_i \mid \theta_1 \ldots \theta_k)$, then the likelihood function is defined as:[3]

$$L(\theta \mid x) = L(\theta_1 \ldots \theta_k \mid x_1 \ldots x_n) = \prod_{i=1}^{n} f(x_i \mid \theta_1 \ldots \theta_k). \tag{1}$$

For each sample point x, let $\hat{\theta}$ be a parameter value at which $L(\theta \mid x)$ attains its maximum as a function of θ with x held fixed. A maximum likelihood estimator of the parameters θ based on sample X is $\hat{\theta}(X)$.

Intuitively, this is a reasonable choice for an estimator as it is the set of parameters for which the observed sample is most likely.

One can think of L describing a k-dimensional surface with the optimum being the global peak. The maximization of L can be performed analytically if the system of equations is twice differentiable, using the standard first- and second-order conditions. To facilitate analytic derivation of the maximum, it is common to take the logarithm of the likelihood function, converting the multiplicative function in (1) to an additive function:

$$LL \equiv \ln L = \sum_i f(x_i \theta_1 ... \theta_k) \ . \tag{2}$$

Maximizing the log-likelihood (LL, or l) is the same as maximizing likelihood because the log function is monotonic. Typically, however, the analytical derivation is challenging, requiring use of numerical methods.

In appendix A1.1, we derive the basic properties of MLE and illustrate this with the case of a simple model with normally distributed i.i.d. errors. For more details on the underlying theory, see Adda and Cooper (2003), Casella and Berger (2001), and Eliason (1993), while sources such as Train (2003) provide some excellent intuition.

MLE Properties

MLE has many desirable properties: sufficiency (it embodies all the useful information in the sample so that knowing the MLE provides as much information about the true value of the unknown parameters θ as the sample of data itself; Fisher 1922); consistency (under plausible conditions the MLE converges—in probability—to the unknown value of θ as the sample size $n \rightarrow \infty$; that is, the law of large numbers applies for sufficiently large samples); asymptotic efficiency (the lowest possible variance of parameter estimates is achieved in the large sample limit); and parameter invariance (if $\hat{\theta}$ is the maximum likelihood estimator of θ, then any function g($\hat{\theta}$) is the maximum likelihood estimator of g($\hat{\theta}$)). Finally, many inference methods in statistics are developed based on MLE (Efron and Tibshirani 1986; Myung 2003). For example, MLE is a prerequisite for the χ^2 test, the G^2 test, Bayesian methods, inference with missing data, modeling of random effects, and many model-selection criteria. MLE is for those reasons well suited for hypothesis testing and is widely used in situations where the standard maintained hypotheses of i.i.d. errors are plausible. Indeed, in the extreme case of models that are linear in parameters, and under the standard assumptions of i.i.d. normally distributed errors, MLE is equivalent to traditional, less robust, ordinary least squares (OLS) estimation (see appendix A2 for more details).

However, a number of challenges arise in the use of MLE in complex (nonlinear, feedback-rich) systems where the standard maintained hypotheses are violated. Analytical derivation of the maximum likelihood estimator can be difficult or impossible, or the likelihood function L may be difficult to specify altogether. Endogeneity may render maximum likelihood estimators inconsistent (Li and Maddala 1996). Multiple (local) optima of the log-likelihood function may exist, and finding the global maximum may be difficult. Finally, the results may be numerically fragile—that is, a slightly different sample might yield quite different outcomes. In such situations, one can begin with simplifying assumptions and test whether they are reasonable, or use more advanced methods (Train 2003). In this chapter, we focus on the former.

MLE Confidence Intervals

Modelers must not only estimate parameter values but also the uncertainty in the estimates so that they and others can determine how much confidence to place in an estimate and select appropriate ranges for sensitivity analysis to assess the robustness of conclusions. Errors in parameter estimates can arise from sampling error (any data set is a potentially unrepresentative sample from the real system), measurement error (the data may be corrupted by random error or systematic bias), violations of maintained hypotheses (the estimation method might yield biased results given the properties of the model and data) or specification error (the model is wrong in ways that matter). Formal confidence interval estimation addresses the sampling error problem assuming the data, estimation method, and model are correct. Other techniques are available to assess (some) of the other types of error, such as the Hausman (1978) specification test.

Estimating confidence intervals can be thought of as finding the shape of an inverted bowl, as illustrated in figure 1.1: the log-likelihood function reaches a maximum for the parameter $\hat{\theta}$. The horizontal dotted line indicates qualitatively the confidence region; that is, the region around the MLE estimate whose likelihood is nearly as high as that of the estimate so that we cannot reject the hypothesis that the true value of θ lies within that region (at some probability, of being wrong, typically 5%). If, for a given data set, the likelihood function for a set of parameters falls off very steeply for even small departures from the

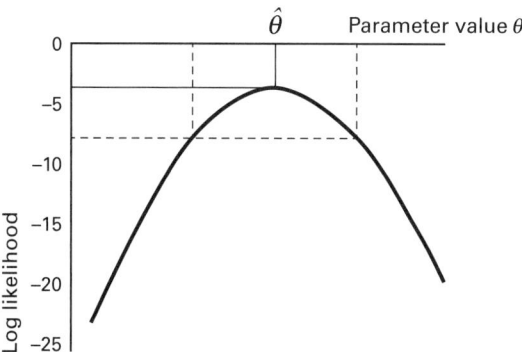

Figure 1.1
Log-likelihood for hypothetical model and data.

best estimates, then one can have confidence that the true parameters are close to the estimated values (as always, assuming the model is correctly specified and other maintained hypotheses are satisfied). If the likelihood falls off only slowly, other values of the parameters are nearly as likely as the best estimates and one cannot have much confidence in the estimated values.

MLE methods provide two major approaches to constructing confidence intervals or confidence regions (Venzon and Moolgavkar 1988). The first is the asymptotic (AS) method, which assumes that the likelihood function can be approximated by a parabola around the estimated parameters, an assumption that is valid for very large samples. The second is the likelihood ratio (LR) method, which involves searching the actual likelihood surface to find values of the likelihood function that yield a particular ratio of the likelihood for a given set of parameter values to the one for the maximum likelihood estimator values.

When the likelihood function is not known, one can use the AS method of confidence interval estimation. The asymptotic method assumes that the log-likelihood function is parabolic in the neighborhood of the best estimates. The asymptotic method makes use of information about the curvature of the log-likelihood function (from the second derivative) at the estimated parameter values.[4] The simplest asymptotic method, the Wald test, further assumes that difference between the maximum likelihood estimate and an alternative parameter value associated with a hypothesis to be tested (e.g., that the parameter is zero) is normally distributed. The Wald test can be executed with the basic summary statistics provided by MLE software, while the AS intervals require the Hessian of the likelihood function (the matrix of partial second derivatives of the log-likelihood over the parameters, $\partial^2 LL / \partial \theta_i^2$) evaluated at the estimated maximum of the likelihood function. Hence, the Hessian describes the curvature of the likelihood function at the maximum likelihood estimator.

The LR method of confidence interval estimation compares the likelihood for the estimated parameters $\hat{\theta}$ with that of an alternative set θ^*. The likelihood ratio is defined as $R = L(\hat{\theta}) / L(\theta^*)$. Asymptotically, the likelihood ratio follows a χ^2 distribution:

$$-2\left[LL(\hat{\theta}) - LL(\theta^*)\right] \sim \chi^2_{df}, \tag{3}$$

where the degrees of freedom df for the χ^2 test are determined by the number of independent parameters involved in the confidence region estimation, compared to those involved in the MLE (Engle 1984; Wilks 1943). The confidence interval yielding a certain probability that the true parameters lie within the interval is determined by the value of the parameters that yield a likelihood ratio corresponding to the relevant cutoff value for the χ^2 distribution. Specifically, for confidence level 1-α the parameter region θ_0 spans those values for which $2\ln R \leq q_k(1-\alpha)$, with $q_k(1-\alpha)$ being the $(1-\alpha)th$ quantile of the χ^2 distribution with k degrees of freedom. Then, the critical parameter value results in a (negative) log-likelihood equal to $LL(\hat{\theta}) - \chi_{df,1-\alpha}$.[5] One should interpret the associated confidence interval for the parameters of interest as follows: if $1 - \alpha$ is the confidence level, we expect that if we collect multiple independent samples, each with the same number of data points, the confidence region, estimated for each sample independently, should contain the true parameters θ in $(1-\alpha)*100\%$ of the samples. Three common LR-based alternatives for confidence interval estimation include: (1) univariate likelihood interval (*LR-univariate*) estimation, which takes a slice of the LL for each of the parameters of interest, holding the others at their estimate; (2) surface likelihood *(LR-surface)* interval estimation, which makes full use of the actual curvature of the inverted multidimensional bowl to estimate the uncertainty in the estimates due to sampling error; and (3) the profile likelihood (*LR-profile*) interval estimation, which re-estimates the LL as one varies each individual parameter of interest over its range.

Comparing the confidence interval approaches, the various LR and AS approaches are asymptotically equivalent and tend to yield similar intervals for large samples. The LR approximation of a χ^2 distribution is also good even for small samples. However, the asymptotic method is often not reliable for small samples. Violations of standard assumptions of asymptotic methods make confidence interval estimation and hypothesis testing difficult (Dogan 2007). The asymptotic methods impose strong assumptions about the geometry of the likelihood surface; specifically, that it is smooth and has only one maximum. In addition, although MLE estimates are invariant under monotonic transformations $f(\theta)$, standard errors derived from the Wald test do not in general transform in the same way.

LR approaches do not suffer from such distributional limitations. A major advantage of AS methods is that they do not require the construction of an explicit likelihood function. However, if the AS interval is quite different from the LR interval, the appropriateness of the normal approximation is doubtful, and the LR approximation is likely to be better. On the other hand, construction of the likelihood function, as required for the LR approach, can be difficult, while approximations may not always be appropriate. Various computational MLE methods are available to address these problems (e.g., Train 2003), but these are beyond the scope of this chapter.

Bootstrapping Confidence Intervals

Bootstrapping (Crawley 2007; Efron and Tibshirani 1986) provides a different approach to confidence interval estimation that does not require knowledge of or assumptions about the likelihood function or error structure, a particular advantage in nonlinear dynamic models. In broad terms, bootstrapping creates "new" data sets by resampling the original data set using the observed residuals, then estimating the parameter value(s) for each of these "new" bootstrap samples. The confidence interval is obtained from the distribution of parameter values in a sufficiently large sample of "new" data sets.

We discuss three main bootstrapping methods (Bates and Watts 1988; Dogan 2007; Efron and Tibshirani 1986; Li and Maddala 1996). The first involves *direct resampling* of the data (x,y), with replacement. The second approach, involving *resampling of the errors,* estimates the model with the data, derives the residuals r, resamples those randomly, and allocates those to the fitted $\mathbf{y}$'s to generate a reconstructed response variable, $\mathbf{y}$, which one regresses with the original explanatory variables x. This second method is generally preferred in dynamic models where the temporal order of the observations cannot be changed. However, one must then take care to test for and, if necessary, correct any autocorrelation. As a third approach, also involving error resampling, if justified, one may specify an error distribution of a particular parametric form (say, i.i.d. normal). This *parametric method* (in contrast to the first two nonparametric methods) fits the probability distribution to the error terms by estimating the required parameters for the specific distribution, then fabricates new error terms according to this probability distribution using a pseudo-random number generator. The parametric method is applicable if the error terms fit a known probability distribution well enough for the purpose at hand. If not, the nonparametric method should be used.

Once the new samples with parameter estimates have been created, there are different methods available to construct the confidence intervals, including the *normal, percentile, studentized, basic,* and *bias-adjusted bootstrap percentile* (BCa) methods (DiCiccio and Efron 1996). The normal method assumes the distribution of the estimated parameters is normal. The percentile interval method assumes symmetric bounds on the left and right. The *studentized* bootstrap interval produces symmetric endpoints like the confidence intervals based on Student's t-statistic, except that the data are used to estimate a distribution for the t-statistic based on the distribution of the estimated parameter for each resampling. The basic bootstrap interval uses separately the upper quantile of the bootstrap distribution to calculate the lower confidence bound and the lower quantile to calculate the upper bound. When the distribution is skewed, results will differ from the percentile interval method. The BCa interval corrects for systemic biases if the interval is not symmetric. The BCa is often the preferred method because one typically decides to use bootstrapping when the normality assumption that would allow other methods to be used is violated. DiCiccio and Efron (1996) provide a more detailed overview about these and other methods. Davison and Kuonen (2002) demonstrate their application in R.

Confidence Interval Estimation for Dynamic Models

In the case of a linear model, the various methods produce similar intervals. Whether any differences between approaches are substantively significant, for example in hypothesis testing or for setting bounds for sensitivity analysis, will depend on the purpose of the model and the analyst's goals. Estimation in the case of complex dynamic models provides contexts where the distributional assumptions for several of the methods discussed above may be violated. For this reason, bootstrapping has been growing in popularity in the statistics and econometrics literature as computation has become cheaper and faster, and because it is applicable to general problems (Croson et al. 2014; Dogan 2007; Li and Maddala 1996; Sterman 1989). The disadvantages of bootstrapping are (1) the relatively elaborate procedure and (2) the computational burden. Regarding the first, the researcher must examine the simulated error terms, regardless of how they are generated, to ensure that they have the correct properties, including centering them around zero and ensuring they have the same variance, distribution and other properties as the original set. These issues commonly arise when the model is nonlinear (Delignette-Muller and Baty 2013; Freedman 1981). Correcting these errors involves judgment.

Computation time can become a prohibitive burden as well. The execution time for bootstrapping grows with the number of replications. For example, the literature suggests 1,000 replications as a minimum sample size for confidence interval estimation (Efron and Tibshirani 1986). While trivial for simple situations, for large models with large data sets and many parameters to be estimated, finding confidence intervals can become prohibitively time consuming and limit structural sensitivity tests that explore alternative specifications. Moreover, alternatives such as flexible MLE and other methods are increasingly employed to estimate parameters and confidence intervals (Imbens and Spady 2002; Train 2003; Wooldrige 2002). Most general-purpose statistical software programs support MLE in some form. Since MLE is essentially an optimization problem, this type of capability is particularly common in mathematical software programs.

Remember that confidence intervals, regardless of the approach, capture only the uncertainty due to sampling error and do not address other sources of error, including measurement error, violations of maintained hypotheses, and specification error. Modelers should select wider ranges for sensitivity analysis than the confidence bounds derived from any of the methods above to assess the robustness of inferences and conclusions regarding the impact of policies (see Sterman 2000, Ch. 21). Consequently, rather than a single preferred method, one needs a process that facilitates evaluation of the various methods and assesses their advantages and disadvantages for any specific context.

A Process for Confidence Interval Estimation for Dynamic Models
We now explore the process for parameter and confidence interval estimation using MLE and bootstrapping with real examples (figure 1.2). The steps include:

1. Define the model, considering whether it is desirable and feasible to construct a likelihood function.
2. Estimate the parameters, and if possible derive maximum likelihood estimators.
3. Assess the asymptotic method and, if the likelihood function is defined, likelihood ratio confidence intervals.
4. Examine the appropriateness of the assumptions required for the asymptotic method (this is relevant for bootstrapping, but also for the interpretation of confidence interval results).
5. Perform, if feasible, bootstrapping; adjust the error terms in case of violation of asymptotic assumptions.
6. Analyze and interpret confidence interval results.

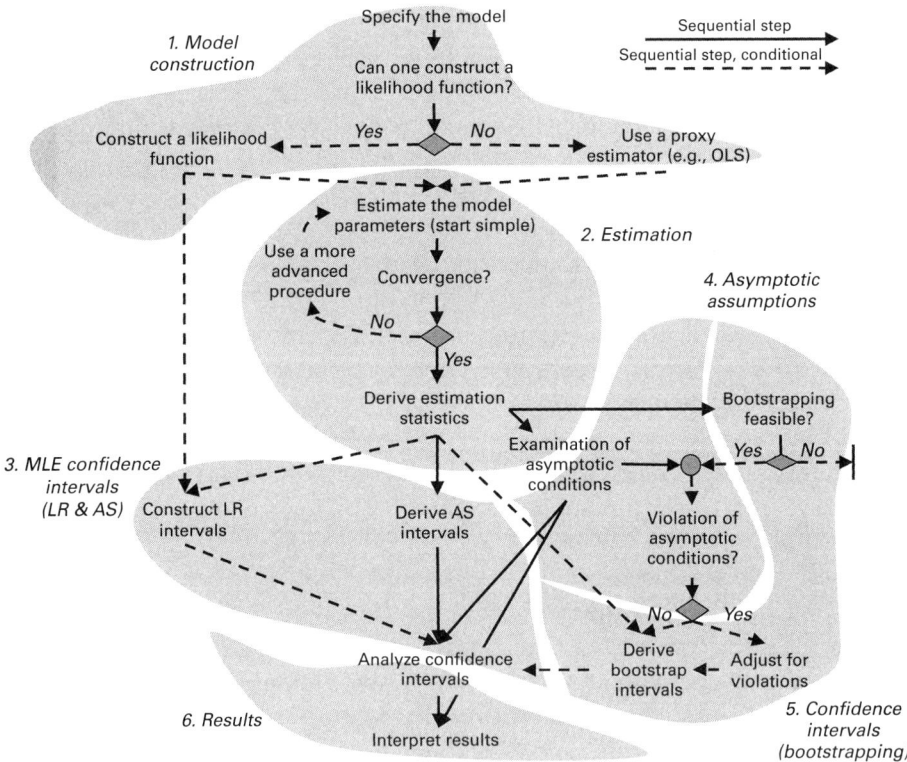

Figure 1.2
Process for confidence interval estimation.

The example below highlights different challenges one may encounter during the confidence interval estimation process. The example follows the steps above and will confront you with issues commonly encountered in projects. The case demonstrates that the likelihood function, while often necessarily approximate due to nonlinearities or feedback, may nevertheless yield appropriate confidence intervals. A subsequent challenge provides the opportunity to explore some similar as well as different considerations you may encounter in applying these tools. The key steps in R are provided in the main text. For details on how to replicate the application and challenge solutions in its entirety, we refer to the R-scripts provided in the electronic supplement. The script "CH1_MLE_BOOT Application.R" contains all the instructions for the application, following the flow and numbering of the sections below, and refers to subordinate scripts for underlying functions. Appendix A1.0 provides more detailed instructions.

Application: Feedback and Nonlinearities in Parameters

Typically, the estimation problem requires finding parameters that best fit the data in a setting where the inputs to the formulation to be estimated are generated by an endogenous feedback system constituting some or all of the model (see Homer 2012).[6]

To illustrate, we examine a model of service quality dynamics (Oliva and Sterman 2001, hereafter OS01; see also Oliva 2001 and Oliva and Sterman 2010, and the replication of OS01 in Dogan 2007, hereafter D07).

The model was developed to examine why so many organizations are unable to deliver consistently high-quality service, despite aspirations to do so, and to explore policies for quality improvement. The model was tested against time-series data from a large bank, including workflows and labor, archival data on policies and procedures, and qualitative data from direct observation. As is common, the data, though extensive, were not sufficient to estimate all model parameters below the level of aggregation of the model. OS01 therefore estimated the critical parameters statistically. Certain key variables needed for the estimation are not directly observable. Thus the estimation problem involves simulating components of the model that include nontrivial endogenous dynamics.

Model

OS01 provide the full model and context. Here we examine the structure representing the time each server spends on each task, time per order, T_t. Time per order depends on the pressure to process the work in a timely manner (work pressure, w), pressure to maintain the quality of the work (quality pressure, q), and the servers' norm for the appropriate time to spend on each order (desired time per order, T^*_t). Further, work pressure and the norm for time per order are endogenous. Work pressure is the ratio of required order processing capacity, C^*_t, to the capacity available, C_t. Required capacity is the capacity needed to process incoming orders, O_t, given the servers' target for the time spent on each task, T^*_t. Actual capacity depends on the size of the staff and the normal workweek:

$$T_t = \max(\tau_f, qwT^*_t)$$

$$w = \left(\frac{C^*_t}{C_t}\right)^\alpha \; ; C^*_t = O_t T^*_t \; ,$$

(4)

where τ_f is the minimum order processing time. Desired time per order, T^*_t, is a stock, which adjusts to actual time per order with a time constant given by τ_o,

$$dT^*_t / dt = (T_t - T^*_t) / \tau_o.$$

(5)

In essence, OS01 hypothesized that when the volume of work to be done exceeded capacity, the bankers would "cut corners" by reducing the time they spent with each customer and on each task. Further, because there were no strong external standards for how long service encounters should take, servers' norms for how much time is appropriate for each customer interaction or task would adjust to actual experience, a classic "floating goal" situation. OS01 further hypothesized that the adjustment time for the norm for time per order is asymmetric. Prior information and observation suggested that desired time per task would adjust downward faster than upward: because corner-cutting and subsequent harm to customer satisfaction are hard to observe, downward adjustments are interpreted by management as improvements in productivity, and would be supported, while upward adjustments would be viewed as declines in productivity that would increase costs. Workers would be reluctant to increase their norms without explicit pressure to raise quality (captured by the management policy q). Hence,

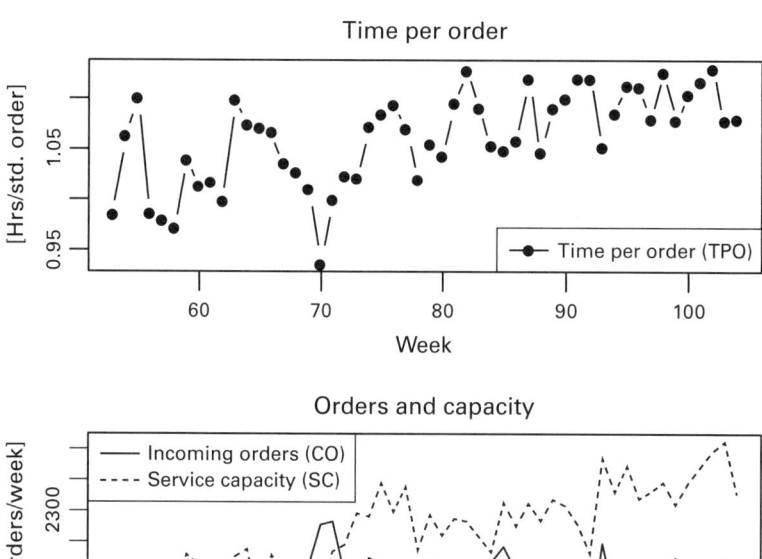

Figure 1.3
Plot of service quality data.

$$\tau_o \begin{cases} \tau_i & T_t > T_t^* \\ \tau_d & \text{otherwise}, \end{cases} \tag{6}$$

where it is expected that $\tau_d < \tau_i$. The parameters to be estimated for this partial model are α (the effect of work pressure on time per order), τ_i (time constant for increasing time per order), τ_d (time constant for decreasing time per order), and T^*_0 (the initial desired time per order, analogous to the constant term in a standard regression).

Figure 1.3 shows the actual time per order, T_t and other input data.

As can be seen from equations (4) and (6), the estimation function $f(\boldsymbol{\theta}, x)$ is nonlinear in parameters and cannot be solved analytically. Further, the large range and variable pattern of time per order suggest the possibility of heteroscedasticity. Specifically, the model errors are likely to be smaller when actual time per order is smaller and larger when orders are larger. The likely violations of standard assumptions suggest least squares and asymptotic confidence interval methods may not be appropriate. Finally, either or both incoming orders and absenteeism may be autocorrelated.

Our first task involves constructing an approximate likelihood function and then estimating it via nonlinear regression. To do this, we first write the model in R. The model *os.model()* produces an estimated time per order based on input parameters (alpha,tau.td,tau.ti,TPOD0) and data (CO,SC):[7, 8]

```
os.model <- function(alpha,tau.td,tau.ti,TPOD0,CO,SC){
        # creation of matrices populated with zero
        TPOt<-TPODt<-as.matrix(0*TPO)
        TPODt[1]<-TPOD0
      # we loop from time = 1 to the number of time steps and
      derive desired and actual time per order
      for(t in 1:N2){
          sc<-as.matrix(CO[t]*TPODt[t])
          wp<-sc/SC[t]
          t.wp<-wp^alpha ; t.wp[which(t.wp==Inf)]<-1e6
          ITPO<-tau.p*t.wp*TPODt[t]
          TPOt[t]<-max(ITPO,tau.f)
          # now we setup the data for the next round
          ifelse(TPOt[t]>TPODt[t],tau.to<-tau.ti,tau.to<-tau.td)
          if(t<N2){TPODt[t+1]<-TPODt[t]+(TPOt[t]-TPODt[t])*(1-exp(-1/
      tau.to))}
      }


      # globally saving the desired and actual time per order
        TPODt<<-TPODt
        TPOt<<-TPOt
      # return the time per order to the global environment
      return(TPOt)
  }
```

As a first approximation, we estimate the model assuming the error terms are i.i.d. normally distributed, in which case we can specify a convenient likelihood function. Because of these assumptions, the likelihood function corresponds to the OLS regression (see appendix A1.2). To estimate the decision rule using R, we created the function *os.model.*

ll(), which converts the R-model into a likelihood function. (All the MLE functions referred to in "CH1_MLE_BOOT_ Application.R" can be found in the R-script "CH1_MLE_Functions.R.") The function reads parameters *par* and data *y* as input. The input *sw.mod.par,* with potential values for each parameter $\in \{0,1\}$, controls which parameters need to be varied and which need to remain constant during an estimation procedure (this is important for the confidence interval examination later, as well as for basic hypothesis testing). Next, the function calculates the time per order *TPOt* given current parameters, using our basic function *os.model()*. The log-likelihood is derived from the result. For flexibility of use, depending on external settings, the function *os.model.ll()* returns either the likelihood *logl* or the *TPOt*.

Estimation and Standard Statistics

The parameter estimation problem is dynamic: find the parameters that maximize the likelihood of the data conditioned on the dynamic model given by equations (4–5). Further, $f(\boldsymbol{\theta}|x)$ is nonlinear in parameters and in x. Estimation of the parameters should follow a nonlinear approach.

R offers a standard nonlinear least squares function *nls* (based on Bates and Watts 1988). For the standard procedure, one provides the data series to be estimated (here *ACT.TPO*), the model with parameters (*os.nls.ll(alpha,tau.td,tau.ti,TPOD0),* specified in the R-script), the start points for the parameters (*start_l*), the data set to be utilized (*reg.data.nls*), the lower and upper bounds for the parameters (lower_b and upper_b), as well as additional convergence criteria:

```
nls(ACT.TPO ~ os.nls.ll(alpha,tau.td,tau.ti,TPOD0),
start=start_l,data=reg.data.nls,
algorithm="port," lower=lower_b, upper=upper_b, trace=T, nls.
control(maxiter=2000) )
```

While the existence of an extensive set of ancillary R analysis tools based on *nls* often makes it a good first choice, a potential issue is that it uses a simple optimization method, Newton-Raphson (NR). NR derives the curvature of the parameter surface at $\hat{\theta}_n$ and uses this to determine the size of the step to $\hat{\theta}_{n+1}$. The curvature information is captured by the matrix of second-order partial derivatives, called the Hessian. The NR and similar methods are fast and work well when the log-likelihood function is close to quadratic. However, the Hessian captures the curvature at the current point, $\hat{\theta}_n$, which may not be a good approximation to the actual shape of the likelihood surface.[9]

R offers several other optimization procedures to handle convergence in more complex models. For example, the Broyden-Fletcher-Goldfarb-Shanno (BFGS) method, which we used here, calculates the approximate Hessian using information at more than one point on the likelihood function (Greene 2006; Oliva 2003; Train 2003). The standard function *optim()* is flexible to handle multiple procedures. Running *optim()* requires similar inputs as *nls()*, but in addition allows specifying the optimization procedure ("method") as well as whether a Hessian is to be returned, which we shall see becomes important later.

While one can run *optim()* directly, we specify an intermediate function *os.mle.ll()* to code for flexibility in varying parameters to be estimated. We call optim() within this *os.mle.ll()* function:

```
est.mle.ll<-optim(par=par.ll.mle,os.model.ll,hessian=TRUE,
method=method, y=ACT.TPO,lower=lowerb,upper=upperb)
```

While an initial estimate with "arbitrary" starting points leads to convergence, we cannot be confident that this is the global maximum likelihood estimator. In general, one must consider the possibility of multiple local optima. To test for this possibility, one needs to carry out the estimation using multiple starting points in the parameter space, including some close to and others far from the expected estimates. The optimization procedures allow for multiple starts. It is easy to automate multiple starts (see the R-script, function *est.os.ll.mult.starts*), for example by setting up a Latin hypercube or similar sampling design. In our example, the estimation procedure identifies multiple optima. Sensitivity to initial parameters is not surprising given the nonlinearity of the model in parameters. The maximum of the individual maxima found by the optimization at each starting point is taken to be the global maximum of the likelihood function.[10]

The next step is to generate summary statistics for the MLE. Standard summary statistics (standard errors and t-values for each parameter) are provided by the package nls. We also derive basic residual statistics using a custom-built function. Table 1.1 reports the results and also compares them to those reported in OS01 and D07. The estimates of α, τ_d, τ_i, and T^*_0 are quite similar across all three studies. Appendix A1.3 compares the estimated time per order for all three papers.

The standard statistics can be used to carry out hypothesis tests, including the usual test of the null hypothesis that the parameters are zero (table 1.1). For example, the null hypothesis that the true value of α given the data is zero is convincingly rejected ($t(48) = -9.5$, $p < 0.0000$). Other tests can be performed just as easily (see the custom-built function *t.test.p.val.fromSummaryStats*): the hypothesis that the true value of α given the data is -1 is also strongly rejected ($t(48) = 5.96$, $p < 0.0000$). Hence, we have strong confidence that the effect of under/overcapacity on work pressure exists ($\alpha<0$) and exhibits diminishing returns ($\alpha>1$).

An intriguing result of the partial model estimation in OS01 is the asymmetry of the adjustment process for desired time per order. When work pressure forces actual time per order to fall below the normal level, the normal level erodes quickly, with an estimated time constant (τ_d) of about 19 weeks. But OS01 find no evidence of any upward revisions in desired time per order when work pressure is low (in OS01 τ_i is effectively infinite, despite the fact that actual time per order exceeded desired time per order in more than half the data set). Our results are consistent with this (table 1.1): the estimates of τ_d and τ_i are 19.76 and 9.90e5 weeks, respectively, very close to the OS01 estimates of $\tau_d=18.83$ and $\tau_i=8.14e5$. Figure 1.4 shows time per order (actual and estimated) and the estimated desired time per order (see also OS01 for comparison). However, the summary statistics suggest that the confidence intervals around these estimates are large: we cannot reject the hypothesis that the two adjustment time constants τ_d and τ_i are equal ($t(48)=$ 4.64e-5, $p>0.999$). However, we can reject that τ_d is effectively infinite (for $\tau_d =1e3$, $p<0.000$), although these results assume asymptotic normality, implying a parabolic approximation of the likelihood function in the neighborhood of the best estimates. To further investigate, we now turn to explicit consideration of the confidence intervals around the MLE estimates.

Confidence Intervals

Turning to estimates of the confidence intervals, the top row of table 1.2 summarizes the default Wald-test based on OLS and assumptions of i.i.d. normally distributed errors as well as normality of the MLE. These results can be directly inferred from table 1.1, which reports the standard error. (They can also be derived by running the *overview()* function on the nls estimate; see the R-script.) Note that the assumption of normally distributed errors means the confidence intervals

Table 1.1
Estimation results (95% confidence level)

Parameter	Estimate	Std. Error	t value	Pr(>|t|)	OS01	D07
α	-0.6151	0.0646	-9.522	1.23e-12 ***	-0.64	-0.60
τ_d	19.76	23.00	0.859	0.395	18.83	12.48
τ_i	99,000	2.131e+08	0.000	1.000	814,000	85,600
T^*_0	1.063	0.02749	38.657	< 2e-16 ***	1.08	1.07
SSE[i]	0.0214[ii]				0.0226	0.0222
SER[i]	0.0211				0.0217[iii]	0.0215
ESS[i]	0.0872				0.0948	0.0788

Significance codes: 0 '***' 0.001 '**' 0.01 '*' 0.05 '.' 0.1 ' ' 1.
[i] SSE = sum of squared errors; SER = standard error of regression; ESS = explained sum of squares. We derived OS01 and D07 values using their parameter estimates.
[ii] On 48 degrees of freedom.
[iii] OS01 report an SER of 0.019.

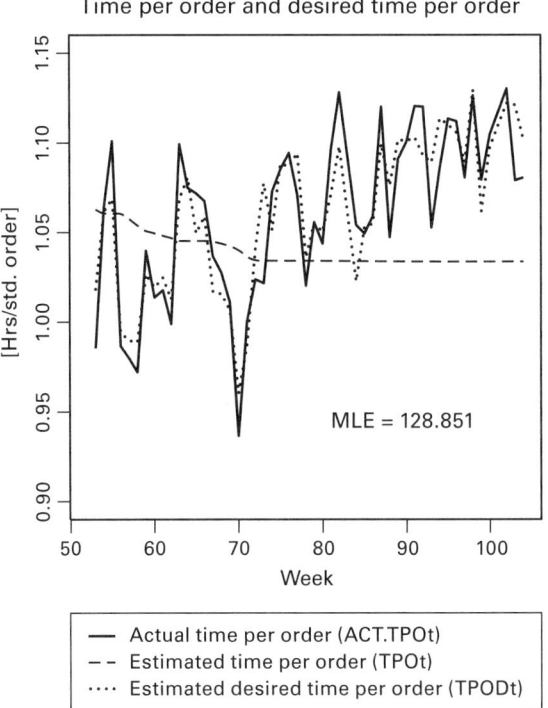

Figure 1.4
Estimation results.

are symmetric around the estimates. We also estimate the confidence intervals using maximum likelihood methods, using both AS- and LR-based intervals.

AS Intervals   We derived the manual Hessian-based AS interval using the Hessian produced by the *optim* function in R. To derive the asymptotic parabola around the MLE, we develop the customized function *anal.conf* (see the R-script). Note that this is also an asymptotic method yielding confidence intervals that are symmetric around the estimate, even for a model that is nonlinear in parameters (table 1.2). The Hessian-based confidence intervals are quite close to those generated by the Wald method.

LR Intervals   Because of the possibility of local optima and resulting sensitivity of results to the initial parameter values used in the optimization, it is important to consider the different approaches to finding LR confidence intervals. First, one can find the likelihood surface by grid search, evaluating the likelihood function for closely spaced points in the parameter space. The number of calculations required grows exponentially with the number of parameters. For k=4 parameters and, say, $p=20$ points for each, will require $p^k=20^4 =160,000$ likelihood values. Using these values, one can then identify, for each parameter, which regions fall within the desired confidence level. Despite the number of parameters, this approach could be feasible because no new optimizations are necessary. However, grid search to obtain the multidimensional likelihood surface may not be effective on large or rugged surfaces, since accuracy would require a large yet fine grid, increasing the computational burden.

Alternatively, we can calculate the likelihood profile. To do so, fix each parameter, one at a time, at its estimated value and obtain its confidence interval by optimizing over the other k-1 parameters. Thus, the required number of estimations grows linearly with the number of parameters. The result is the outer bounds of a k-1 dimensional sphere. The profile method is robust for rugged and flat surfaces. Still faster is a univariate LR approach in which all other parameters are held fixed at their best estimates when assessing the confidence interval for one parameter. While fast (no new estimates are necessary), this approach may be inaccurate (this is equivalent to estimating the size of the sphere at the axes).

Here we first derive the likelihood profile and univariate intervals. To do so, we developed the custom function *LL.multipar.plot()*. The function requires specification of the model (*os.model.ll*), parameter ranges (*par.1d*), the tolerance for the confidence bounds (*alpha.conf=0.5*), degrees of freedom for the χ^2 tests (*df.1p=1* for both univariate and profile intervals), and selection of the profile or univariate method through sw.profile. The function also asks for parameters for a plot of the results:

```
LL.multipar.plot(os.model.ll,par.est,y,sigma2,par.1d,par.1d.
range,par.2d,par.2d.range,alpha.conf,df.1p,df.2p,sw.same.LL.range,LL.
range.rel,sw.2dplot=0)
```

The profile likelihood intervals resulted from 80 estimates for each parameter, which required about two minutes.[11, 12] Figure 1.5 shows the results for univariate (top four) and profile (bottom four) graphs, and table 1.2 reports the confidence intervals. For α and T^*_0, the confidence regions were consistent across all MLE-based approaches, all suggesting a steep likelihood surface and narrow confidence bounds. For the time constants, the function is quite flat: it is hard to reject any plausible value above 12 weeks for either increasing or decreasing the norm for time per task. (More on the shorter time constant below.)

Unlike the asymptotic method, the likelihood functions and resulting confidence intervals are not symmetric and, in some cases, clearly not parabolic (e.g., the likelihood function for τ). Also note that, while the MLE is constrained to the admissible regions (0.1–1e6), the LR intervals are not. The example shows that LR methods, in contrast to AS methods, can handle more complex parameter relations. For example, for both τ's the likelihood function drops off steeply for values below, but not above their respective MLE. The example illustrates that the LR method can detect and correctly capture important nonlinearities in parameters that the AS method, which assumes the symmetric parabolic approximation around the MLE by construction, does not. In this case, the consequences for confidence interval estimation are large.

Validity of Assumptions

So far we have, optimistically, assumed normality and independence of the errors (no autocorrelation in the errors). We now assess the appropriateness of these assumptions by analyzing the residuals. The R function *nlsResiduals* provides, by default, four classic plots of residuals (Delignette-Muller and Baty 2013): nontransformed residuals against fitted values, standardized residuals against fitted values, autocorrelation plot of residuals, and QQ-plot of the residuals. Of direct interest are normality and autocorrelation (figure 1.6).

Regarding normality, the histogram (figure 1.6, left) does not obviously deviate from normality, an impression confirmed by the Shapiro-Wilk normality test: the hypothesis that the residuals are normally distributed cannot be rejected (Shapiro-Wilk normality W = 0.9784, p = 0.33). Next, to assess whether the residuals show significant autocorrelation, we plot the autocorrelation by lag. Autocorrelation can be extracted using the standard function *acf()* (figure 1.6). Visually, we cannot draw any strong conclusion. To test for autocorrelation formally, we use the custom function *autcov.var.test()*. The test shows that autocorrelation is not significant at the 5% level. Both results are consistent with those of D07. Additional standard tests available in R support these findings (appendix A4).

Table 1.2
Comparative confidence intervals (95% level)

Confidence Interval Method	Parameter				Duration
	α	τd	τi	T*0	
MLE (SSE=0.0214)					
	-0.62	19.76	99,000	1.06	2 min
AS, Wald test	(-0.75, -0.49)	(-27, 66)	(-4.3e8, 4.3e8)	(1.01, 1.12)	negligible
AS, manual (Hessian)	(-0.71, -0.52)	(-18, 58)	(87,000, 111,000)	(1.02, 1.11)	negligible
LR, Univariate	(-0.69, -0.54)	(11, 53)	(54, Inf)	(1.04, 1.08)	20 sec
LR, Profile	(-0.71, -0.52)	(8.1, Inf)	(32, Inf)	(1.03, 1.26)	2 min
Boot[*], percentile	(-0.70, -0.49)	(5.8, Inf)	(19, Inf)	(1.03, 1.19)	6.0 hrs
Boot[*], BCa	(-0.74, -0.52)	(6.34, Inf)	(1000, Inf)	(1.03, 120)	
Boot manual[*], percentile	(-0.70, -0.50)	(6, Inf)	(20, Inf)	(1.03, 1.13)	6.3 hrs
OS01 (SSE=0.0226)					
	-0.64	**18.83**	**814,000**	**1.08**	**na**
In OS01, univariate	(-0.7, -0.59)	(13.30, 28.95)	(33,000, Inf)	(1.06, 1.09)	na
AS, Wald test	(-0.77, -0.51)	(-17, 54)	(-3e12, 3e12)	(1.01, 1.15)	negligible
LR, Univariate	(-0.66, -0.56)	(8, 22)	(216, Inf)	(1.04, 1.09)	20 sec
LR, Profile	(-0.71, -0.52)	(6, Inf)	(35, Inf)	(1.03, 1.13)	2 min
D07 (SSE=0.0222)					
	-0.60	**12.48**	**8.56e4**	**1.07**	**NA**
In D07, boot, percentile	(-0.68, -0.49)	(4.65, 5.00e4)	(25.02, 2.5e8)	(1.03, 1.14)	na
In D07, boot, BCI	(-0.71, -0.52)	(5.22, 8.85e4)	(25.53, 3.43e8)	(1.03, 1.15)	na

[*] 600 resamplings.

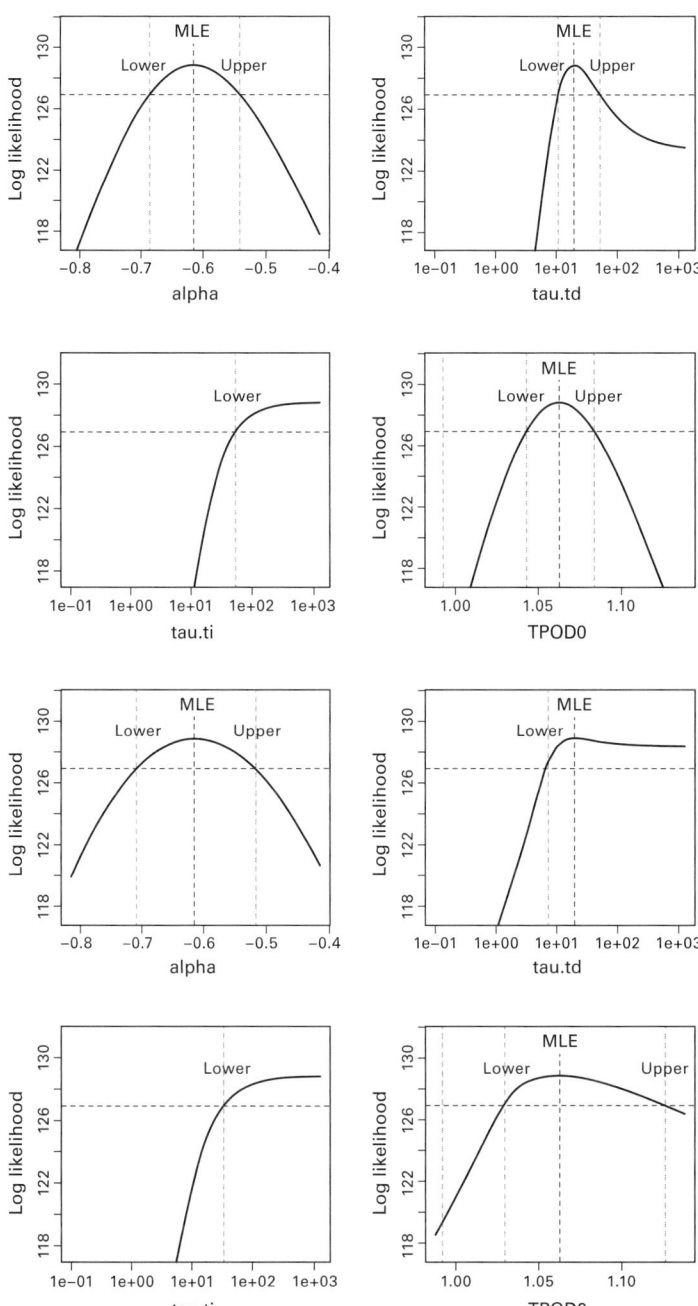

Figure 1.5
Log-likelihood function and confidence bounds for the four parameters (95% confidence level) univariate (top) and profile (bottom).

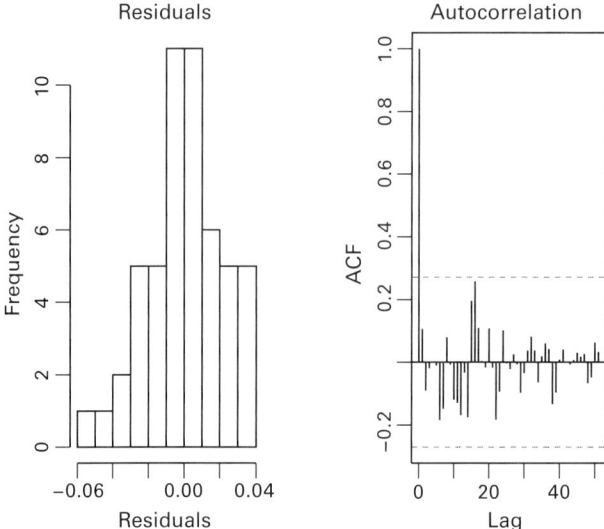

Figure 1.6
Histogram (left) and autocorrelation function values (right) of the maximum likelihood estimator error terms for OS01 data.

Bootstrapping

Given the characteristics of the error distribution, specifically that we cannot reject the hypothesis that the residuals are i.i.d. normal, it would be acceptable to use nonparametric residual-based bootstrapping. Reshuffling the observed error terms is inappropriate in the presence of autocorrelation unless that is also corrected for. In our example, the bootstrapping process is straightforward because error-term autocorrelation is not statistically significant.

Multiple predefined bootstrapping functions exist in R. First, *nlsboot* is specifically designed for nonlinear least squares and uses nonparametric bootstrapping with mean-centered residuals. The bootstrap estimate distributions of this function can be visualized, using the function *plot.nlsBoot*, either by plotting the bootstrap sample for each pair of parameters or by displaying the boxplot representation of the bootstrap sample for each parameter. However, in this case nlsboot does not work because Newton-Raphson fails to converge. Other bootstrap packages and functions differ in their ease of use, flexibility, and effectiveness depending on the estimation procedure and visualization. In particular, the package *boot,* with functions *boot* () and *boot.ci* to resample and then to extract the intervals, is powerful (Crawley 2007). One needs to supply the function *boot* with: (1) the actual data; (2) a "statistic function," providing information on whether the data or the residuals are to be resampled; and (3) the estimation method to be used (e.g., *optim*). Once the bootstrapped data sets have been generated, the *boot.ci()* function generates the confidence intervals.[13]

Regardless of the bootstrapping method, one must use multiple starts for each resampled bootstrap estimate, b. Because of this one needs to perform b*k^s bootstrapping optimizations, with k the number of parameters and s the number of start values for each parameter. We constructed bootstrapping intervals using manual reshuffling of error terms. In addition, we report results from boot.ci. Details are provided in the R-script. Using two starts for each parameter and b=600, we performed 10,000 optimizations, taking about six hours of processing time for each bootstrap. For the manual bootstrap, we implemented the percentile method. The bootstrap-generated intervals (figure 1.7, table 1.2) are consistent with each other.

Interpreting Results and Additional Analysis

We now revisit the result of the partial model estimation in OS01 for which the adjustment process for desired time per order is asymmetric. The small confidence intervals OS01 find for both τ_d and τ_i support this notion.

However, table 1.2 shows that different methods for estimating the confidence intervals for the time constants yield different results (figure 1.8). The asymptotic and univariate methods (except Wald for τ_i) yield relatively narrow intervals, consistent with OS01. In contrast, the MLE LR-profile and bootstrap methods show large upper bounds for τ_d.[14] Note that the curvature around the MLE drops as τ_d rises (figure 1.5), explaining the differences between the asymptotic and nonasymptotic methods. The bootstrap results are consistent with the LR-profile interval method, and, based on this information, we cannot reject the hypothesis that τ_d and τ_i are *not* different.

More detailed analysis using MLE and bootstrapping results can be performed to improve intuition about the parameter estimates and intervals. We can explore interaction between variables by examining bivariate intervals, keeping the other k-2 parameters at their estimated values (df=2). Thus, the bivariate confidence interval is an intermediate form between univariate and surface plots showing the effect of interactions between two (but not all) parameter estimates on their confidence intervals. Figure 1.9 shows the results for the case of α and τ_d.

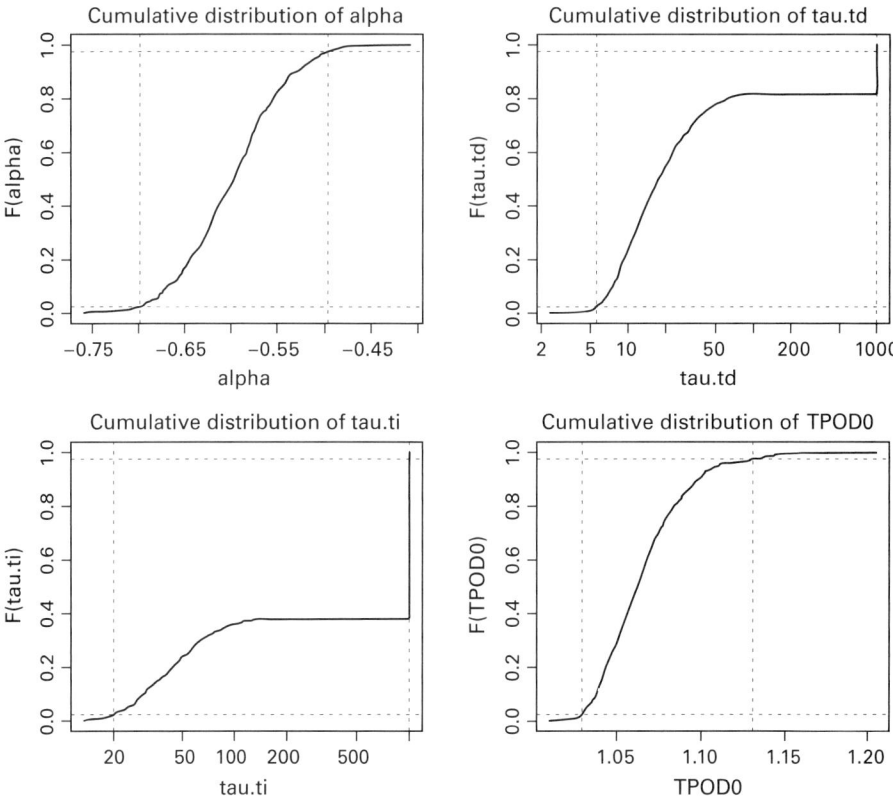

Figure 1.7
(Manually derived) percentile bootstrap confidence interval limits of the parameters.

Their bivariate intervals are, respectively, *(-0.72, -0.51)* and *(7.5, 118)*, larger than for the univariate case. However, one can also see that parameter estimates interact: for example, given a stronger effect of work pressure (larger $|\alpha|$), we can reach greater confidence in a smaller τ_d. A faster downward adjustment τ_d of the normal time per order "makes up" for a larger responsiveness to work pressure. Hence, conditional upon more information about α, we might not reject τ_d being smaller than τ_i. In addition, one can use the MLE LR for alternative hypothesis testing, which tests whether, when comparing models, some can be rejected (appendix A5).

More detailed examination of the bootstrap results does provide some stronger support for the finding in OS01 that $\tau_d < \tau_i$. First we perform the nonparametric Mann-Whitney test (e.g., Conover 1980),

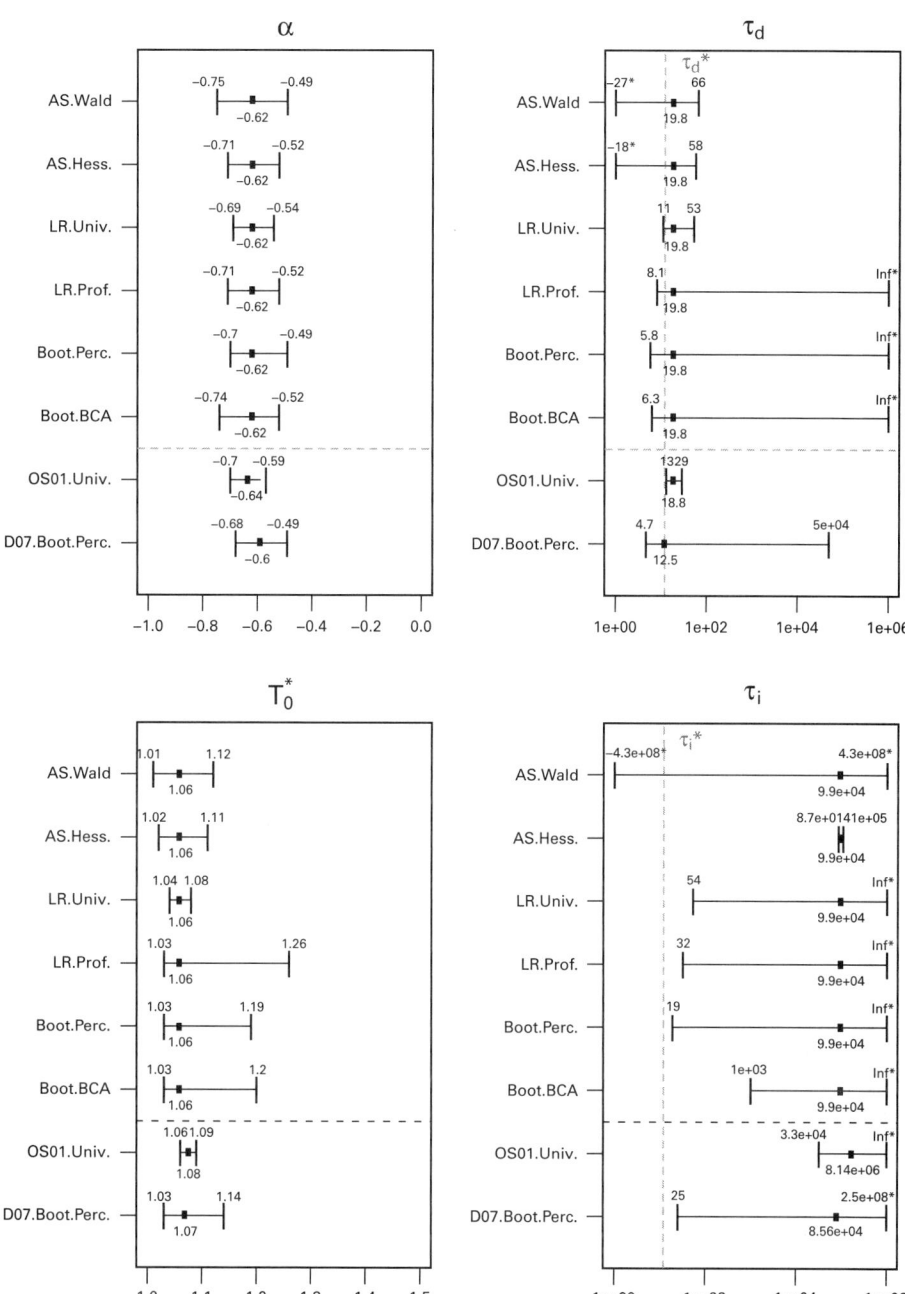

Figure 1.8
Visual comparison of parameter confidence intervals across methods.

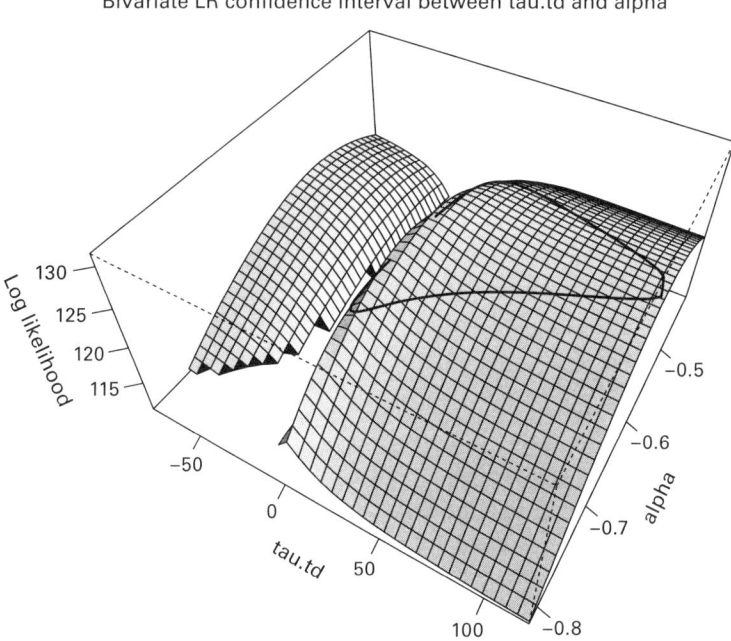

Figure 1.9
Bivariate confidence interval for τ_d and α. (Other parameters remain at their estimated value.)

designed to compare sample means, to test for $\tau_d < \tau_i$. With p<0.000, we must reject equality of τ_d and τ_i. These results are also supported by inspection of the bootstrap data: in 78% of the bootstrap sample cases, $\tau_d < 52$ weeks (one year), while this is true for τ_i in only 25% of the cases. Further, in 41% of the cases $\tau_d < 0.5$ years (26 weeks), while at the same time $\tau_i > 2$ years. The opposite ($\tau_i < 0.5$ year and $\tau_d > 2$ years) occurs in less than 1% of the cases.

Conclusion

Selecting the best method to estimate model parameters and confidence intervals is difficult and presents the modeler with strong trade-offs. The relative benefits of asymptotic approaches, likelihood ratio methods, and bootstrapping depend on the context and characteristics of the model and the data. In this chapter, we demonstrated how to estimate parameters in nonlinear dynamic models by maximum likelihood, and how to estimate confidence intervals using likelihood or bootstrapping methods. We highlighted some of the main challenges and trade-offs one encounters in this process, and discussed how to select appropriate methods.

If the maintained assumptions of the likelihood method are valid, bootstrap results are almost identical to likelihood ratio results. If tests show that these assumptions are appropriate, both methods are viable, though likelihood methods are computationally more efficient.

However, for nonlinear models, confidence regions and intervals constructed using the covariance matrix (or Hessian), following asymptotic assumptions, are only approximate and may not give results consistent with exact, though computationally more intensive, methods. In contrast, likelihood ratio approaches are only based on the asymptotic approximation regarding the χ^2 distribution of the log-likelihood ratio test. LR methods often behave well for smaller samples. Further, LR-based approaches provide insight into the true curvature of the LL surface, and respect the allowable ranges for the parameters. In addition simulation-assisted estimation procedures such as maximum simulated likelihood are available to provide more flexibility in estimation of the parameters of a model.

Alternatively, if the maintained assumptions of MLE are not valid, bootstrapping may be superior. However, although bootstrapping is fairly straightforward, the process involves several manual steps that require judgment and can be computationally demanding.

To make these trade-offs appropriately, researchers need to be able to use and compare the different methods quickly and easily. Affordable and widely used software such as R includes a wide range of statistical methods and reflect the latest developments for rigorous model testing. Packages such as R go well beyond the built-in statistical and estimation capabilities of most dynamic simulation software. On the other hand, while R has dynamic simulation capabilities, they are not as advanced as those of specialized dynamic simulation software and more cumbersome to use. In the future, linking simulation and statistical software would avoid the need for modelers to develop their models in one environment only to recode them in a statistical package. Whatever approach one uses, however, it is essential that the parameters and uncertainties in dynamic models be estimated formally and rigorously, using the best available methods.

Challenge

In this challenge you get the chance to explore MLE and bootstrap techniques by estimating the parameters for a well-analyzed model of decision making in the Beer Distribution Game (Sterman 1989). Appendix A1.6 summarizes the model and provides additional details and tips on the questions.

Question 1. Data and replication model and estimation
  a. Import the data into R, replicate the model, and reproduce the data and estimated orders reported in Dogan (2007).

Question 2. Estimation
  b. Develop the model in R. Estimate the orders using the model. How do you get confidence that you have found a global optimum? Explain.
  c. Derive parameter estimates. Produce summary statistics. Discuss the results.

Question 3. MLE confidence intervals
  a. Develop a simple log-likelihood function. Assume normal i.i.d. error terms.
  b. Suppose you estimated the model parameters using MLE. Do you expect the results to be the same as or different from those in question 2? Why? Now, re-estimate parameters using MLE. Interpret your results.
  c. Construct AS confidence intervals. Examine at least two types. Tabulate and plot the results.
  d. Construct LR confidence intervals. Examine at least two types. Tabulate and plot the results.
  e. Compare and interpret the confidence intervals.

Question 4. Validity of asymptotic assumptions
  a. Examine the validity of the asymptotic assumptions. Perform graphical and formal analysis of normality and autocovariance.
  b. Comment on your results. Were your assumptions under 3a justified? (If not, under question 6 below, comment on whether this is a problem or not.)
  c. If needed, adjust the data to correct for autocorrelation and rerun the analysis under 3.

Question 5. Bootstrap-generated confidence intervals
  a. Decide on the appropriate bootstrapping method to use. Can you reshuffle the error terms?

    b. Perform the bootstrap analysis. Consider the appropriateness of the various automated and custom-defined bootstrap functions discussed above and provided in the supplement.

    c. Compare and interpret the results.

Question 6. Conclusions

Comment on your results. In particular, are basic assumptions violated? If so, is it a problem? How do you interpret any differences between AS-confidence, LR-confidence, and bootstrapping results?

Notes

1. But only one such test. Tests of robustness under extreme conditions and many others are also necessary to build confidence in any model (see Sterman 2000, Ch. 21).

2. In addition, extensive help for use of predefined R functions is available within R, with many other references available online. For good general introductions to R, see Venables and Smith (2012) and Crawley (2007).

3. Notationally, $X_1 \ldots X_n$, are the random variables of which the estimator is a function, while the $x_1 \ldots x_n$ represent realized values. Note that the likelihood function is expressed as a density function for the parameters θ and not a function of the variables, given the parameters, even though L and f may be identical in form.

4. To do this one constructs the so-called Fisher information (FI) matrix, the asymptotic variance matrix of the MLE. The inverse of the FI determines the asymptotic performance of the maximum-likelihood estimator as the number of data points goes to infinity.

5. This illustrates that likelihood ratio confidence estimation is fundamentally a form of likelihood-ratio hypothesis testing. The confidence interval specifies those parameter sets for which we are unable to reject the hypothesis that the estimated parameter ("the null") given the desired confidence level α. For values of the difference in equation (4) exceeding the critical value of the χ^2 distribution, the "null" hypothesis is rejected.

6. The need to infer parameters governing processes for which there are no directly observable data is common not only in social science but in physical systems. A large literature in control theory addresses the estimation challenges in such complex dynamic systems (Stengel 1994; Bryson 2002; Crassidis and Junkins 2011).

7. Variable names for the same variables differ between D07 and OS01. Our R-script follows D07. Conversions are as follows (D07/OS01): required service capacity $(sc \, / \, C_t^*)$; service capacity $(SC \, / \, C_t)$; desired time per order $(TPODt \, / \, T_t^*)$; incoming orders $(CO \, / \, O_t)$; time per order $(TPOt \, / \, T_t)$; initial time per order $(TPOD01/ \, T_0^*)$.

8. The code that follows can be more compact (e.g., using R's "*apply()*" function instead of the classic "*for[ ]*" operation), but that would compromise tractability of the model.

9. If the function is perfectly quadratic, then curvature information at one point on the function provides all the information that is needed about the shape of the function. This is why AS approaches perform well in models that conform to the maintained hypotheses of standard linear models but may fail in more complex ones with nonlinearities.

10. Examination of the likelihood surface using a formal Latin hypercube approach can provide further evidence of having found the global optimum.

11. This ignores implementation of multiple starts beyond those generated by the estimation procedure.

12. For reference, estimation times reported in the tables are on a 2012 MacBook Pro with a 2.6 GHz Intel Core i7 processor and 8 GB 1600 MHz DDR3 memory.

13. *Boot()* can handle more advanced procedures as well, including parametric residual resampling. Further, for nonparametric bootstrapping, multiple resampling strategies are supported. Resampling strategies are useful, for example, when particular ranges of explanatory variables are important, but underrepresented in the original data. Sampling strategies include ordinary, balanced, antithetic, permutation and stratified. Resampling weights can also be specified.

14. The manual AS likelihood approach did not work for the Oliva and Sterman estimates because the Hessian, necessary to derive the second derivative, can only be derived in *optim* for an optimum, which was not the case.

References

Adda, J., and R. Cooper. 2003. *Dynamic economics: Quantitative methods and applications.* Cambridge, MA: MIT Press.

Aldrich, J. 1997. R. A. Fisher and the making of maximum likelihood 1912–1922. *Statistical Science* 12 (3): 162–176.

Barlas, Y. 1989. Multiple tests for validation of system dynamics type of simulation models. *European Journal of Operational Research* 42 (1): 59–87.

Barlas, Y. 1996. Formal aspects of model validity and validation in system dynamics. *System Dynamics Review* 12 (3): 183–210.

Bates, D. M., and D. G. Watts. 1988. *Nonlinear regression analysis and its applications.* New York: Wiley.

Bryson, A. E. 2002. *Applied linear optimal control: Examples and algorithms.* Cambridge, UK: Cambridge University Press.

Casella, G., and R. L. Berger. 2001. *Statistical inference.* Pacific Grove: Duxbury.

Conover, W. J. 1980. *Practical nonparametric statistics.* 2nd ed. New York: Wiley.

Crassidis, J. L., and J. L. Junkins. 2011. *Optimal estimation of dynamic systems.* Boca Raton: CRC Press.

Crawley, M. J. 2007. *The R book.* Chichester: Wiley.

Croson, R., K. Donohue, E. Katok, and J. Sterman. 2014. Order stability in supply chains: Coordination risk and the role of coordination stock. *Production and Operations Management* 23 (2): 176–196.

Davison, A. C., and D. Kuonen. 2002. An introduction to the bootstrap with applications in R. *Statistical Computing and Statistical Graphic Newsletter* 13 (1): 1–11.

Delignette-Muller, M.-L., and F. Baty. 2013. Use of the package nlstools to help the fit and assess the quality of fit of a Gaussian nonlinear model. Available at: http://www .icesi.edu.co/CRAN/web/packages/nlstools/vignettes/nlstools_vignette.pdf.

DiCiccio, T. J., and B. Efron. 1996. Bootstrap confidence intervals. *Statistical Science* 11 (3): 189–212.

Dogan, G. 2007. Bootstrapping for confidence interval estimation and hypothesis testing for parameters of system dynamics models. *System Dynamics Review* 23 (4): 415–436.

Efron, B., and R. Tibshirani. 1986. Bootstrap methods for standard errors, confidence intervals, and other measures of statistical accuracy. *Statistical Science* 1 (1): 54–75.

Eliason, S. R. 1993. *Maximum likelihood estimation: Logic and practice*. Quantitative applications in the social sciences, vol. 96. Newbury Park: Sage Publications.

Engle, R. F. 1984. Wald, likelihood ratio, and Lagrange multiplier tests in econometrics. In *Handbook of Econometrics*, vol. 2, 775–826. Amsterdam: Elsevier.

Fisher, R. A. 1920. A mathematical examination of the methods of determining the accuracy of an observation by the mean error, and by the mean square error. *Monthly Notices of the Royal Astronomical Society* 80 (8): 758–770.

Fisher, R. A. 1922. On the mathematical foundations of theoretical statistics. *Philosophical Transactions of the Royal Society of London. Series A, Mathematical and Physical Sciences* 222: 309–368.

Forrester, J. W., and P. M. Senge. 1978. *Tests for building confidence in system dynamics models*. Cambridge, MA: System Dynamics Group, Sloan School of Management, Massachusetts Institute of Technology.

Freedman, D. A. 1981. Bootstrapping regression models. *Annals of Statistics* 9 (6): 1218–1228.

Graham, A. K. 1980. Parameter estimation in system dynamic modeling. In *Elements of the system dynamics method*, ed. J. Randers, 143–161. Waltham, MA: Pegasus Communications.

Greene, W. H. 2006. *Econometric analysis*. 5th ed. New Jersey: Prentice Hall.

Hausman, J. A. 1978. Specification tests in econometrics. *Econometrica* 46 (6): 1251–1271.

Homer, J. B. 2012. Partial-model testing as a validation tool for system dynamics. *System Dynamics Review* 28 (3): 281–294.

Imbens, G. W., and R. Spady. 2002. Confidence intervals in generalized method of moments models. *Journal of Econometrics* 107 (1): 87–98.

Keith, D. 2012. Essays on the dynamics of alternative fuel vehicle adoption: Insights from the market for hybrid-electric vehicles in the United States. PhD thesis. Massachusetts Institute of Technology.

Li, H., and G. S. Maddala. 1996. Bootstrapping time series models. *Econometric Reviews* 15 (2): 115–158.

McAfee, A., and E. Brynjolfsson. 2012. Big data: The management revolution. *Harvard Business Review* 90 (10): 61–67.

Meadows, D. H., J. Richardson, and G. Bruckmann. 1982. *Groping in the dark: The first decade of global modelling*. New York: Wiley.

Meeker, W. Q., and L. A. Escobar. 1995. Teaching about approximate confidence regions based on maximum likelihood estimation. *The American Statistician* 49 (1): 48–53.

Millar, R. B. 2011. Some widely used applications of maximum likelihood. In *Maximum likelihood estimation and inference: With examples in R, SAS and ADMB*. Chichester: Wiley.

Myung, I. J. 2003. Tutorial on maximum likelihood estimation. *Journal of Mathematical Psychology* 47 (1): 90–100.

Oliva, R. 2001. Tradeoffs in responses to work pressure in the service industry. *California Management Review* 43 (4): 26–43.

Oliva, R. 2003. Model calibration as a testing strategy for system dynamics models. *European Journal of Operational Research* 151 (3): 552–568.

Oliva, R., and J. D. Sterman. 2010. Death spirals and virtuous cycles: Human resource dynamics in knowledge-based services. In *Handbook of service science*, ed. P. Maglio, C. Kieliszewski, and J. Spohrer, 321–358. New York: Springer.

Oliva, R., and J. Sterman. 2001. Cutting corners and working overtime: Quality erosion in the service industry. *Management Science* 47 (7): 894–914.

R Core Team. 2012. R: A language and environment for statistical computing. R Foundation for Statistical Computing, Vienna, Austria. ISBN 3-900051-07-0. URL http://www.R-project.org/.

SAS Institute Inc. 2011. *Base SAS® 9.3 Procedures Guide*. Cary, NC: SAS Institute Inc.

Simini, F., M. C. González, A. Maritan, and A.-L. Barabási. 2012. A universal model for mobility and migration patterns. *Nature* 484 (7392): 96–100.

Stengel, R. F. 1994. *Optimal control and estimation*. New York: Dover Publications.

Sterman, J. D. 1989. Modeling managerial behavior: Misperceptions of feedback in a dynamic decision making experiment. *Management Science* 35 (3): 321–339.

Sterman, J. D. 2000. *Business dynamics: Systems thinking and modeling for a complex world*. Boston: Irwin McGraw-Hill.

Stigler, S. M. 2007. The epic story of maximum likelihood. *Statistical Science* 22 (4): 598–620.

Train, K. 2003. *Discrete choice methods and simulation*. Cambridge, UK: Cambridge University Press.

Venables, W., and D. Smith. 2012. An introduction to R. Available at http://cran.r-project.org/doc/manuals/R-intro.pdf, 1–109.

Venzon, D. J., and S. H. Moolgavkar. 1988. A method for computing profile-likelihood-based confidence intervals. *Journal of the Royal Statistical Society, Series C (Applied Statistics)* 37 (1): 87–94.

Wilks, S. S. 1943. *Mathematical statistics*. New York: Wiley.

Wooldrige, J. 2002. *Econometric analysis of cross section and panel data*. Cambridge, MA: MIT Press.

2 Using the Method of Simulated Moments for System Identification

Mohammad S. Jalali, Hazhir Rahmandad, and
Hamed Ghoddusi

Increasingly, dynamic modelers face problems where estimating model parameters from numerical empirical data is a requirement. This trend is partly motivated by increasing availability of numerical data from a large number of ongoing and one-off data collection projects that survey different concepts of interest to dynamic modelers, from individuals and firms to disease incidences and measures of economic performance, just to name a few. For example, as of February 2015, the Data.gov portal contains data on over 138,000 machine-readable data sets. Another driver of this trend is the increasing application of dynamic models, beyond case-specific corporate projects, to theoretical and academic problems (Repenning 2003; Sterman 2006). In these cases, generic models for a category of objects (e.g., individuals, firms, and countries) are desired. Parameterizing such models requires specifying the different parameters that quantify similarities and differences across different objects, a goal often dependent on using robust and replicable parameter-estimation procedures. In fact, in light of the rapid growth and dissemination of improved parameter-estimation methods for model calibration, hypothesis testing, and policy recommendation in social and behavioral disciplines, continued relevance of any modeling subdiscipline may partially be tied to its ability to remain up to date with the best available tools in this domain.

Closely tied to the advances of the digital computer revolution, the field of system dynamics (SD) at its inception was ahead of many approaches available in social sciences in using advanced analytical methods (Forrester 1961), and kept this edge for many years. For example, advanced filtering and estimation methods were introduced into the SD literature in the 1970s (Peterson 1975). However, over the last three decades the research in SD has largely focused on diverse applications of the original toolbox, with limited methodological expansions in the parameter-estimation domain. In contrast, research in econometrics and other related fields has provided many relevant tools over this period (Greene 2012). As a result, many dynamic modeling studies do not currently report formal parameter estimates common in social science research, or, when calibration is pursued, typical measures of confidence in estimated parameter values are not reported. While formal parameter estimation may not be feasible for many modeling problems, expert dynamic modelers should be equipped with the relevant tools when numerical data is available, when model purpose requires reliable parameter estimates, or when the audience requires formally estimated parameters and confidence intervals.

Hand-calibration is commonly practiced for assigning parameter values (Lyneis and Pugh 1996). When formal estimation procedures are used, modelers typically compare time-series data against the same variables in a model, and minimize the weighted sum of a function of the error term by changing the uncertain parameters until best-fitting estimates are found through a nonlinear optimization algorithm (Oliva 2003). The error function is frequently defined as the squared error, but absolute error and absolute percent error terms are also common (Sterman 2000). Weights for different data points are given based on the confidence the researcher has in the accuracy of the data and its relevance to the problem at hand. When reported, confidence intervals are calculated using normality and independence assumptions for error terms, which, with weights proportional to the reciprocal of error variance, would turn least-squared error estimates into maximum-likelihood estimates (MLE). Bootstrapping methods are also sometimes used for estimating confidence intervals (Dogan 2007). While these approaches cover many important estimation challenges, they each include some shortcomings. Ad hoc selection of the error term and the weights for different data points reduces the consistency of the methods and their ability to provide confidence intervals. Normality and independence may regularly be violated, which negate the benefits of MLE when using squared errors. Bootstrapping, while flexible, increases the computational costs significantly and as a result may prove infeasible for many realistically sized problems. Finally, the majority of these methods rely on having time-series data, and cannot extract from distributions in cross-sectional data the dynamics that have led to those distributions. In general, the estimation procedures ideal for dynamic models have the following characteristics.

Model independence. Given that most dynamic models do not follow
a fixed structural form (e.g., linearity), estimation procedures that
are independent of model structure are most beneficial.

Analytical confidence intervals. The ability to find confidence intervals analytically is important because of the computational costs of
optimizing nonlinear dynamic models and replications needed for
bootstrapping methods.

Assumption-free error terms. Independence and distributional assumptions on error terms for dynamic models are not always easy
to justify, so methods with fewer such assumptions are preferred.

Applicability to diverse data types. Both time-series and cross-sectional data are included.

No single method fully satisfies all these requirements. Therefore, modelers need to choose from a menu of available estimation methods to
match their problem requirements.

In this chapter, we offer an introduction to the method of simulated
moments (MSM) for application to dynamic modeling problems. The
basic idea of this method is to define appropriate moments of data and,
by changing uncertain parameters, minimize the difference between
those moments and their simulated counterparts resulting from the
model simulations. Moments are statistics that are calculated using
some function that gets as input the empirical data points available and
provides as output a single numerical value. For example, population
mean could be a moment, getting the individual data points as input
and generating the population mean as the output. While any function
can be used for generating moments, for analytical confidence intervals
to be available, one needs these moments to be normally distributed,
often meaning that each moment is an average across a function of
multiple independent observations coming from the same underlying
distribution (then normality follows from the central limit theorem). In
practice, those observations (that feed into the moments calculations)
are picked either from time-series data when a system is in steady state
(e.g., stock prices over time), or at similar points in the life of similar
units of observation (e.g., all five-year-old individuals in a country).
Typically, randomness plays an important role in how these units have
ended up with different observed values (e.g., different weights for
similarly aged individuals). As a result, the MSM is best fitted for
dynamic modeling problems when some of the following problem
characteristics are present:

- Population data. The MSM is suitable for estimation of generic models to population data. Different units of data such as individuals, firms, and countries could be available. For each unit one or more data items (e.g., weight, height, and age for individual data) could be available.
- Role of random processes. The MSM could be a good choice when models include stochastic processes that drive the model, and their impact on the model behavior is reflected in the data against which the model is to be calibrated (e.g., when we are trying to match the variance observed across multiple units).
- Cross-sectional data. The MSM applies to both cross-sectional and time-series data. Whereas time-series data include multiple data points for the same unit over time, cross-sectional data includes data points for multiple units at the same time. The MSM may be the only viable choice for estimating dynamic models when data is cross-sectional, since it allows us to extract the information about the historical trajectories of units hidden in their cross-sectional distributions.
- Confidence intervals. The MSM would be a suitable choice when analytical confidence intervals are sought.

The material in this chapter complements the chapter on MLE. Both are parameter-estimation methods. However, MLE requires distributional assumptions on error terms. When such assumptions are justified (either empirically, based on the observed error, or based on theoretical considerations), MLE provides a more efficient estimation method than the MSM. When error terms do not follow any well-established distribution, the MSM may be a better choice. An alternative method, maximum simulated likelihood, extends MLE's basic premise to situations when error terms do not follow predefined distributions. We do not discuss the similarities and differences between these methods, but refer interested readers to Adda and Cooper (2003) and Gourieroux and Monfort (1996).

Consider an example from the dynamic modeling literature. While the MSM has become a major econometrics tool in the past two decades, it has rarely been applied in the system dynamics literature. Barlas (2006), in the design of the behavior pattern testing (BTS II) approach and software, uses some of the basic ideas of the MSM to match moments of the model against data, but does not draw on the MSM literature or discuss issues related to confidence levels. Rahmandad and Sabounchi (2011) adapt the MSM to estimating the parameters of an individual weight gain and loss model. In this section we provide a brief overview of their application to provide a more concrete example of use of the MSM. A simple model of an individual's body mass, consisting of fat mass and fat-free mass, was developed. The model included a few uncertain parameters. In the absence of time-series data, those parameters were estimated from cross-sectional data on individual weights from the National Health and Nutrition Examination Survey (NHANES). The NHANES 2005–2006 population of 5,971 subjects was categorized into 110 subpopulations based on different ethnicities (five ethnicities), genders (two genders), and ages (11 age groups). For each population group, two moments, average body weight and variance of body weight, were calculated as the moments to be matched, leading to a total of 220 moments to match.

On the other hand, the model was replicated (using subscripts in Vensim[1] software) for 5,971 instances that matched the demographic characteristics (age, gender, and ethnicity) of the NHANES sample of year 2006. Initial body weight and fat fraction for these individuals were drawn from distributions of another NHANES sample of 1999–2000. Note that each round of NHANES uses a sample different from other rounds; thus, we cannot track the same individual over time, and the data is cross-sectional. The model was then simulated to grow this synthetic population from their initial age in year 2000 to their final age (consistent with the NHANES sample) in year 2006. Mean and variance of weight for different subpopulations in the simulated population were calculated for year 2006 and compared against the 220 moments coming from the data. The weighted sum of squared errors was calculated using weights of the reciprocal of variance in each moment, itself calculated using variance and kurtosis of different moments. This error was minimized by changing 17 uncertain parameters using the Vensim internal optimization engine. The estimated parameters provided the minimum error. As a result, the authors were able to estimate a dynamic model, including individual growth mechanisms, from cross-sectional data with individuals in different age groups.

While this application follows the basic ideas of the MSM, it has some differences from the canonical MSM procedure. First, in this application the number of moments (220) was larger than many typical applications, in which the number of moments and parameters to be estimated are of the same order of magnitude. Second, given the computational costs in this setting, each moment was only simulated once, whereas typically multiple simulations, using different noise seeds, provide the estimation for the moment before it is compared with data. Finally, confidence intervals were not reported in this application.

Historical Background

The MSM is an offspring of the method of moments. Here we provide a quick review of this method and its basics, and refer interested readers to common econometric handbooks for further details.

Method of Moments

As a classical estimation method in statistics, the method of moments (MM) is based on finding unknown parameters of a certain distribution by relating these parameters to the moments of the distribution and then using empirical moments (obtained from data) to back up the unknown distribution parameters. We explain this using a few examples.

Example 1: Normal Distribution

The most convenient (and a straightforward) example for MM estimation can be expressed using a case for the normal distribution. Suppose you have collected a large sample of independent and identically distributed (i.i.d) observations for an experiment (e.g., a sample of heights of individuals in a country, y_i). Let us assume that we are confident the true (or the best-fitting) functional form for the distribution is normal, $N(\mu, \sigma^2)$. However, we do not know the values for the mean (μ) and variance (σ^2) of this distribution to fully characterize it.

Estimating the mean is easy since we can rely on the law of large numbers (LLN), which suggests that the mean of a large sample of trials will converge to its true mean. So we can simply calculate $\mu_s = \dfrac{\sum_{i=1}^{n} y_i}{n}$ and use it as the best estimator of μ. Now we need to estimate the variance. Remember the formula for variance: $\sigma^2(X) = E(X^2) - (E(X))^2$. Here we know $E(X)$ from our estimation of mean. Moreover, we can calculate $E(X^2)$ using our sample $E(X^2) \approx \dfrac{\sum_{i=1}^{n} (y_i^2)}{n}$. Plugging these two variables back into the variance equation, we obtain $\sigma^2 = \dfrac{\sum_{i=1}^{n} (y_i^2)}{n} - \left(\dfrac{\sum_{i=1}^{n} y_i}{n} \right)^2$. Note that our ability to estimate the two parameters here is dependent on knowing the analytical formulas that specify the unknown parameters (μ, σ^2) as a function of quantities that can be directly measured.

Example 2: Binomial Model
Now suppose that our data points are drawn from a binomial distribution $B(n,p)$ in which p is the probability of success and n is the number of trials. Let us assume we have several observations from this distribution, but do not know the value of parameters p and n. We know that the mean and variance of a binomial distribution is given by $\mu = np$ and $\sigma^2 = np(1-p)$. Similar to the previous example, we can calculate the first and second moments of the data ($E(X)$ and $E(X^2)$). Using the formula for the variance, $\sigma^2(X) = E(X^2) - (E(X))^2$, we express the second moment of data using the parameters of the binomial model, $E(X^2) = np(1-p) + (np)^2$. Therefore, we can now use the first- and second-moment equations together to provide a system of two equations and two unknowns that can be solved to recover $(n$ and $p)$. Specifically, $p = 1 - \dfrac{E(X^2) - E(X)^2}{E(X)}$ and $n = \dfrac{E(X)}{p}$, and equations for $E(X)$ and $E(X^2)$ are given above.

These two examples provide some intuition concerning the merits and difficulties of MM techniques. While for certain probability distributions MM can be used to recover parameter values through analytical expressions, it faces two major challenges. First, we need to know the true functional form of the distribution of outcomes. Second, we should be able to express the parameters of the distribution in terms of the data moments, a task only feasible for a small set of probability distributions. For many distributions, we cannot find an analytical (closed-form) solution to relate moments to parameters. Realistic dynamic models usually do not have an analytical solution to relate the output of the model to its structural parameters. Therefore, the classical method of moments discussed above is not directly applicable to these models.

From Method of Moments to Method of Simulated Moments

Mcfadden (1989) was the first to propose using simulation to find moments as a function of model parameters, instead of trying to solve the moment conditions analytically. His paper was focused on discrete-response models (multinomial Probit); however, he provided theoretical foundations for more general models. Mcfadden (1989) works on the basis that an unbiased simulator can be used to generate a sample of moments given a set of parameters, that the simulation errors are independent across observations, and that the variance in estimates of moments will be normally distributed due to the law of large numbers operating across simulations of those moments. Lee and Ingram (1991) and Duffie and Singleton (1993) extended the framework and provided a rigorous treatment of the MSM estimators for time-series and panel-data cases and provided relevant statistics for making tests. Duffie and Singleton (1993) showed that the MSM estimator is, under regularity conditions, consistent and asymptotically normal.

Since then, the MSM has been widely used in various subfields of economics such as finance (both asset-pricing and corporate finance), macroeconomics, industrial organization (IO), international trade, and labor economics. Novales (2000) and Ruge-Murcia (2012) provide a useful review of MSM methods for estimating macroeconomic and dynamic stochastic general equilibrium (DSGE) models and show various examples step by step.

Basics of the Method of Simulated Moments

Let us go back to the body weight example discussed in the introduction. Suppose you have built a model that captures the dynamics of people's body weight as a function of their initial weight, eating and physical exercise habits, genetics, age, gender, and other fixed and time-varying characteristics.

People differ both in terms of their idiosyncratic characteristics (genetics, initial weight, etc.) and environmental factors (e.g., quality of food, cost to exercise, social eating habits, etc.). By changing initial conditions and model parameters, one will get different dynamic paths for an individual's weight as a function of her age. Suppose we have data on the weight of several children of age 10 (our initial value) as well as samples of children of ages 11 and 12, one and two years later. Further, assume that we are interested in estimating a structural parameter (e.g., average weight growth per year), which determines the weight path as a function of initial weight.

By fixing this (unknown) parameter to an initial value and simulating the model with all empirical values for the initial weight (age 10), we will generate different paths of weight-age for a simulated population the same size as the number of subjects in our data set. Now we can compare the distribution of model-predicted weight profiles at ages 11 and 12 against the empirical distributions. Specifically, we can compare the mean and variance of weight for a simulated population at ages 11 and 12 against the mean and variance at the same ages observed in the data. It is likely that our initial choice for the structural parameters leads to mean and variance weights different from those observed in the data. However, these simulated moments are a function of the parameter. By changing the structural parameter of the model, we will change both the mean and the variance of simulated weight values. We can therefore use an optimization method to search for the parameter value that minimizes the difference between model-generated mean and variance and their empirical values over all available moments (i.e., mean and variance at ages 11 and 12). This is the core idea behind the method of simulated moments: we simulate the moments of the model to find simulated counterparts for observed data, then change the structural parameters until the simulated moments match the observations as closely as possible.

Formal Definitions

The core of the MSM is to minimize the (weighted) difference between the empirical and simulated moments by changing the unknown parameters. In general, we can assume that our empirical data $\{y_i\}_t, i \in \{1,...,N\}, t \in \{1,...,T\}$ are observed for N different agents in T conditions or instances (e.g., T different times). To apply the MSM, first we need to calculate the empirical moments. For those empirical data in each population group of agents (e.g., profits of N firms each three years old, or body weights in a sample of 11-year-old children in the body weight example), different moment conditions (such as mean, variance, etc.) can be calculated. Those moment functions (sometimes called descriptive statistics) can be put together as different elements of a vector of empirical moments M_D. M_D is a vector $<p \times 1>$, where p is the number of moment functions and the pth element of M_D vector would be calculated using function m_p: $m_p(\{y\}_t)$.

For instance, let us say that mean is the first moment condition (average) for N agents at time t, hence the respective element in the vector of empirical moments M_D is $m_p(\{y\}_t) = \dfrac{1}{N}\sum_{i=1}^{N}\{y_i\}_t$. Step one in the recipe for MSM discusses how to choose moment functions. Since we only have access to a sample of data for estimating moments, the true moments of population from which the data sample is collected are approximated by empirical moments M_D.

Now we need to determine simulated moments $\tilde{M}_S$ to be matched against the M_D. Consider a fully specified model—that is, a model that can be simulated given a set of parameter values. Assume there are d unknown parameters (θ)we are interested in estimating; that is, θ is a $<d\times1>$ vector. The true functional form of the system's dynamics that leads to output $\{y_i\}_t = g(.)$ is approximated by the model's output $\{\hat{y}_i\}_t = \hat{g}(.)$. The output of the model is a function of known parameters vector C, unknown parameters vector θ (to be estimated) and random inputs ξ. Choosing different values for ξ will generate different values for $\hat{g}(.)$. We assume that the model is correctly specified so that $\hat{g}(.)$ is an unbiased estimator of the true model $g(.)$, such that:

$$E(\hat{g}(C,\theta,\xi)) = E(g(.)) . \tag{1}$$

This ensures that if we generate a large enough (K) sample of outputs using a true random stream of inputs ξ, the arithmetic average of the model output should generate a reasonable approximation of the real-world processes that generate the observations:

$$m_p(\hat{g}(.)) \approx \frac{1}{K}\sum_{n=1}^{K} m_p(\hat{g}(C,\theta,\xi^{(n)})) \tag{2}$$

$\tilde{M}_S$ is then the vector of simulated moments, elements of which are calculated using equation (2), in parallel with the calculations for the empirical moments (M_D). The averaging component in equation (2) is necessary to make sure that the moments we estimate from the model are least affected by sampling bias. Ruge-Murcia (2012) shows that relatively small K values are sufficient for accurate estimates; however, we need to use a much larger number of replications to estimate asymptotic properties of the MSM (e.g., to calculate confidence intervals; see step 2 in the recipe for MSM).

A necessary (but not sufficient) condition for being able to identify the model is $p \geq d$. Otherwise, we will have an unidentified model with free parameters and the unknowns will not be uniquely determined. Ideally, we need more moment functions than the unknown parameters ($p >> d$), which is called overidentification. Overidentification allows us not only to estimate the parameters but also to evaluate the model's overall goodness of fit. A test of overidentification is discussed in step 5 of the recipe for MSM.

The core of the MSM is to minimize the (weighted) difference between the empirical and simulated moments by changing the unknown parameters. The estimated parameter set is the value of parameters that minimizes this difference. Specifically, with vector of simulated moments $\tilde{M}_S$ consisting of $m_p(\hat{g}(.))$ elements and the $< p \times p >$ matrix W for weighting the moment conditions, unknown parameters are estimated as:

$$\hat{\theta} = \text{argmin}(\tilde{M}_S - M_D)'W(\tilde{M}_S - M_D). \tag{3}$$

Note that $\tilde{M}_S$ and M_D are both vectors $< p \times 1 >$. Calculation of W is discussed in step 2 of the recipe for MSM.

Recipe for MSM

Step 1: Choose the Moment Conditions

The first step is to determine which moment conditions to use. Usually, the first and second moments of a model's outcomes (mean and variance) are good candidates. Remember that the number of moment conditions (p) should be (equal to or) larger than the number of unknown parameters (d). Thus, depending on the number of parameters you should decide to use informative moment conditions. The most informative moments are the ones that: 1) are sensitive to at least one of the unknown parameters (i.e., if we do a sensitivity analysis on unknown parameters, the moment changes significantly when changing at least one of the parameters); and 2) have rather small variances, since the larger the variance of a moment (across multiple simulations), the more noisy, and thus less informative, it is for estimation. For more discussions on which moments to match, see Gallant and Tauchen (1996).

In addition to single-variable moments (e.g., mean and variance of one variable), one can also try cross-variable moments such as the correlation/covariance between two output variables, autocorrelation of a variable with itself, skewness of a variable, and so on. For instance, Nikolov and Whited (2009) use the MSM to quantify the magnitude of agency conflicts in corporate finance. They use 16 moments to identify the model, which includes first and second moments as well as first-order autoregression of firm's investment, profits, and cash holdings. Franke (2009) also uses the MSM to estimate an agent-based asset pricing model. He uses mean and variance of returns and several short, medium, and long lags of autocorrelations between returns. Moments that are asymptotically normally distributed (typically due to the central limit theorem, when a moment can be seen as a summation or average over a large number of statistics) are more useful due to their asymptotic properties, which allow us to calculate the confidence intervals directly. If moments are not normally distributed, then calculation of confidence intervals will require bootstrapping or other nonparametric methods.

Conditions for Identification

The right choice of moment conditions is the most crucial step in identifying the model and recovering model parameters. Identification of a model using the MSM requires that the model-generated moment conditions fit their empirical counterparts if and only if the structural parameters equal their true values. Otherwise, the model will generate spurious results. Furthermore, the sufficient condition for identification is a one-to-one mapping between the structural parameters and a subset of the moment restrictions of the same dimension. Because dynamic models often do not yield such a closed-form mapping, to help ensure an identified model, one should choose moments that are sensitive to variations in the structural parameters. One way to check this is to run one-dimensional sensitivity analyses on each parameter being estimated, and check if selected moment conditions vary substantially with changes in parameter values. If a moment condition does not vary much or if its response to changes in parameters is not smooth and monotonic, then using that moment may not be very informative and may even cause the optimization engine to stop in a local optimum or never converge to an optimal solution.

Figure 2.1a shows examples of informative and noninformative moment conditions. The moment specified by the solid line moves smoothly as the unknown parameter changes and has a unique, well-defined value for each parameter value. Therefore, minimizing the distance between this function and the empirical moment will generate a unique parameter value. On the other hand, the moment represented by the dashed line is not informative. It is not very sensitive to changes in parameter value. We cannot even be sure that these small changes are due to a true response of the model to various parameter values or are the artifact of computational or sampling errors (though if the graph is smooth, these conjectures will be less valid). Moreover, the moment shows multiple extreme values and thus minimizing its distance from the empirical value will not identify a unique value.

Figure 2.1b shows an example of a noisy (oversensitive) moment. In this case, the value of the moment changes very abruptly as the

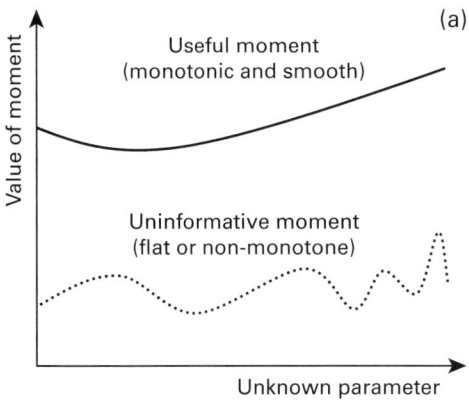

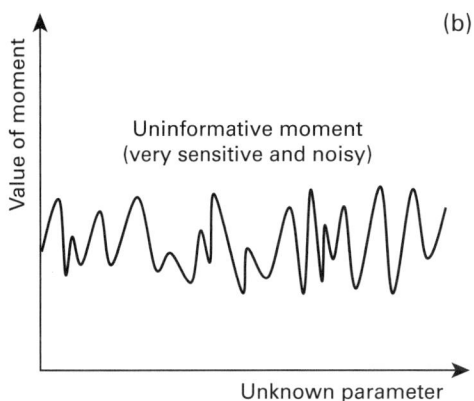

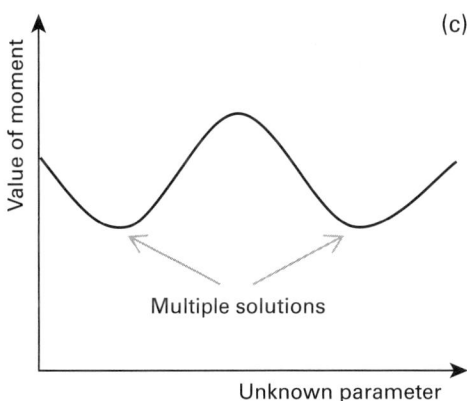

Figure 2.1
Relationship between moments and parameters. (a) Monotone versus nonmonotone moments. (b) Oversensitive nonmonotone moment. (c) Smooth but nonmonotone moment.

parameter value changes. From a theoretical perspective, it is harder to argue that a socioeconomic system will have such an oversensitive moment. Nonmonotone oversensitive moments also make it difficult to search for parameter values using numerical methods. Another issue with the choice of moments is the multiplicity of parameter values that make a good fit with the data. A sufficient condition for identifying parameters is the existence of a one-to-one relationship between parameter values and the moments, so if multiple parameter values fit the moment condition well, we may have a challenge with identifying the model (figure 2.1c).

Step 2: Weighting Matrix

Suppose you have d unknown parameters and exactly the same number of moment functions p. Under regularity conditions, we expect that these equations will yield the exact values of d unknowns. However, in the more common case when we have more moment functions than unknowns, it is very unlikely that all moment conditions will be precisely satisfied. This case is called an overidentification condition. In such a case, rather than solving a system of d unknowns and p moment functions, we need to conduct a minimization to get the simulated moments as close to empirical ones as possible. This raises the issue of weighting those moments; specifically, we need to use a symmetric, positive-definite weighting matrix W in equation (3).

A two-step procedure should be used to calculate the optimal weighting matrix. The right choice of weighting matrix ensures that we get the smallest asymptotic variance for our estimated values and thus the tightest confidence intervals possible. In the first step a simple matrix is usually used as the initial value of the weighting matrix (e.g., $W = I$, the identity matrix). Using an identity matrix for weighting provides a consistent estimate but is not optimal. Not being optimal means that this weight will not generate the estimation results with the lowest possible variance. While popular in econometric applications, using an identity matrix may cause problems if the scales of different moments differ significantly. An identity matrix implies the application of the same error weight to all moments, which may unrealistically increase the importance of some moments over others. For instance, if one moment has uncertainty in the order of thousands and the other in the order of fractions, the first moment will dominate the optimization and the second moment will be practically ignored, which is not a good outcome. One way to overcome this problem is to use a scaling factor in the first step. We have found that weighting the error terms for each moment by the reciprocal of corresponding empirical moments—that is, using percentage error of moments instead of absolute error—provides a good alternative first step moment. To get percentage error, diagonal elements of W should be $1/(M_D)^2$ and the other elements should be zero.

Once the model is estimated using this initial W (first step estimates), the estimation can be repeated for a second step using a more efficient weighting matrix that is calculated using the estimated model from step 1. To estimate the optimal weighting matrix for the second step, we first need to estimate the variance-covariance matrix of simulated moments, $\hat{S}$:

$$\hat{S} = \frac{1}{L_1} \sum_{l_1=1}^{L_1} \left[m^{l_1}(\hat{g}(C,\hat{\theta},\xi_1^{l_1})) - \frac{1}{L_2} \sum_{l_2=1}^{L_2} m^{l_2}(\hat{g}(C,\hat{\theta},\xi_2^{l_2})) \right].$$

$$\left[m^{l_1}(\hat{g}(C,\hat{\theta},\xi_1^{l_1})) - \frac{1}{L_2} \sum_{l_2=1}^{L_2} m^{l_2}(\hat{g}(C,\hat{\theta},\xi_2^{l_2})) \right]' \tag{4}$$

Note that in equation (4) the function m^l includes the p dimensional output vector of simulated moments for replication l. To estimate $\hat{S}$ in practice, we do the first-round estimation using equation (3) in which W can be a diagonal matrix with elements $1/(M_D)^2$ (or alternative weighting options). Then a large number of replications are generated; the moments of interest are calculated for each replication, and the variance-covariance matrix of these moments are calculated using equation (4). Note that for a good estimation of $\hat{S}$, we need a very large number of replications for L_1 and L_2 (practically, a number over a thousand is often large enough; Adda and Cooper 2003; Gourieroux and Monfort 1996). This large number of replications is used only once and does not significantly increase the computational costs of the method. The optimal weighting matrix W^* will be the inverse of the variance-covariance matrix $\hat{S}$, $W^* = \hat{S}^{-1}$. By choosing this W^*, a new round of optimization using equation (3) can take place to achieve a better estimate for $\hat{\theta}$.

The intuition behind this choice of weighting matrix is that the more noisy and uncertain a moment, the less weight we want to put on matching it. If an estimated moment is very uncertain (i.e., sensitive to the random inputs we do not control), its value in the empirical sample is likely to be far from the true population level value (i.e., if we had a very large sample). The weight of this moment should therefore be smaller in the optimization problem, as reflected in the larger corresponding element in the diagonal of matrix $\hat{S}$. On the other hand, those moments that are more robust against the choice of the sample will show smaller dispersion and will have small corresponding $\hat{S}$ elements and (thus) larger W^* elements. Essentially, using the inverse of variance-covariance $\hat{S}$ as W^* helps us give more weight to more robust moments and reduce the importance of those that change a lot from one simulation to another. Ruge-Murcia (2012) shows that the efficiency gained with the use of optimal W^* increases with the nonlinearity of the model. In fact, if you find that parameter estimates from step 2 are significantly different from those coming from step 1, it is recommended that you continue with additional iterations (three or more steps) until the estimated parameters converge.

For example, suppose that our data is for weights of 50 individuals ($N = 50$) and we want to use two moments, mean and variance. Running the model for 50 similarly parameterized individuals gives us 50 simulated observations for weight. We call this one round of simulation. Using these 50 model-generated data points, we can calculate the moments of interest: that is, the mean and the variance of simulated weights. Saving these moments and repeating the previous step, we can get a new set of model-generated moments (simulated moments). Notice that the only factor that changes between rounds of simulation is the random input ξ fed into the model; the structure of the model embedded in $\hat{g}(.)$, the initial conditions, and the parameters C and $\hat{\theta}$ remain unchanged. After repeating the simulation step for, for example, 2,000 rounds ($L_1 = L_2 = 1,000$), we have generated 2,000 different observations for our moments, the mean and the variance of weight. We then calculate $\hat{S}$, the variance-covariance matrix of the estimated moments, which includes elements such as variance of mean, variance of variance, and covariance between mean and variance (a 2x2 matrix). Finally, the inverse of $\hat{S}$ gives us the optimal weighting matrix W^*.

Step 3: Initial Value and Simulation
For nonequilibrium models, we want to start the simulations from initial stock/state values as close to the empirical sample as possible. This reduces the error in simulated moments (compared to the empirical ones) that is due to differences in the initialization of the model from the real cases. In such models, we can only use the data points at times that correspond to the empirical sample. For example, we can only pick

the data for 10-year-olds in our simulation model to compare with 10-year-olds in the empirical sample. For steady-state models, where all of the dynamics are in the steady state, we can be more efficient in the use of data generated by simulation. Essentially, rather than running the model N different times, we could use data from N subsequent time points coming from a single simulation, because the model is in a stochastic steady state and the differences across different points in time represent the steady-state distribution of the outcomes. If this process is pursued, we should discard early observations (in the time-series sense), since these observations are sensitive to the initial value of the model and are not in the steady state. If we need n observations over time, it is recommended to generate a vector of $2n$ observations over time and discard the first n observations. Note that these considerations are only relevant if the empirical sample is from a system that can be assumed to be in a steady state.

Step 4: Optimization Routine and Iteration
This is the most computationally intense step of the MSM procedure. We need to minimize the weighted distance of model-generated moments from empirical moments. More formally:

$$\theta^* = \operatorname{argmin} F = (\tilde{M}_S - M_D)' W^* (\tilde{M}_S - M_D) \tag{5}$$

We need to use numerical optimization routines to find the minimum of the total error function. A smart choice of initial values for parameters may significantly facilitate quicker convergence of the optimization routine. Any numerical optimization method requires a tolerance rule to stop. This is given as the error tolerance for the objective function $\|F^i - F^{i-1}\| \varepsilon_F$ as well as for the parameters $\|\hat{\theta}^i - \hat{\theta}^{i-1}\| \varepsilon_\theta$.

Similar to any nonlinear optimization routine, the MSM estimator may fall into the trap of a local optimum. Moreover, if some of the moment conditions are not very informative, they will have low sensitivity to parameter values and the problem may face a flat value function, which makes it very difficult to progress and converge. The MSM uses numerical methods to find the minimum of the objective function. Therefore, the results might be sensitive to initial values, the precision of the search algorithm (the level of error tolerance), and the quality of the algorithm to distinguish local and global extreme points. We recommend rerunning the optimization using distant initial values to check whether the results are sensitive to the choice of initial values for parameters.

Another important implementation concern is consistency in random number streams. To avoid introducing sampling error into rounds of simulation-optimization, we should work with the same sample of random numbers in each simulation-optimization step. This ensures that changes in results are due to changes in structural parameters and not the random sample.[2]

Step 5: Interpretation and Making Inferences
The results we get from the previous steps are consistent point estimates of true parameter values. Confidence intervals of estimated values allow us to assess the accuracy of those estimates and the range of potential errors.

Variance-Covariance Matrix In order to estimate the confidence interval of the estimated parameters, we need their distribution form and the parameters of those distributions. The simulated moments estimator is (asymptotically) normal when the original moments are (asymptotically) normal. This can be achieved through selecting moments that are calculated by summing some statistic over the N instances of agents (e.g., mean, variance, autocorrelation) and following the central limit theorem (normality follows as $N \rightarrow \infty$). In practice, much smaller samples (e.g., with $N>30$) lead to results that are adequately close to normal. In this case, the asymptotic distribution of estimated parameters (θ^*) is normally distributed and given by:

$$(\theta^* - \theta) : \text{Normal}(0, Q). \tag{6}$$

The asymptotic variance-covariance matrix of parameters (Q) is given by the following formula:

$$Q = \left(1 + \frac{1}{K}\right)[\hat{D}'\hat{S}^{-1}\hat{D}]^{-1}. \tag{7}$$

$\hat{D}$ measures the sensitivity of the moments to small changes of unknown parameter values, $\hat{D} = \mathsf{E}_0\left(\dfrac{\partial m}{\partial \theta}\right)$, and can be calculated numerically by changing different model parameters with very small increments ($\partial \theta$) and measuring the impact on the moments (∂m). This procedure should be repeated multiple times and $\dfrac{\partial m}{\partial \theta}$ should be averaged over that sample; however, only a small number of replications is typically enough to precisely estimate the $\frac{\partial m}{\partial \theta}$ (Gourieroux and Monfort 1996). In equation (7), $\hat{S}$ is the variance-covariance matrix of moments introduced in step 2 of the recipe for MSM for which large numbers of replications (L_1 & L_2) are required. Once the variance-covariance matrix for the estimated parameters (Q) is found, it is easy to find confidence intervals at different levels of confidence: for example, $\theta_i^* \pm 1.96 q_{i,i}$ for the 95%[3] confidence interval for parameter θ^*_i, using $q_{i,i}$, the ith diagonal element of Q matrix. The variance-covariance matrix remains valid even if the moments are not asymptotically normal; in that case, calculation of confidence bounds using normality assumptions may include too much error, and bootstrap methods may be preferred.

Overidentification Test

When the model is overidentified ($p > d$), some of the moment conditions will be different from zero. This provides us with an opportunity to assess how well the estimated model matches the data. For a good model whose structure and parameter values are close to the true system, the value of these nonzero moment conditions should only be different from zero to the extent allowed by random error in the values of the empirical moments. If the difference between the empirical and simulated moments is larger than what is expected due only to randomness, then we can claim that the model has some systematic error (i.e., it does not match the true data-generating process very well). The J-test is used to quantify and assess the significance of this error. The J-statistic is given by:

$$J = \frac{K}{1+K}((\tilde{M}_m - M_D)'W^*(\tilde{M}_m - M_D)) \sim \chi^2_{p-d} . \tag{8}$$

This statistic is distributed with chi-square distribution with $p-d$ degrees of freedom under the null hypothesis that the true data-generating process (g(.)) is not different from the estimated model $\hat{g}(.)$.

Other Methodological Considerations

The MSM is seen by some as a "black box" estimation method. It does not require the detailed consideration of model structure and error terms called for by likelihood-based methods. Moreover, it provides a

large degree of freedom for choosing the moment conditions. By imposing different moment conditions, one may recover different structural
parameters, and as long as there are enough meaningful moment conditions, the researcher has the luxury of using an arbitrary subset of
moments. This is a risk if the researcher picks the moments that best
serve his biases. Providing graphs showing the sensitivity of the
moments to parameter values is one way to ensure that the moments
are indeed informative.

Adding an additional parameter to be estimated is costly, in terms
of both computation time and identification strategy. Thus, we suggest
trying to estimate as many parameters as possible from other sources
(e.g., review of literature, regressions, etc.) and leaving the minimum
number of parameters for the MSM technique. Finally, following the
reporting guidelines for simulation-based research (Rahmandad and
Sterman 2012) is strongly recommended to ensure transparency and
replicability of the results using the MSM.

An Applied Example

In this section, we provide a simplified example of using the MSM to
estimate a dynamic model.[4] Assume we have observations from a
sample of $I=200$ firms. Our modeling project has allowed us to build
a simple dynamic model of these firms. The firms have similar structures and parameters, but are exposed to different random shocks.
Firms require resources (R) to produce outputs (O). The relationship
between resources and outputs is given by a concave production function, $f(R)$. Concavity ensures that the rate of production grows more
slowly than the rate of resource accumulation. Thus, no firm grows
forever, and there is a steady-state point for each firm to stop growing.

The existing resources depreciate at a proportional rate (depreciation
rate, D). Firms devote a fraction (β) of their output to investment
(I), which increases their resources. Moreover, resource accumulation
and depreciation are also subject to random shocks (ξ). Random
shocks capture events such as sudden technological shocks, failures in
adopting new technology inside the firm, shifts in customer tastes, and
inherent risks of the production technology, among others. Random
shocks are pink noise distributed normally, $\xi \sim N\left(0, \sigma_\xi^2\right)$, and with correlation time ρ_ξ as a half-year. Figure 2.2 shows this simple model of
firms that go through a process of expansion and stabilization.

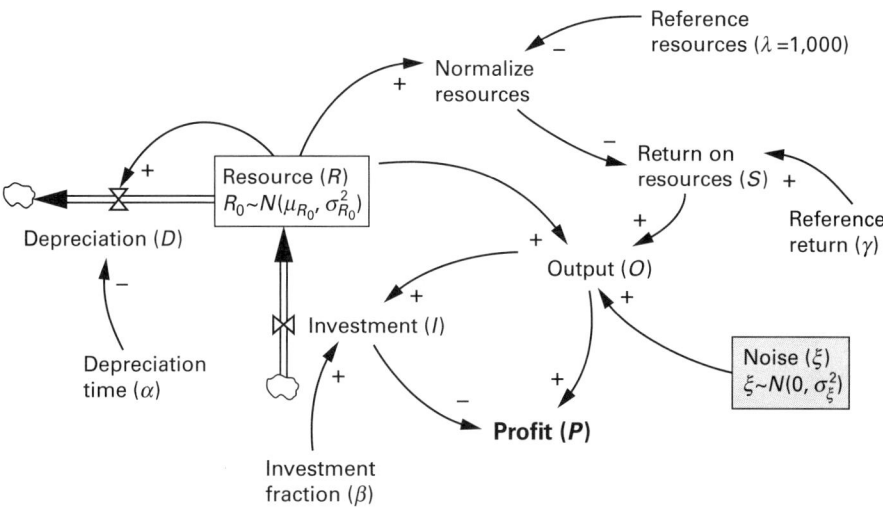

Figure 2.2
Dynamic model of firms (unknown parameters are gray). Vensim file is available in the online appendix.

Let us assume that we have data on a sample of 200 firms, all starting from $R_0 \sim N\left(\mu_{R_0}, \sigma_{R_0}^2\right)$ initial resources and growing in this market. Data include annual profit values for all the firms in this market; however, the specific firms are not identified, so we cannot connect the profit from one year to that in the next year. We impose this additional restriction to increase the complexity of the estimation problem: if time-series data were available for the full panel of 200 firms, one could use calibration methods in which the behavior of a simulated firm is matched against the data from each firm, either multiple times or simultaneously for all the firms. Without firm IDs, all we have is the cross-sectional data on profits for each year, but we cannot tell if a firm that made a lot of money one year did well during the next. Given this data, we want to estimate the unknown parameters of our model, identified in figure 2.2 and listed in table 2.1.

Equations that identify the model are summarized in table 2.2.

To ensure that we know the true model parameters, we generate our data by simulating the model once for a population of 200 firms for 10 years (a panel of 2,000 observations) using the parameters in table 2.3. We then treat this data as input into an MSM estimation procedure, in which we treat these six parameter values as unknown and estimate them.

Table 2.1
Unknown parameters in the model

Unknown Parameters (θ)	Notation	Unit
Depreciation time	α	Year
Investment fraction	β	Dmnl*
Noise—standard deviation	σ_ξ	Dmnl
Reference return	γ	1/Year
Resource—mean (for initial resource)	μ_{R_0}	M$
Resource—standard deviation (for initial resource)	σ_{R_0}	M$

* Dimensionless.

Table 2.2
Model's equations

Variables	Function (i^{th} firm)	Unit
Depreciation	$D_i = R_i / \alpha$	M$/Year
Investment	$I_i = \beta O_i$	M$/Year
Resources	$R_i = R_{0i} + (I_i - D_i)d_t$	M$
Normalized resources	$Z_i = R_i / \lambda$	Dmnl*
Return on resources	$S_i = \gamma(1 - (Z - 1)^2)$	1/Year
Output	$O_i = R_i S_i (1 + \xi_i)$	M$/Year
Profit	$P_i = O_i - I_i$	M$/Year

* Dimensionless.

Table 2.3
Parameters to generate synthetic data

Parameter	True Value
α	5
β	0.8
σ_ξ	0.1
γ	1.5
μ_{R_0}	200
σ_{R_0}	20

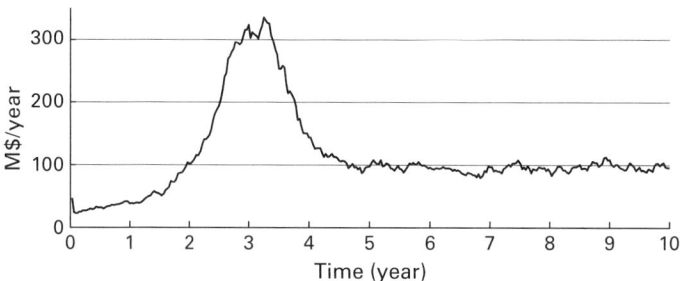

Figure 2.3
Profits of one of the 200 firms. Note that the actual data used for estimation does not include firm IDs; therefore, we cannot put together a similar time path for the growth of any single firm. The graph is provided only to offer an intuition about the type of behavior generated by this model.

Step 1: Choose the Moment Conditions
Following the recipe for MSM, we need to choose appropriate moments. Here we choose the first moment $\mu_t = \frac{1}{200}\sum_{i=1}^{200}(P_i)_t$ and function

$$\vartheta_t = \frac{\frac{1}{200}\sum_{i=1}^{200}(P_i^2)_t}{\mu_t}$$

of profits of 200 firms per year t, for 10 years.

Putting all 10 μ_t and 10 ϑ_t in one vector, M_D would be a vector of 20 elements. The reason that ϑ is selected instead of the second central moment (variance) is to scale both μ and ϑ in a reasonably similar range (compare the vertical axes in figure 2.4); however, it is not necessary to do so as long as a proper choice of W is selected—see the next step for more discussion. Note that the number of moments (p=20) is larger than the number of unknown parameters (d=6). Figure 2.4 shows selected moments of profits of 200 firms per year.

Step 2: Weighting Matrix
In the first round of estimation, the $<20\times20>$ weighting matrix is assumed diagonal with none-zero elements $W_{ii} = 1/(M_i)^2$. This W essentially minimizes the squared percentage error of moments, which is a good starting point but is not optimal. After running the first round of optimization and estimating $\hat{\theta}$, the W^* is approximated based on the inverse of the variance-covariance matrix of simulated moments, equation (4). Note that the number of replications we use for estimation is N=10, while the number we use in equation (4) is $L_1 = L_2 = 1,000$. This

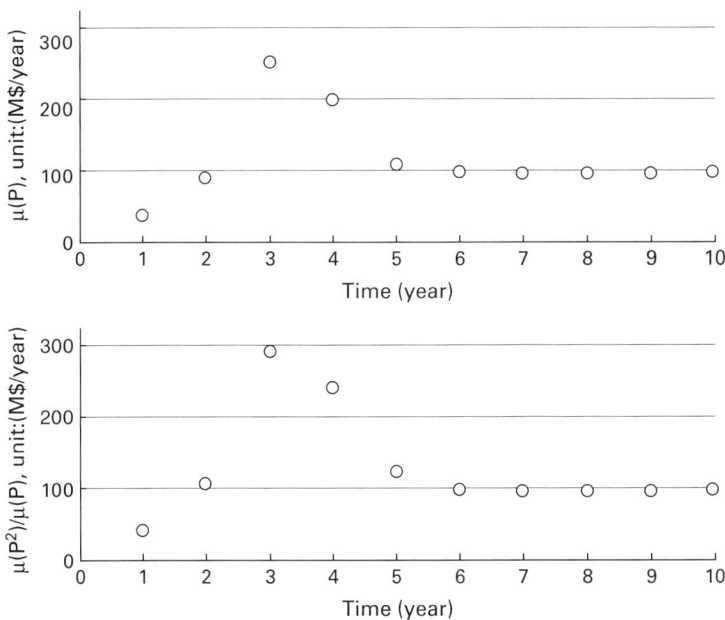

Figure 2.4
μ and $\vartheta = \mu(P^2)/\mu(P)$ of profits of 200 firms (each dot represents a moment in year i).

larger number is required for getting reliable estimates for the covariance matrix.

Step 3: Initial Value and Simulation
Initial values of the stock resource in figure 2.2 (R_0) are unknown. We estimate the distribution of R_0 by assuming that initial resources come from a normal distribution with two unknown parameters. Those parameters, μ_{R_0} and σ_{R_0}, are estimated as part of the estimation process. For any given values for these parameters, we draw 200 random numbers from the resulting distribution to initialize the simulated firms.

Step 4: Optimization and Iteration
Initial values of parameters to start the optimization could be arbitrary. Yet, if optimization restarts are limited due to computational costs, a reasonable set of initial values can help optimization not fall into local optima and will speed up convergence. We use the following initial

parameters, which are not too far off the true value and simplify the optimization step:

$$\theta_0 = \{\alpha, \beta, \sigma_\xi, \gamma, \mu_{R_0}, \sigma_{R_0}\} = \{2, 0.5, 0.5, 5, 200, 10\}.$$

We use *fmincon* solver in MATLAB to estimate the unknown parameters θ. The same random sample shocks are used in each optimization iteration to avoid sampling errors. Given W^* estimated in step 2, the optimization solver starts at θ_0 and, using numerical techniques, searches over the feasible space to find a minimum of the payoff, F, function of the MSM, presented in equation (5).

Step 5: Interpretation and Making Inferences

The MSM matches simulated moments $\tilde{M}_S$ against empirical moments M_D by changing the unknown parameters θ. Estimated parameters in the first round of optimization (based on $W = 1/(M_D)^2$) and the second round of optimization (based on W^*) are presented in table 2.4. Comparing estimated values with the true values in table 2.4, the first round of optimization estimates $\alpha, \sigma_\xi, \gamma$ and μ_{R_0} very close to the true values, but it fails to estimate σ_{R_0} and also does not provide a good estimate for σ_ξ. The second round of optimization provides relatively close estimates for all parameters. Figure 2.5 shows how simulated and empirical moments are matched at $\hat{\theta}^*$ after the second round.

To interpret the findings, we estimate the confidence intervals of estimated parameters $\hat{\theta}^*$. First we calculate the variance-covariance matrix Q of parameters using equation (7). Given $\hat{\theta}^*$ and variances of $\hat{\theta}^*$ (which is the diagonal of the Q matrix), estimated parameters and the true values of those parameters are summarized in table 2.5. Notice

Table 2.4
Estimated values and true values

Parameter	First optimization results	Second optimization results ($\hat{\theta}^*$)	True Value
α	5.013	5.035	5
β	0.800	0.799	0.8
σ_ξ	0.209	0.100	0.1
γ	1.478	1.527	1.5
μ_{R_0}	201.213	196.140	200
σ_{R_0}	0.0003	19.702	20

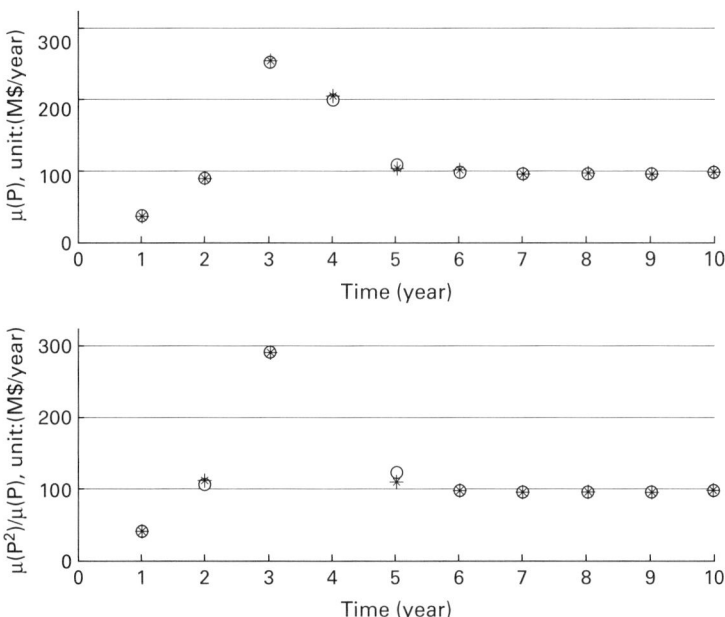

Figure 2.5
Empirical moments (○) and simulated moments (*).

that 95% confidence intervals for all six parameters include the true values, and the estimated values are also very close to the true values. The J-test is calculated to be 5.95, which is significantly smaller than the 99% cut-off value for a chi-square distribution with $p - d = 20 - 6 = 14$ degrees of freedom ($J = 5.95 < \chi^2_{16} = 29.14$), suggesting that the estimated model is not statistically distinguishable at a 99% confidence level from the true data-generating process (if $J > \chi^2_{p-d}$, the estimated model is systematically biased). This should not come as a surprise in this setting; after all, we have used the exact model that generated the "empirical" data to estimate the parameters in this exercise.

Finally, we test the reliability of the calculated confidence intervals by repeating the above procedure 100 times (using 100 different sets of data points, all generated by the same underlying dynamic model and only varying in their random noise streams). The results are summarized in table 2.6.

The values in table 2.6 represent the percentage of 100 replications where the true value of the parameter lies in the estimated confidence interval. For instance, 79 runs out of 100 estimated an 80% confidence

Table 2.5
Estimated values and true values

Parameter	Estimated Value (std*)	95% Confidence Interval	True Value
α	5.035 (0.064)	[4.953, 5.117]	5
β	0.799 (0.002)	[0.797, 0.801]	0.8
σ_ξ	0.100 (0.014)	[0.082, 0.118]	0.1
γ	1.527 (0.023)	[1.498, 1.556]	1.5
μ_{R_0}	196.140 (3.132)	[192.127, 200.153]	200
σ_{R_0}	19.702 (0.980)	[18.446, 20.959]	20

std*: Standard deviation.

Table 2.6
Percentage of estimated 80% and 95% confidence intervals including the true value over 100 estimations

Parameter	80% Confidence Interval	95% Confidence Interval
α	77%	90%
β	79%	92%
σ_ξ	84%	90%
γ	79%	89%
μ_{R_0}	81%	91%
σ_{R_0}	77%	90%

interval for β that included its true value. As shown in table 2.6, the estimated 80% and 95% confidence intervals include approximately the same percentages of the true values over the 100 replications (i.e., 80% and 95%). The results are more precise for the 80% confidence intervals, compared to the 95% intervals. This indicates that the estimated parameters slightly diverge from the normal distribution and have somewhat fatter tails, which lead to tighter-than-real confidence intervals when using the normality assumption.

Exercise

In this section, we provide a simple exercise for practicing the use of the MSM in a dynamic setting. Assume we have data on a population of 1,000 individuals who are starting in a study and their initial body weights are normally distributed, with mean of 80 kg and standard deviation of 5 kg. The individuals are then monitored as they gain weight over time and the distribution of their weight changes. Both mean and standard deviation of the distribution increase over time, and the distribution may no longer be normal. In general, body weight changes when energy intake and energy expenditure are not balanced. When energy intake (kcal/year) from food and drink exceeds energy expenditure (kcal/year), body weight (kg) increases; otherwise, people lose weight (or remain at their current weight if energy intake and energy expenditure are equal). Energy expenditure depends on physical activity, basal metabolic rate (the energy consumed at complete rest), and digestion, but for simplicity in this exercise, we assume that energy expenditure is a fixed number (100,000 kcal/year) + weight (kg) × energy cost of weight (10,000 kcal/kg/year). We may not have direct measures of energy intake, but can approximate it by multiplying the energy expenditure by a dimensionless fluctuation factor. We can then quantify the fluctuation factor as a function of different inputs, such as random shocks, starvation or overfeeding feedback, and general trends in energy intake. Let us use the following assumptions in formulating the fluctuations of energy intake:

- For biological robustness, assume that energy intake never goes below the fixed energy expenditure (100,000 kcal/year) that we assume is the minimum for survival.
- Fluctuation factor changes around 1, and have the following components added to value 1.

  - A random shock factor that is a pink noise for each individual distributed normally with mean and standard deviation of 0 and 0.007, respectively, and with correlation time of 5 (years). Essentially, this noise factor recognizes that people eat more/less than their energy expenditure by chance, and there is some autocorrelation in how they change their eating behavior.
  - Energy intake balance (energy intake/energy expenditure), after a short first-order delay (delay time=0.033 year), generates either starvation or overfeeding feedback, such that:
    - if (delayed energy intake balance − 1) ≤ 1, there is an overfeeding feedback, so the fluctuation rate is changed by adding (delayed energy intake balance − 1) × overfeeding.
    - if (delayed energy intake balance − 1) > 1, there is a starvation feedback, so the fluctuation rate is changed by adding (delayed energy intake balance − 1) × starvation.
  - Overfeeding and starvation are two unknown parameters.

- Extra energy intake trend is assumed to be constant and the third unknown parameter.

We have generated data using the true model parameters by simulating the model once for a population of 1,000 individuals for 20 years, and we have saved only weights at years 1, 5, 10, 15, and 20 (you are provided with a panel of 5,000 observations in *MSM_Exercise_Data.xls* file). Note that the specific individuals are not identified, so you cannot connect the weights between the given years.

Based on the information above, the true model (which is discussed above and is also available in *MSM_Exercise_Model.mdl* Vensim file), and the data in *MSM_Exercise_Data.xls* file, leverage the MSM estimation procedure to find the three unknown parameters (overfeeding, starvation, and extra energy intake trend are unknown). See the Vensim model for more details of the model equations. Estimate the unknown parameters and report confidence intervals for estimated parameters along with J-test results for the validity of the model. In the online appendix, MATLAB codes and a sample solution are provided for this exercise.

Conclusion

Over the last three decades, the research in system dynamics has largely focused on diverse applications of the original toolbox, with limited methodological expansions in the parameter-estimation domain. In this chapter, we offer an introduction to the MSM for application to dynamic modeling problems. The basic idea of this method is to define appropriate moments of data and, by changing uncertain parameters, minimize the difference between those moments and their simulated counterparts coming from the model.

Given that most dynamic models do not follow a fixed structural form (e.g., linearity), estimation procedures such as the MSM that are independent of model structure are most beneficial. Moreover, independence and distributional assumptions on error terms for dynamic models are not always easy to justify, so the MSM, which has fewer such assumptions, is preferred. The MSM is especially useful when error terms do not follow any well-established distribution. It could also be a good tool when models include stochastic processes that drive the model, and their impact on the model behavior is reflected in the data against which the model is to be calibrated; for example, when we are trying to match the variance observed across multiple units.

The MSM is also applicable to diverse data types, including both time-series and cross-sectional data. When likelihood functions cannot be calculated or are too expensive to calculate numerically, the MSM may be the only viable choice for estimating dynamic models with cross-sectional data, since it allows us to extract the information about the historical trajectories of units hidden in their cross-sectional distributions.

Notes

1. Vensim software, ©2014 Ventana Systems, Inc.

2. The concept of "seed" in software packages such as MATLAB and Vensim controls random number generation. Using the same seeds across optimization iterations helps draw the same random samples.

3. The 2.5% and 97.5% critical values for standardized normal distribution are 1.96 standard deviation away from the mean.

4. The analysis is conducted in MATLAB, but could be also fully implemented within Vensim. The MATLAB implementation simplifies the optimization and estimation steps, while the Vensim implementation is easier for model building and presentation. Overall, we recommend MATLAB or other general computational packages for similar applications due to the efficiencies gained in the optimization step, which is the computational bottleneck for applying the MSM. Alternatively, one can use the Vensim or other simulation software's connectivity features to couple them with MATLAB or other general computational software, conducting the simulations in the customized environment while running the MSM steps in general software.

References

Adda, J., and R. W. Cooper. 2003. *Dynamic economics: Quantitative methods and applications.* Cambridge, MA: MIT Press.

Barlas, Y. 2006. Model validity and testing in system dynamics: Two specific tools. 24th International Conference of the System Dynamics Society, Nijmegen, The Netherlands.

Data.gov. 2013. United States Government. Accessed Nov. 1, 2013. http://www.data.gov.

Dogan, G. 2007. Bootstrapping for confidence interval estimation and hypothesis testing for parameters of system dynamics models. *System Dynamics Review* 23 (4): 415–436.

Duffie, D., and K. J. Singleton. 1993. Simulated moments estimation of Markov-models of asset prices. *Econometrica* 61 (4): 929–952.

Forrester, J. W. 1961. *Industrial dynamics.* Cambridge, MA: MIT Press.

Franke, R. 2009. Applying the method of simulated moments to estimate a small agent-based asset pricing model. *Journal of Empirical Finance* 16 (5): 804–815.

Gallant, A. R., and G. Tauchen. 1996. Which moments to match? *Econometric Theory* 12 (4): 657–681.

Gourieroux, C., and A. Monfort. 1996. *Simulation-based econometric methods, CORE lectures.* Oxford: Oxford University Press.

Greene, W. H. 2012. *Econometric analysis.* 7th ed. Boston: Prentice Hall.

Lee, B.-S., and B. F. Ingram. 1991. Simulation estimation of time-series models. *Journal of Econometrics* 47 (2–3): 197–205.

Lyneis, J. M., and A. L. Pugh. 1996. Automated vs. "hand" calibration of system dynamics models: An experiment with a sample project model. 14th International Conference of the System Dynamics Society, Cambridge, MA.

Mcfadden, D. 1989. A method of simulated moments for estimation of discrete response models without numerical-integration. *Econometrica* 57 (5): 995–1026.

Nikolov, B., and T. Whited. 2009. Agency conflicts and cash: Estimates from a structural model.

Novales, A. 2000. The role of simulation methods in macroeconomics. *Spanish Economic Review* 2 (3): 155–181.

Oliva, R. 2003. Model calibration as a testing strategy for system dynamics models. *European Journal of Operational Research* 151 (3): 552–568.

Peterson, D. W. 1975. Hypothesis, estimation, and validation of dynamic social models: Energy demand modeling. Ph.D., Massachusetts Institute of Technology.

Rahmandad, H., and N. Sabounchi. 2011. Building and estimating a dynamic model of weight gain and loss for individuals and populations. 29th International Conference of the System Dynamics Society, Washington, DC.

Rahmandad, H., and J. D. Sterman. 2012. Reporting guidelines for simulation-based research in social sciences. *System Dynamics Review* 28 (4): 396–411.

Repenning, N. P. 2003. Selling system dynamics to (other) social scientists. *System Dynamics Review* 19 (4): 303–327.

Ruge-Murcia, F. 2012. Estimating nonlinear DSGE models by the simulated method of moments: With an application to business cycles. *Journal of Economic Dynamics & Control* 36 (6): 914–938.

Sterman, J. D. 2006. Learning from evidence in a complex world. *American Journal of Public Health* 96 (3): 505–514.

Sterman, J. 2000. *Business dynamics: Systems thinking and modeling for a complex world.* Boston: McGraw-Hill/Irwin.

3 Simultaneous Linear Estimation Using Structural Equation Modeling

Peter S. Hovmand and Nishesh Chalise

Statistics seems to concentrate on the deviation of processes from the mean. The approach through continuous variables lays first emphasis on the causal structure that controls the mean and, when this is understood, adds randomness to determine the influence of uncertainty on the system.

(Forrester 1968, 414)

Structural equation modeling (SEM) is a multivariate statistical method for testing hypothesized causal relations between variables that focuses on how variables deviate from their means. That is, SEM is a method for the analysis of covariance, and hence, sometimes also called analysis of covariance (ANCOVA). What distinguishes SEM from other statistical techniques is its ability to (1) include both directly observed variables and unobserved or latent variables, (2) simultaneously estimate parameters and measurement error, (3) control for spurious correlations between variables and measurement error, and (4) include reciprocal or feedback relationships (Bollen 1989). It is this last feature—the inclusion of reciprocal or feedback relationships in a statistical model—that has arguably motivated the exploration and comparison of SEM and system dynamics (Levine and Lodwick 1992; Levine, Sell, and Rubin 1992; Hovmand 2003). Moreover, recent advances in using SEM to study dynamical systems (Bisconti, Bergeman, and Boker 2004; Boker and Nesselroade 2002; Boker and Wenger 2007) have raised expectations around the possibilities of using SEM as a method for estimating model parameters and testing models.

The overarching aim of this chapter is to provide a basic introduction to SEM and how SEM can be used in a productive way to support dynamic modeling, along with giving specific attention to the limitations of SEM from this perspective. In particular, SEM is a standard and effective tool for partial model testing of causal relationships involving intangible or latent variables. One can also readily use SEM to estimate parameters for structures involving feedback relationships involving simultaneous equations (e.g., market price as a function of supply and demand). In both cases, SEM excels at being able to include relationships between latent or unobserved variables, and measurement models that relate latent variables to observable indicators with measurement error. However, SEM might not be as helpful as one might initially expect when modeling nonlinear feedback systems, as is commonly done when working with systems of nonlinear differential equations. This is not to say there aren't ways to use SEM effectively or test dynamic models using SEM, but the approaches required are less attractive than other, more widely accepted approaches from control theory and system dynamics modeling (e.g., Kalman filtering).

Background

SEM is a general modeling approach that encompasses a number of methods, including path analysis, multiple regression, factor analysis, analysis of variance, and analysis of covariance. SEM as we know it today draws on a number of different developments across the history of statistics (Bollen 1989), originating in the basic idea that structural equations specifying the causal relationships between variables mathematically imply a specific pattern of variances and covariances.

For example, the expression $y = a x + b + e$, where x is a cause, y is the effect, a and b are the slope and interpretation of this relationship, e is an error term, and y, x, and e are all taken to be random variables, implies specific formulas for variances and covariances of x and y, written as $\text{VAR}(x)$, $\text{VAR}(y)$, and $\text{COV}(x,y)$, respectively. In this case,

$$\text{COV}(x, y) = \text{COV}(x, a x + b + e) = \text{COV}(x, a x) + \text{COV}(x, e) = a \, \text{VAR}(x)$$

and

$$\text{VAR}(y) = \text{VAR}(a x + b + e) = \text{VAR}(a x + e) = a^2 \, \text{VAR}(x) + \text{VAR}(e).$$

These equations therefore *imply* the following covariance matrix:

$$\begin{pmatrix} VAR(x) & aVAR(x) \\ aVAR(x) & a^2VAR(x) + VAR(e) \end{pmatrix}.$$

This expression has two unknown terms, a and $\text{VAR}(e)$, involving three equations that can be estimated by comparing the implied covariance from above with the observed sample variance; for example, by estimating a and VAR(e) such that:

$$\begin{pmatrix} VAR(x) & COV(x,y) \\ COV(x,y) & VAR(y) \end{pmatrix} - \begin{pmatrix} VAR(x) & aVAR(x) \\ aVAR(x) & a^2VAR(x)+VAR(e) \end{pmatrix} \approx 0 . \quad (1)$$

There are a variety of estimators that can be used for equation (1), including ordinary least squares, generalized least squares, and maximum likelihood. Ordinary least squares assesses the sums of squares of the differences between the observed and implied covariance matrices. Generalized least squares is similar, but scales the expression in terms of the observed moments. Finally, maximum likelihood estimates the difference in the log-likelihoods.

The essence of SEM is that one can use the implied covariance matrix, $\Sigma(S)$, for one or more structural equations involving a set of unknown parameters and compare it with the sample covariance matrix, Σ, to estimate the unknown parameters, S, under the hypothesis that $\Sigma(S)=\Sigma$ for a given model. Building on this idea has led to a number of extensions, including the estimation of feedback relationships, although it is important to realize that the notion of feedback in SEM differs somewhat from the usual use of the term "feedback" among dynamic modelers, which will be discussed later in this chapter.[1]

SEM allows for the inclusion of unobserved or latent variables. That is, instead of just estimating the strength of association between observed variables, SEM allows one to estimate the strength of associations between variables that are not directly observed. For example, many psychological and sociological variables of interest, including concepts such as trust, frustration, stress, political affiliation, social capital, and discrimination, are considered latent variables. SEM therefore provides the social scientist with more advanced techniques for identifying constructs and developing more formal and testable theories of social and psychological phenomena. It also addresses a historical limitation and general skepticism by scholars oriented more toward the natural sciences about the appropriateness of applying statistical methods to the study of unobserved social and psychological variables and their relationships.

Most importantly, SEM allows social scientists to address issues related with multiple regression techniques such as association between error terms, spurious correlations between variables and error terms, and reciprocal relationships. Most of the multivariate techniques offered in graduate courses force research questions and designs to explicitly exclude consideration of colinearity between the variables, despite the fact that most social systems have many variables interacting that give rise to precisely this situation. In contrast to physical systems where variables can be neatly isolated in experiments, social systems are characterized by "everything connected to everything," which presents a host of methodological issues for social scientists (Meehl 1990) that are familiar to and addressed by system dynamicists.

The introduction of the LISREL notation for defining linear structural relations by Karl Jöreskog in the 1970s and the basis for the LISREL software made SEM widely available to social and behavioral scientists for testing more sophisticated theories. SEM provided an efficient approach to conduct both exploratory theory development and confirmatory theory testing. It allowed scientists to routinely consider both the measurement model and underlying causal relationships simultaneously. By the mid- to late 1990s, SEM had become a common advanced statistical method taught at the graduate level in social sciences with numerous software implementations (e.g., LISREL, AMOS, EQS, MPlus, PROC CALIS in SAS, SEM package in R). For most social scientists, SEM was also the first introduction to formal mathematical modeling of social theories (Bollen 1989).

The next sections will focus first on introducing SEM through a series of examples that build up to a feedback relationship, then discuss the limitations of trying to apply SEM to the general case of SD modeling, and lastly suggest ways that these limitations can be addressed by other methods and potential extensions of SEM. To do this, the examples use simulated data generated in Vensim that has been exported to R, where the SEM package in R is used to specify and test the structural equation models. This will provide an opportunity to introduce a variety of structural equation models, notation for path diagrams, the output typical of SEM software packages, and, lastly, provide a generic template for readers exploring the correspondence between SD models in Vensim and structural equation models. All of the examples and data are available as part of the electronic supplement for this handbook.

Steps to Specifying, Estimating, and Testing Structural Equation Models

The general procedure to applying SEM begins with a correlation or covariance matrix from some empirical data. It is important to realize that generally the only information needed about the data is the aggregate covariance matrix along with the number of observations; that is, the individual data are often not needed. This means that one can, in principle, begin an SEM analysis with published summaries of data where the covariance matrices are included in the publication. This is important, because it essentially makes the process of developing and testing structural equation models transparent and replicable. For example, it is possible for two or more researchers to be debating a given theory and testing their theories using various structural equation models against the same data.

After acquiring a correlation or covariance matrix, the next step is to specify a structural equation model. This can be done in a variety of ways depending on the software package. Some implementations, such as AMOS and LISREL, allow one to specify the model using graphical conventions, while other implementations such as the R SEM package require models to be specified in the reticular action model (RAM) path notation. In addition to defining the variables, the SEM specification also defines the error terms and which parameters are "fixed" and which are "free." Free parameters are estimated by attempting to fit the equations describing the implied covariance matrix to the observed covariance matrix, whereas fixed parameters are not estimated but are assigned a value by the modeler. Generally fixed parameters are assigned a value of zero, which means there is no relationship between the two variables unless the modeler explicitly chooses a nonzero value for the fixed parameter. One of the advantages of SEM over other statistical methods is the ability to set specific patterns of fixed and free parameters. For example, it is possible to set all the error terms to the same value and thereby build in the assumption that the variances of all measurement errors are the same.

Once the model has been specified, SEM packages iteratively try to estimate the parameters that maximize the fit. This can be done using any number of algorithms, including two-stage ordinary least squares (2-SLS) and full-information maximum likelihood. The specific approach depends on the package, normality of distributions, and sample size. If the algorithm converges, the results will be a set of estimates for the parameters and error terms along with fit statistics.

Convergence is generally not guaranteed, and typically depends on the number of observed variables in relation to the number of latent variables. Failures to converge are often due to models being underdetermined and overdetermined. Underdetermination occurs when there are more parameters to be estimated than data available in the covariance matrix. Overdetermination occurs when there are more instrumental variables than predictors, and the model is said to be overidentified. This is usually a consequence of sampling error and can be addressed by using 2-SLS (Fox 1997).

In SEM, models are primarily evaluated by (1) their underlying logic, and (2) fit statistics. The former requires a sound theory and is evaluated in a way not unlike SD models. The latter involves employing one or more of the existing fit indices, comparing the observed against the implied covariance matrix. Models that are statistically significantly different from the observed covariance matrix are rejected.

Examples

For the first example, consider a structural equation model of academic achievement where (1) hours studied per week is thought to influence weekly homework grade and weekly quiz grade, and (2) attendance influences the weekly quiz grade. All four variables are directly observable, so we use the graphical conventions most common with path diagrams, where boxes represent observed variables. The exogenous variables are hours studied per week, x_1, and attendance, x_2. The endogenous variables are homework grade, y_1, and quiz grade, y_2. Figure 3.1 shows the hypothesized causal relationships, where y_1 (homework grade) is influenced by x_1 (hours studied per week) and an error term δ_1, and y_2 (quiz grade) is influenced by x_1 (hours studied per week) and x_2 (attendance) and an error term δ_2.

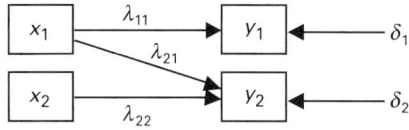

Figure 3.1
Conventional path diagram of a structural equation model with four observed variables of weekly academic achievement relating hours studied per week (x_1), attendance (x_2), homework grade (y_1), and quiz grade (y_2).

The parameters λ_{11}, λ_{21}, and λ_{22} determine the strength of association between the causes and effects, where $y_1 = \lambda_{11}x_1 + \delta_1$ and $y_2 = \lambda_{21}x_1 + \lambda_{22}x_2 + \delta_{12}$. For this system, the covariance matrix is

$$\begin{pmatrix} VAR(x_1) & & & \\ COV(x_1,x_2) & VAR(x_2) & & \\ COV(x_1,y_1) & COV(x_2,y_1) & VAR(y_1) & \\ COV(x_1,y_2) & COV(x_2,y_2) & COV(y_1,y_2) & VAR(y_2) \end{pmatrix}.$$

The equations can then be used to derive the elements of the implied covariance matrix:

$$VAR(y_1) = VAR(\lambda_{11}x_1 + \delta_1) = \lambda_{11}^2 VAR(x_1),$$

$$VAR(y_2) = VAR(\lambda_{21}x_1 + \lambda_{22}x_2 + \delta_2) = \lambda_{21}^2 VAR(x_1) + \lambda_{22}^2 VAR(x_2),$$

$$COV(x_1,y_1) = \lambda_{11}VAR(x_1),$$

$$COV(x_2,y_1) = \lambda_{11}COV(x_1,x_2),$$

$$COV(x_1,y_2) = \lambda_{21}VAR(x_1) + \lambda_{22}COV(x_1,x_2),$$

$$COV(y_1,y_2) = \lambda_{11}\lambda_{21}VAR(x_1) + \lambda_{11}\lambda_{22}COV(x_1,x_2), \text{ and}$$

$$COV(x_2,y_2) = \lambda_{21}COV(x_1,x_2) + \lambda_{22}VAR(x_1).$$

With a slight modification to the conventions of path diagrams, the structural equation model can be represented in Vensim software by (1) adding causal paths from the parameters to the variable, and (2) using boxes to represent observed variables (see figure 3.2). It is important to note that the structural model we have just specified in figure 3.1 is not dynamic. That is, the values of the outputs (y's) are strictly a function of the value of the inputs (x's). To generate a synthetic data set for SEM analysis, we simply iterate the calculations once for each case. The easiest way to do this in Vensim is to let the model step through 100 cases. Note that in doing so, *Time* in Vensim no longer represents time, but the iteration number. This is merely for the convenience of using Vensim to draw diagrams with notation that correspond to structural equation models and generating synthetic cross-sectional data corresponding to the equations generating the data.

Table 3.1 shows the Vensim equations, which can be used to generate a synthetic data set for SEM analysis. The Vensim model shown in figure 3.2 is run from 0 to 100 with a time step of 1, and therefore

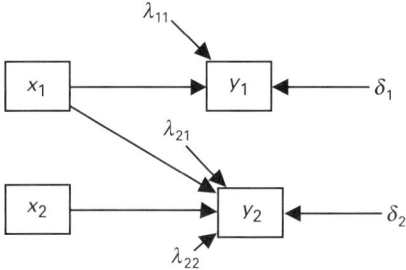

Figure 3.2
Vensim path diagram for a structural equation model with four observed variables.

Table 3.1
Vensim equations for structural equation model in figure 3.2.

```
x1=RANDOM NORMAL(-100, 100, 0, 20, 0)
x2=RANDOM NORMAL(-100, 100, 0, 20, 0)
y1=λ11*x1 + RANDOM NORMAL(-100,100,0, δ1, 0)
y2=λ21*x1 + λ22*x2 + RANDOM NORMAL(-100,100,0, δ2, 0)
δ1=5
δ2=3
λ11=5
λ21=3
λ22=8
```

generates 100 cases for analysis. This data set can be exported to a tab-delimited file, then imported into the R software package.

The R code for testing the model is shown in table 3.2. The first step is to specify the model using RAM notation (see appendix to Fox 1997). RAM notation has three columns. The first column specifies the relationship between two variables. The second specifies the name of the corresponding parameter. If the relationship is to be fixed at a nonzero value instead of being estimated, then the name in the second column is NA or missing. The third column specifies the start value for the estimation. If the parameter is fixed,[2] then a value must be specified. For example, in the expression "x1->y1, lam11, NA" from table 3.2, x1->y1 means X influences Y, lam11 names the parameter $\lambda 11$ to be estimated as a free parameter, and NA means that no initial value has been specified. Meanwhile, in the expression "x1 <-> x1, NA, 1" from table 3.2, X<->Y is the unobserved covariance between X and Y, NA means that parameter is fixed, and 1 gives the value at which this covariance is fixed.

The output from running the R code is shown in table 3.3, which provides a set of statistics for assessing the fit of the model. Table 3.4 provides an overview of the different fit statistics. A more thorough review of fit indices used in SEM is beyond the scope of this chapter, and readers are encouraged to see standard texts for a more complete treatment (Bollen 1989) where the pros and cons of different measures are covered in a manner comparable to the way that model fit is discussed in system dynamics (Sterman 2000). The general idea behind all of these measures, however, is to assess the fit by comparing the implied or fitted covariance matrix against the observed or empirical covariance matrix. Hence, the null hypothesis is that the implied and observed covariances are the same, and a statistically significant result is a rejection of the structural equation model.

Table 3.2
Sample R code for structural equation model in figure 3.2 with explanations

R Code	Explanation

sem1 <- specify.model()
  x1->y1, lam11, NA
  x1->y2, lam21, NA
  x2->y2, lam22, NA
  x1 <-> x1, NA, 1
  x2 <-> x2, NA, 1
  y1 <-> y1, del1, NA
  y2 <-> y2, del2, NA
	This specifies the model, sem1, defining the variables and parameters along with their relationships, whether or not they are fixed or variable, and the initial value used during the estimation procedure.

sem1.rslt <- sem(sem1,cov(df),n,colnames(df))	This calls the SEM function and estimates the model. The call takes the arguments of the SEM specification, the covariance matrix, and a list of observed variables in the covariance matrix.
summary(sem1.rslt)	This provides a summary of the results, including parameter estimates and fit statistics.

Table 3.3
Results from running the R code shown in table 3.1 estimating the parameters in the
implied covariance matrix for the structural equation model shown in figure 3.2.

```
Model Chisquare =      794563      Df =    5 Pr(>Chisq) = 0
 Chisquare (null model) =    14108      Df =   6
 Goodness-of-fit index =    0.0049393
 Adjusted goodness-of-fit index =    -0.99012
 RMSEA index =    12.612      90% CI: (NA, NA)
 Bentler-Bonnett NFI =    -55.322
 Tucker-Lewis NNFI =    -66.615
 Bentler CFI =    0
 SRMR =    0.78152
 BIC =    794529


 Normalized Residuals
  Min. 1st Qu.   Median     Mean 3rd Qu.        Max.
   178     2340     4330     4870     8320      9130


 Parameter Estimates
         Estimate Std Error z value   Pr(>|z|)
lam11    5.0055   0.0078931    634.165 0           y1 <--- x1
lam21    2.9971   0.0047423    631.986 0           y2 <--- x1
lam22    7.9952   0.0046853   1706.421 0           y2 <--- x2
del1    24.8823   1.1135413     22.345 0           y1 <--> y1
del2     8.9781   0.4017716     22.346 0           y2 <--> y2
Iterations =   0
```

While SEM can adequately handle situations such as the model
shown in figure 3.2, SEM excels at representing models involving
hypothesized causal relations between latent or unobserved variables
of the type shown in figure 3.3. Following the convention of path diagrams in SEM, observed or indicator variables are shown as boxes and
unobserved or latent variables are shown as circles.

Consider, then, an example in which one has hypothesized a causal
relationship where organizational culture influences job satisfaction.
Organizational culture (ξ_1) and job satisfaction (η_1) are unobserved or
latent variables. To measure them, we have indicators (e.g., responses
to items on a questionnaire). For example, we might have three questions on a 1 to 10 scale from strongly disagree to strongly agree, reflecting the exogenous variable organizational culture (x_1, x_2, and x_3), and

Table 3.4
Goodness of fit for SEM models (adapted from Schumacker and Lomax 1996)

Fit Statistic	Range	Significance
Goodness of fit index (GFI)	0 = no fit; 1= perfect fit	Value below .90 is poor.
Adjusted goodness if fit (AGFI)	0 = no fit; 1= perfect fit	Adjusted for degrees of freedom; value below .90 is poor.
Root mean square error of approximation (RMSEA) index		Model with values less than 0.05 is a good fit.
Bentler Bonnet or normed fit index (NFI)	0 = no fit; 1= perfect fit	Value below .90 is poor. No penalty for adding parameters.
Tucker Lewis or non-normed fit index (NNFI)	0 = no fit; 1= perfect fit	Value below .90 is poor. Penalty for adding parameters.
Bentler comparative fit index (CFI)	0 = no fit; 1= perfect fit	Value below .90 is poor.
Standarized root mean square residual (SRMR)		Model with values less than 0.08 is a good fit.
Bayesian information criterion (BIC)	Not an absolute measure of fit. Used to compare two models.	The model with smaller BIC has a better fit.

three indicators reflecting the endogenous variable job satisfaction (y_1, y_2, and y_3).

In figure 3.3, the causal direction goes from the latent variable to the indicator variables, with an associated measurement error for each indicator variable. By convention, ξ is used to represent latent exogenous variables and η is used to represent latent endogenous variables. A lowercase x is used to represent indicator variables for exogenous variables, and lowercase y is used to represent indicator variables for endogenous variables. Error terms ε, ζ, and δ are used to represent the model specification and measurement error for exogenous indicator variables, latent variables, and exogenous indicator variables, respectively. Table 3.5 shows the Vensim equations of the underlying system shown in figure 3.3 used to generate the synthetic data for this illustration, while table 3.6 lists the specification of the structural equation model that corresponds to the system shown in figure 3.3. Table 3.7 shows the resulting output, including the fit indices.

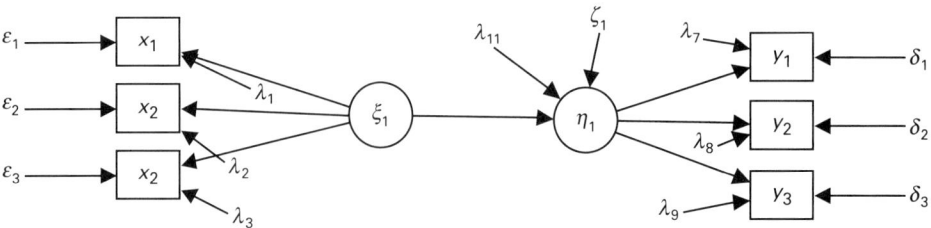

Figure 3.3
Vensim path diagram of structural equation model with two latent variables, organizational culture (ξ_1) and job satisfaction (η_1), with three indicators for organizational culture (x_1, x_2, and x_3) and three indicators for job satisfaction (y_1, y_2, and y_3).

Table 3.5
Structural equation model in Vensim for figure 3.3

```
x1=λ1*ξ1+RANDOM NORMAL(-100,100,0, ε1, 0)
x2=λ2*ξ1+RANDOM NORMAL(-100,100,0, ε2, 0)
x3=λ3*ξ1+RANDOM NORMAL(-100,100,0, ε3, 0)
y1=λ7*η1+RANDOM NORMAL(-100,100,0, δ1, 0)
y2=λ8*η1+RANDOM NORMAL(-100,100,0, δ2, 0)
y3=λ9*η1+RANDOM NORMAL(-100,100,0, δ3, 0)
γ11=3
δ1=5
δ2=5
δ3=5
ε1=5
ε2=5
ε3=5
ζ1=5
η1=γ11*ξ1 +RANDOM NORMAL(-100,100,0, ζ1, 0)
λ1=1
λ2=1
λ3=1
λ7=1
λ8=1
λ9=1
ξ1=RANDOM NORMAL(-100, 100, 0, 20, 0)
```

Table 3.6
Sample R code for figure 3.3

```
sem2 <- specify.model()
xi1->x1, NA, 1
xi1->x2, NA, 1
xi1->x3, NA, 1
xi1->eta1, gam11, NA
eta1->y1, NA, 1
eta1->y2, NA, 1
eta1->y3, NA, 1
x1 <-> x1, eps1, NA
x2 <-> x2, eps2, NA
x3 <-> x3, eps3, NA
y1 <-> y1, del1, NA
y2 <-> y2, del2, NA
y3 <-> y3, del3, NA
eta1 <-> eta1, zeta1, NA
xi1 <-> xi1, NA, 400
```

These examples highlight some of the traditional uses of structural equation models, but much more sophisticated models are possible. One situation arises in dynamical models where one has a feedback relationship involving very short time delays in relation to the time horizon, effectively forming a set of simultaneous equations. This can happen, for example, if one seeks to include a market price mechanism whereby the price of some product or service influences demand, and demand influences the price. In some cases, this could be handled by static table functions that describe the price-demand relationship. However, there are situations where this relationship changes over time, and one wants to model this as a set of simultaneous equations.

Simultaneous equations represent a feedback relationship but, generally speaking, a special type of feedback relationship that is more peripheral to the problem of interest where the dynamics settle around some equilibrium point (e.g., a price for some product or service). In these cases, we are generally more interested in having an adequate description of the behavior and less interested in modeling the underlying causal relationships generating the dynamics. It is this type of situation, involving simultaneous equations, where the notion of a feedback loop in SEM maps nicely onto the concept of a feedback loop

Table 3.7
Results from fitting SEM specification in table 3.5 using data generated by the Vensim model shown in figure 3.3 and table 3.4

```
> summary(sem2.rslt)

  Model Chisquare =    11.141      Df =    13 Pr(>Chisq) = 0.59899
  Chisquare (null model) =    1723.4      Df =    15
  Goodness-of-fit index =    0.96693
  Adjusted goodness-of-fit index =    0.94658
  RMSEA index =    0      90% CI: (NA, 0.086666)
  Bentler-Bonnett NFI =    0.99354
  Tucker-Lewis NNFI =    1.0013
  Bentler CFI =    1
  SRMR =    0.059111
  BIC =    -48.855


  Normalized Residuals
    Min. 1st Qu.   Median      Mean 3rd Qu.      Max.
  -0.016    0.240    0.381    0.400    0.556    0.940


  Parameter Estimates
        Estimate Std Error z value    Pr(>|z|)
gam11    2.9464      0.046698 63.09454 0.0000e+00 eta1 <--- xi1
eps1    27.7308      4.137614  6.70212 2.0541e-11 x1 <--> x1
eps2    33.1949      5.208554  6.37315 1.8518e-10 x2 <--> x2
eps3    30.3530      4.695559  6.46419 1.0184e-10 x3 <--> x3
del1    25.4215      5.104189  4.98052 6.3414e-07 y1 <--> y1
del2    17.9941      4.319475  4.16580 3.1026e-05 y2 <--> y2
del3    27.4854      5.349333  5.13809 2.7755e-07 y3 <--> y3
zeta1-7.0332      15.177526-0.46339  6.430
```

in system dynamics, and SEM can be used to estimate the parameters we would use in a system dynamics model involving simultaneous equations.

For the next example, consider an extension to the previous example where we now assume that job satisfaction (η_1) "feeds back" to influence organizational culture (ξ_1), and that over the time horizon of interest for our dynamical system of an organization of 50 years, the underlying coupled differential equations reach equilibrium relatively

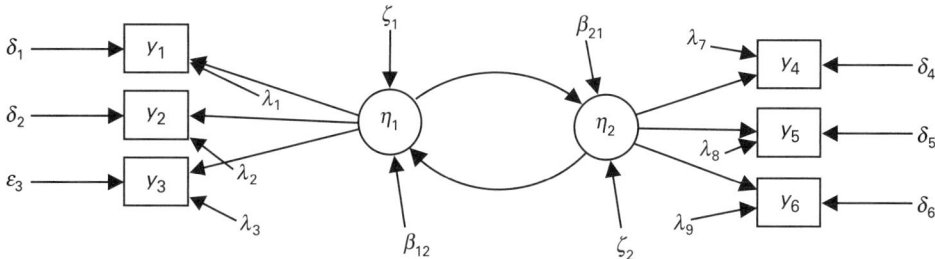

Figure 3.4
Vensim path diagram of structural equation model with two latent variables organizational culture (η_1) and job satisfaction (η_2) with three indicators for organizational culture (y_1, y_2, and y_3) and three indicators for job satisfaction (y_4, y_5, and y_6).

quickly on the order of several months. Although one could model these underlying differential equations explicitly, doing so would add to the computational burden and slow down the overall simulation model. So there are advantages to replacing this underlying system with a simpler description of the resulting behavior.

Figure 3.4 shows an example of a system involving simultaneous equations between two latent variables with multiple indicator variables for each latent variable, in addition to error terms for both the indicator variables and latent variables. Table 3.8 shows the equations in the Vensim model used to generate the synthetic data for fitting the structural equation model in R (table 3.9), and then the results from the exercise (table 3.10). Overall, the structural equation model specified in table 3.8 does a reasonably good job of recovering the parameters from the underlying system of equations shown in table 3.7.

This last example represents a special case where substructures in system dynamics (SD) modeling and SEM overlap in some interesting ways, and has utility when one needs to include simultaneous equations in an SD model with parameters estimated from empirical data.

Discussion

The generality of SEM has allowed a stream of innovations and extensions, including multilevel modeling, latent growth-curve modeling, longitudinal analysis, linear dynamic oscillators, and state-space modeling. Most recent developments have focused on more rigorous approaches to inferring and testing causal relationships (Pearl 2009) and Bayesian approaches to SEM.

Table 3.8
Vensim model of structural equation model in Vensim for figure 3.4

```
y1=λ1*η1+RANDOM NORMAL(-100,100,0, δ1, 0)
y2=λ2*η1+RANDOM NORMAL(-100,100,0, δ2, 0)
y3=λ3*η1+RANDOM NORMAL(-100,100,0, ε3, 0)
y4=λ7*η2+RANDOM NORMAL(-100,100,0, δ4, 0)
y5=λ8*η2+RANDOM NORMAL(-100,100,0, δ5, 0)
y6=λ9*η2+RANDOM NORMAL(-100,100,0, δ6, 0)
β12=0.5
β21=1.2
δ1=25
δ2=25
δ3=25
δ4=25
δ5=25
δ6=25
ζ1=RANDOM NORMAL(0,100, 50, 10,0)
ζ2=RANDOM NORMAL(0,100, 50, 10,0)
η1= SIMULTANEOUS (β12*η2+ζ1, 1)
η2=β21*η1+ζ2
λ1=1
λ2=1
λ3=1
λ7=1
λ8=1
λ9=1
```

As a mathematical modeling approach, a number of cautionary topics also appeared that parallel some of the methodological issues in system dynamics. The problem of assessing model fit has led to a proliferation of indices, while the iterative use of SEM on the same data set has led to a blurring of the role of exploratory modeling and confirmatory testing of theories and a related fixation on statistically fitting the model to the data at the expense of sound development of formal theory. These are, at least in spirit, not unlike the concerns often expressed in system dynamics about model development and testing.

SEM as introduced in this chapter focuses on applications that represent, not the entire system of nonlinear differential equations, but instead subsystems that are essentially static and include both latent

Table 3.9
Sample R code for figure 3.4

```
sem4<-specify.model()
eta1->y1, NA, 1
eta1->y2, NA, 1
eta1->y3, NA, 1
eta2->y4, NA, 1
eta2->y5, NA, 1
eta2->y6, NA, 1
eta1->eta2, beta21, NA
eta2->eta1, beta12, NA
y1<->y1, del1, NA
y2<->y2, del2, NA
y3<->y3, del3, NA
y4<->y4, del4, NA
y5<->y5, del5, NA
y6<->y6, del6, NA
eta1<->eta1, zeta1, 100
eta2<->eta2, zeta2, 100
```

variables with indicators and measurement error. Such applications can readily be used as part of a partial model testing strategy to establish confidence in the underlying causal structure of specific subsystems. This can help rule out formulations that might on the surface appear to be mathematically equivalent from a dynamical systems perspective, and thereby help narrow down the set of possible feedback loops for subsequent analysis.

Additionally, there are important underlying differences between SEM and system dynamics that ultimately limit the application of SEM to system dynamics modeling and disappoint those expecting SEM to be a method that can be used to fully test and estimate parameters in system dynamics models.

Of greatest interest both within the system dynamics literature and statistical approaches to studying dynamical systems has been the possibility of including feedback relationships. However, this is where approaches in SEM and system dynamics modeling generally diverge, in large part because the emphasis in SEM is primarily on understanding linear systems of differential equations, whereas system dynamics generally focuses on nonlinear systems of differential equations. The

Table 3.10
Results from fitting SEM specification in table 3.9 using data generated by the Vensim
model shown in figure 3.4 and table 3.8

```
> summary(sem4.rslt)


 Model Chisquare =    46.654      Df =   11 Pr(>Chisq) = 2.4767e-06
 Chisquare (null model) =    436.98      Df =   15
 Goodness-of-fit index =    0.88971
 Adjusted goodness-of-fit index =    0.78946
 RMSEA index =    0.18004      90% CI: (0.12865, 0.23481)
 Bentler-Bonnett NFI =    0.89323
 Tucker-Lewis NNFI =    0.88478
 Bentler CFI =    0.9155
 SRMR =    0.078109
 BIC =    -4.1121


 Normalized Residuals
    Min. 1st Qu.    Median      Mean 3rd Qu.      Max.
 -1.1000-0.5330    0.2270    0.0467    0.3850    1.2600


 Parameter Estimates
          Estimate    Std Error z value Pr(>|z|)
 beta21     1.21138      0.15554 7.78820 6.8834e-15 eta2 <--- eta1
 beta12     0.52535      0.13609 3.86021 1.1329e-04 eta1 <--- eta2
 del1     454.02836     51.53296 8.81045 0.0000e+00 y1 <--> y1
 del2     437.48978     54.93979 7.96308 1.7764e-15 y2 <--> y2
 del3     431.72997     56.04062 7.70388 1.3101e-14 y3 <--> y3
 del4     411.72035     53.08317 7.75614 8.6597e-15 y4 <--> y4
 del5     299.74605     47.15886 6.35609 2.0695e-10 y5 <--> y5
 del6     426.95266     54.83589 7.78601 6.8834e-15 y6 <--> y6
 zeta1     99.99901    104.29342 0.95882 3.3765e-01 eta1 <--> eta1
 zeta2    100.00206     81.53433 1.22650 2.2001e-01 eta2 <--> eta2


 Iterations =    10
```

implication of this fundamental shift from linear to nonlinear systems of differential equations is often overlooked, especially as it relates to the notions of nonlinearity in a system of differential equations leading to emergent behavior of shifts in feedback loop dominance.

Structural equation models can be used to model the underlying system of causal relations but, generally speaking, only for linear situations where the system of equations is already identified. In earlier work, it was shown that one could recover the relationships from a linear system of differential equations within a given phase of behavior, but structural equation models tended to break down with shifts of dominant feedback loops characteristic of nonlinear systems of differential equations (Hovmand 2003).

Some efforts in SEM (Boker and Nesselroade 2002) begin with assumption of a given system of equations (e.g., simple linear oscillators) and have been able to consistently demonstrate both the soundness of the method and its application to oscillatory phenomena. These efforts represent important extensions to both SEM and social science literature in providing a means to move from static systems to linear dynamic systems. However, these efforts generally presuppose a set of generic differential equations and, in this sense, represent an effort to reduce all oscillatory phenomena to the case of a simple linear oscillator. This is contrary to the general approach to building dynamic theories using system dynamics.

Moreover, there is a tendency to confuse the system of equations describing the *solution* to an underlying system of differential equations with the *underlying system of differential equations* itself. In statistically modeling the time series behavior of a system, one is essentially modeling the equations representing the *solution to the underlying system* and not the system itself. The simplest example of this is the distinction between a simple system where growth, *dx(t)*, is proportional to the population size, *x(t)*,

$$dx(t) = r \cdot x(t) \, , \tag{2}$$

which has the general solution,

$$x(t) = x(0) \cdot e^{r \cdot t} \, . \tag{3}$$

Equations (2) and (3) are not the same equation. Equation (2) is the equation for the underlying system representing the causal mechanism, while equation (3) is the equation describing the solution or trajectory of the system over time. If one hypothesizes (2), one can derive (3) and estimate the parameters for (2). SEM is often used to model the trajectory (3), but it is important to keep in mind this is not the same as modeling (2). Even in this context, however, SEM is still valuable, because it can help identify and establish an empirical basis for the reference modes.

When statistically modeling dynamical systems using SEM, there can also be a tendency to forget what the underlying differential equations mean. For example, it is mathematically impossible to have anything other than the net rate of change for a variable influence the variable. That is,

$$x(t) = \int_0^u dx(u)\,dt,$$

where $dx(u)$ is the net rate of change. Yet one sees examples where there are other direct influences on $x(t)$ in addition to the net rate of change.

To be clear, it is not that there are not appropriate methods of using SEM that avoid these types of errors and can handle nonlinear systems of differential equations (there are), but instead one needs to consider whether or not using SEM in the conventional sense is the most efficient approach *if one already has a mathematical simulation model of a dynamical system available.* In particular, one extension is worth considering.

The major conceptual difficulty in SEM is generally developing a theoretically sound set of hypothesized causal relations to test against empirical data. Recall that the basic idea behind SEM is that the hypothesized causal relations *imply* a specific pattern in a covariance matrix, and that this implied covariance matrix can be used to estimate parameters and test models using an observed covariance matrix. The main limitations we run into with traditional SEM stem from the underlying assumptions required to calculate the implied covariance matrix. However, in system dynamics we already have not only a model but also a simulation model that can be used to generate covariance matrices. The fact that the simulation produced the covariance matrices means that this is the implied covariance matrix from the underlying structure of causal relations. Knowing this, all we need to do then is compare the observed covariance matrix against the implied covariance matrix from our SD model. This is a much more direct way to use the ideas from SEM to test SD models.

This chapter has sought to review the application of SEM to SD through a series of examples, with the goal of clarifying the relationship between the two. Using Vensim models to generate data for SEM analyses, and the R SEM package to implement the SEM, the results highlighted the applicability of SEM to simple linear open loop and feedback systems. Contrary to common expectations that the notion of feedback in SEM can be applied to SD models, the notion of feedback in SEM is generally limited to the specific situation in SD where simultaneous equations are included in a model. In such situations, SEM can be appropriately used to estimate the parameters. However, recognizing the underlying logic of SEM, a more general approach is suggested that covers most cases of SD, which involves using SD simulation models to generate the implied covariance matrices that are compared against one or more observed covariance matrices. While uncommon, such an approach will yield more meaningful tests and comparisons of SD models against data.

Exercise

Consider the following example from Duncan, Haller, and Portes (1968) looking at peer influences on aspiration. Table 3.11 provides the

Table 3.11
Correlation matrix from Duncan, Haller, and Portes (1968) for 329 respondents

		Xa	Xb	Xc	Y1	Y2	Xd	Xe	Xf	Y3	Y4
Respondent:											
Intelligence	Xa		.1839	.2220	.4105	.4043	.3355	.1021	.1861	.2598	.2903
Parental aspiration	Xb			.0489	.2137	.2742	.0782	.1147	.0186	.0839	.1124
Family SES	Xc				3240	.4047	.2302	.0931	.2707	.2786	.3054
Occupational aspiration	Y1					.6247	.2995	.0760	.2930	.4216	.3269
Educational aspiration	Y2						.2863	.0702	.2407	.3275	.3669
Best friend:											
Intelligence	Xd							.2087	.2950	.5007	.5191
Parental aspiration	Xe								-.0438	.1988	.2784
Family SES	Xf									.3607	.4105
Occupational aspiration	Y3										.6404
Educational aspiration	Y4										

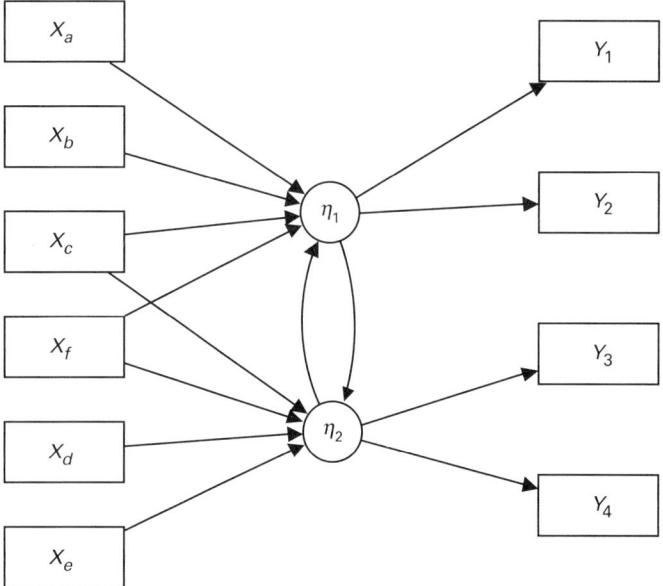

Figure 3.5
Path diagram of peer-influence model.

correlation matrix for the 329 respondents in their study, along with the variable names and variable symbols for each. Specify and test a structural model shown in figure 3.5. Note that not all of the variables from the correlation matrix are used, and that η_1 and η_2 represent the respondent and friend's aspiration as latent variables, respectively, with occupational and educational aspiration as the indicator or observed variables.

Notes

1. Awkwardly, the term "feedback relationship" is also commonly called a "nonrecursive relationship" in SEM.

2. When a covariance is listed *and* fixed using RAM notation, the assumption is that the parameter is fixed to a nonzero number that must be provided by the modeler. All other covariances and relationships that are not listed *are assumed to be fixed* with a value of zero, meaning that there is no relationship between the variables and no covariate relationship.

References

Bisconti, T. L., C. S. Bergeman, and S. M. Boker. 2004. Emotional well-being in recently bereaved widows: A dynamical systems approach. *Journal of Gerontology* 59B (4): 158–167.

Boker, S. M. & Nesselroade, J. R. 2002. A method for modeling the intrinsic dynamics of intraindividual variability: Recovering the parameters of simulated oscillators in multi-wave panel data. *Multivariate Behavioral Research* 37 (1): 127–160.

Boker, S. M., and M. J. Wenger. 2007. *Data analytic techniques for dynamical systems*. New Jersey: Lawrence Erlbaum Associates.

Bollen, K. A. 1989. *Structural equations with latent variables*. New York: Wiley & Sons, Inc.

Duncan, O. D., A. O. Haller, and A. Portes. 1968. Peer influences on aspirations: A reinterpretation. *American Journal of Sociology* 74 (2): 119–137.

Forrester, J. W. 1968. Industrial dynamics-after the first decade. *Management Science* 14 (7): 398–415.

Fox, J. 1997. *Applied regression analysis, linear models, and related methods*. Thousand Oaks, CA: SAGE Publications.

Hovmand, P. S. 2003. Analyzing dynamic systems: A comparison of structural equation modeling and system dynamics modeling. In *Structural equation modeling: Applications in ecological and evolutionary biology*, ed. B. H. Pugesek, A. Tomer, and A. von Eye, 212–234. New York: Cambridge University Press.

Levine, R. L., and W. Lodwick. 1992. Psychological scaling and filtering of errors in empirical systems. In *Analysis of dynamic psychological systems: Methods and applications*, ed. R. L. Levine and H. E. Fitzgerald, 87–117. New York: Plenum Press.

Levine, R. L., Sell, M. V., & Rubin, B. 1992. System dynamics and the analysis of feedback processes in social and behavioral systems. In *Analysis of dynamic psychological systems: Basic approaches to general systems, dynamic systems, and cybernetics*, ed. R. L. Levine and H. E. Fitzgerald, 145–266. New York: Plenum Press.

Meehl, P. E. 1990. Appraising and amending theories: The strategy of Lakatosian defense and two principles that warrant it. *Psychological Inquiry* 1 (2): 108–141.

Pearl, J. 2009. *Causality: Models, reasoning, and inference*. 2nd ed. New York: Cambridge University Press.

Schumacker, R. E., and R. G. Lomax. 1996. *A beginner's guide to structural equation modeling*. Routledge.

Sterman, J. D. 2000. *Business dynamics: Systems thinking and modeling for a complex world*. Irwin McGraw-Hill.

4 Working with Noisy Data: Kalman Filtering and State Resetting

Robert Eberlein

In appendix K of *Industrial Dynamics*, Forrester states: "If the presence of noise is admitted, we must necessarily come to the conclusion that even the perfect model may not be a useful predictor of the specific future state of the system it represents" (Forrester 1961, p.431). This simple sentence has important ramifications not only on the way we think about models and the future, but also on the way we think about models and past observations. The same conclusion about models not being able to predict the future applies to prediction of the past. This means that comparing model behavior with observed behavior is a potentially risky undertaking.

In this chapter, we will use the same model Forrester used in appendix K to demonstrate the same point, then use a simple pendulum model to look at both direct state adjustment and filtering as methods to more meaningfully compare model output with observed behavior. The pendulum, like Forrester's supply chain simulation, is an oscillator, and such systems pose the most complete challenge to making comparisons with data. Systems displaying other patterns of behavior are more amenable to comparisons with data, though all pose challenges, and we will explore these using simple models.

The most important learning in this chapter is that some systems generate behavior that can only be replicated by either coincidence or systematic intervention in the simulation. Other systems generate behavior that can reliably be replicated, with some error expected. Knowing which situation you are dealing with is important in deciding what role replication of history will play in the model-validation process. The specific techniques outlined in this chapter represent some of the ways to compensate for the fact that you are dealing with an uncooperative system. Other solutions, including those outlined in chapters 5 and 8, may also be effective.

The models and ancillary files required to reproduce the examples presented in this chapter are included in the electronic support files for this book. The examples presented here are all done using Vensim. Some of the examples relating to state resetting are easily implemented in other platforms, including both system dynamics packages and, in many cases, spreadsheets. The examples that explicitly use a Kalman filter will require Vensim DSS (or Professional) to be replicated directly, as Vensim uses an idiosyncratic form of the Kalman filter. Largely similar results can be obtained with other filter implementations (in products such as R), though the details of doing that are beyond the scope of this chapter. As of writing, filtering capabilities are not built in to the other available system dynamics software packages.

Background Production Distribution

Anyone who has played the Beer Game (Sterman 2000, ch. 17.4) is familiar with production distribution systems, and we start with a model of such a system. The system has three stages: retail, wholesale and manufacturing. This is the model used to demonstrate how a cascaded order and replenishment system can generate oscillation (Forrester 1961, ch. 2). Figure 4.1 shows the production over time that results when the system is driven by a noisy but patternless order stream at the retail level. Two runs are shown. For the first 50 months, both have the same retail orders. After time 50, the two runs have different retail orders drawn from random sequences with the same statistical properties.

For a short time after the 50th month, the two runs show identical results. By time 60, however, the results begin to diverge. Both runs come from exactly the same model. If we compare the character of the results in terms of mean values, variability, and periodicity—whether by simply looking at the graph, considering phase relationships (Barlas 1989), or performing a statistical analysis such as the Kolmogorov-Smirnov test (Encyclopedia of Mathematics 2012)—we could reasonably conclude they are the same. Point-by-point, however, the curves are very different, only rarely having the same value and often moving in opposite directions over time.

To demonstrate this observed divergence with broader sampling, we can run a sensitivity simulation with a different noise sequence for orders beginning at time 50, as shown in figure 4.2. Here we can see that there is an invariant distribution of values for all times after about

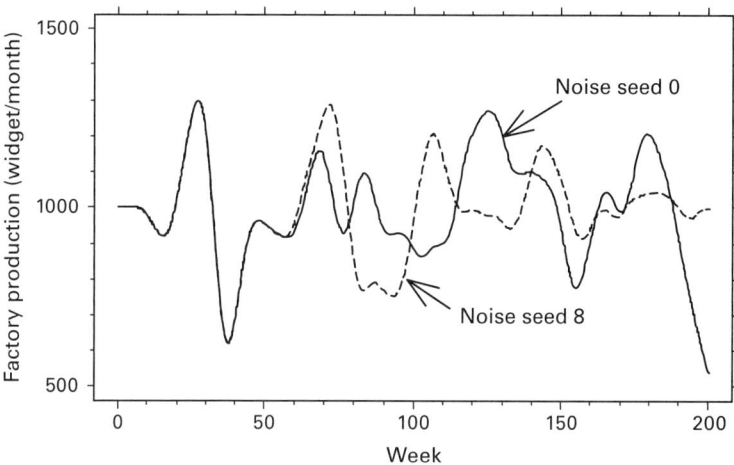

Figure 4.1
Production in a three-stage supply chain with alternative order streams after time 50.

100. That is, after time 100, the previous history of the system no longer places any constraints on factory production (or other model variables, though that is not shown here). Output, at any time, ranges from about 750 to about 1,250 with 90% confidence, and the best point estimate is simply the mean of the distribution (that is, 1,000 widgets per week). That is not to say that the best estimate of next week's production is 1,000. It is clear from this graph that for almost 10 weeks, production is completely predictable based on previous knowledge. This is because 10 weeks is the length of time it takes a perturbation to propagate through the system, from retail orders to actual production.

Figures 4.1 and 4.2 demonstrate the point that Forrester made in our opening quote. The same structure generates completely different results at any given time. We can see the same thing using a much simpler two-stock model, and it is helpful to do so in order to both better understand the problem and see a possible solution.

Damped Pendulum

Consider a damped pendulum half a meter in length for which there is noise affecting the acceleration. The differential equation for this would be:

$$\ddot{\theta} = \sin\theta \cdot \frac{g}{l} - \dot{\theta} \cdot d + \epsilon, \tag{1}$$

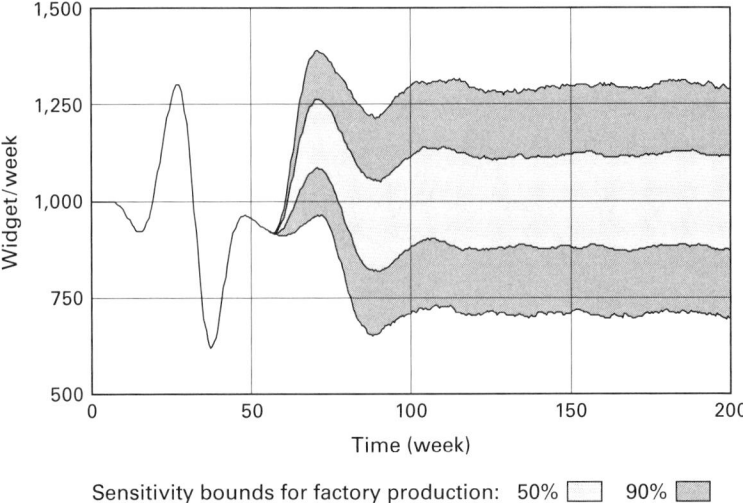

Sensitivity bounds for factory production: 50% ☐ 90% ▨

Figure 4.2
Trajectory profiles of factory production for 20,000 random realizations of retail orders.

where d is the drag coefficient (with value 1 per second) and ϵ is a random disturbance. We start the pendulum at rest and vertical and then watch its behavior for 10 seconds. With no noise input (ϵ *always* 0), the pendulum simply stays where it is. With noise, it swings back and forth in a periodic, though somewhat irregular, fashion. Figure 4.3 shows a simulation of the equilibrium case as well as one using a noise seed of 0 in Vensim.

We are using Euler integration, with a solution interval of 1/32nd second for the simulations. This is to simplify the explication. Euler integration is well known to underdamp oscillatory systems, so the behavior is not reflective of a physical pendulum with the specified damping constant. However, substantially the same results would be obtained for a physical system (or using an integration technique such as Runge-Kutta) by using a smaller drag coefficient (approximately 0.42).

Suppose that we want to get estimates of the length of the pendulum and the damping given this behavior. Clearly, if we use a model with no noise to try to accomplish this, we will fail, since such a model will only produce a straight line, independent of length and damping. If, on the other hand, we add noise to the model, then unless we happen

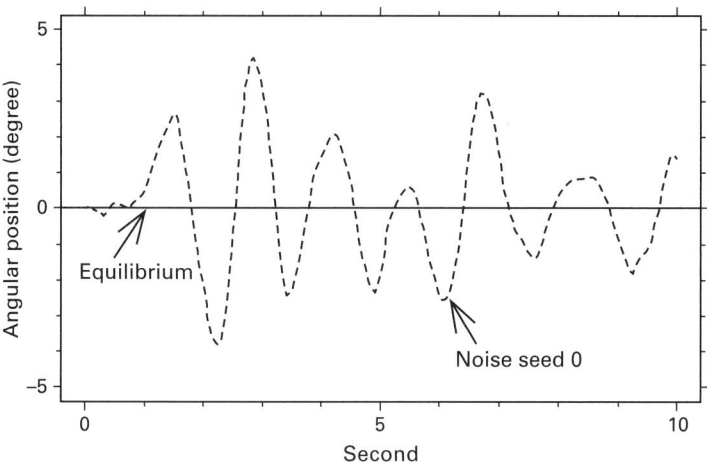

Figure 4.3
Angular position with noise versus no noise (equilibrium).

upon exactly the right noise sequence, we will have trouble. For example, using a random noise seed of 8 and calibrating the model to the sample run shown in figure 4.3 (which used a noise seed of 0) would give a drag coefficient of 62 (instead of 1), a length of 0.2 meters (instead of 0.5), and a graph for position that looks like that in figure 4.4.

The results are nonsensical, and that should not come as a surprise. We are using point comparisons for something that can follow very different tracks depending on influences outside the scope of the model. By using stochastic optimization (see chapter 8), it may be possible to get reasonable parameter estimates, but for this particular model, the number of replications required would likely be in the billions, which presents computational challenges.[1] Looking at the behavior graph, however, it is not that hard to make a rough estimate of length. The behavior is periodic, with an average period of about 1.5, so the length of the pendulum should be about 0.56 meters, using the standard formula for the period of a pendulum.[2] That is reasonably close—certainly a much better result than we obtained from our naïve calibration.

It is clear there is information in the time behavior, but that deterministic calibration is not going to extract it for us. Estimating the period is one alternative, but it is possible to do better than that. In fact,

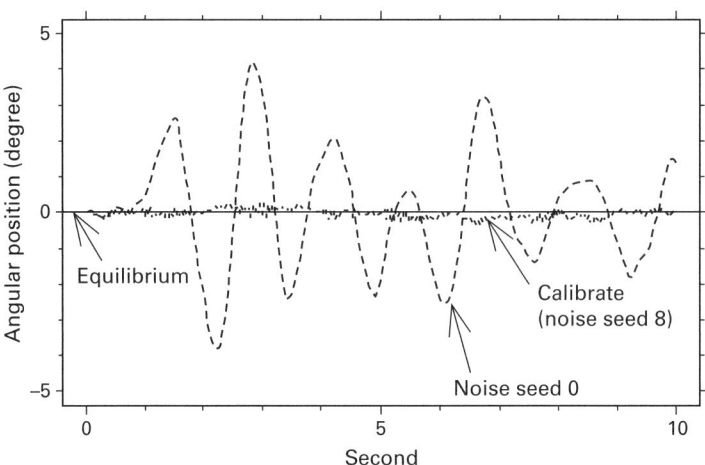

Figure 4.4
Calibration to noise seed 0 run using a noise seed of 8.

for this case we have measurements of position, so we can derive velocity and acceleration and reduce this to a nonlinear least-squares estimation problem—and one in which the assumptions of that form are actually met. This yields estimates of 0.44 meters and 1.3 per second for the length and damping, with wide confidence bounds that easily encompass the true values of 0.5 and 1.0.

It practice, it is rare to have a situation in which you can simply run a regression. Should it happen, however, take advantage of the opportunity. Standard regression techniques yield very good results (consistent, efficient estimators using the language of econometrics) when the conditions they assume are met. Even when you need to stretch those assumptions, the techniques can be of great practical value. The barrier to using regressions is not that they have something inherently wrong with them, but that the data are generally not available.

Challenge 1: In Forrester's production distribution example, he uses variation in retail orders to drive the differences in output. In the beer game, the same set of orders are given to everyone, and yet the results still differ. Introduce noise in different locations in the three-stage supply chain (or a beer game model) and see what kind of variation you get in results with an invariant order stream (using a step input will be easiest). Are the resulting variations as large as those from team to team in the beer game? What else might be changed to get higher variety with the same orders? What does the nature of the variation imply about directly calibrating against recorded inventories?

State Resetting

Another approach to computing the length and damping would be to start not at time 0, but instead at time 1.5, 2.2, 2.8, and so on when the pendulum is changing directions (so we know that the velocity is approximately 0), then follow the observed position for a while and compare results. In fact, we can take this idea further and simply reset position to match up with the data after each observation (this is done in DampedPendulumResetting05.mdl in the material included with this chapter). When we do this, and calibrate, the estimates for length and damping are 0.56 and 1.6. The values are not correct, but clearly they at least make sense. Comparing position is not that informative, since we are continually resetting the model to match the data. It is, however, interesting to compare the resulting model-generated veloc-ity to the data, as shown in figure 4.5.

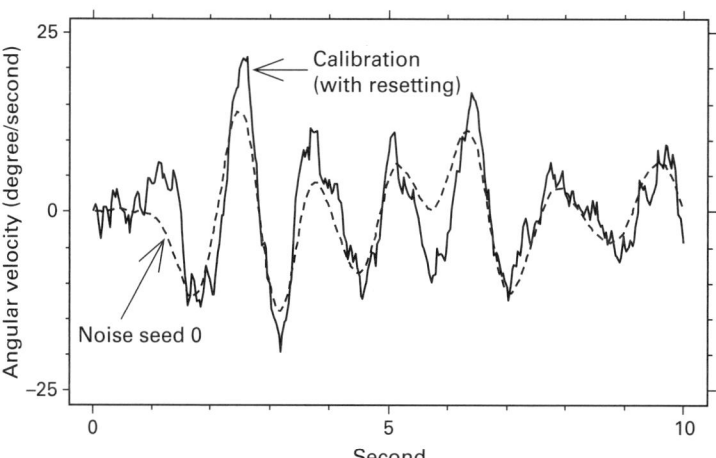

Figure 4.5
Calibration to noise seed 0 run using state resetting.

This is not as good a representation of velocity as we could have obtained by computing it from changes in position, but it is entraining reasonably well. In this case, we are relying on the forward effect of acceleration to change velocity. That is, by changing position, we change the computed acceleration, which, through integration, changes velocity. We are not making an adjustment to velocity based on the observed mismatch in position, but we could. If the model computes a position that is larger than that observed, we could reasonably conclude that this is because the velocity is too large. We could then adjust velocity (one state) based on the observation of position (a different state). This brings us to the idea of Kalman filtering.

Challenge 2: A playground swing is a good example of a damped pendulum. On a windy day, visit a playground and observe the swings. While the swing is moving back and forth, try to see the connection between variations in the wind and the movement of the swing. Now stop the swing at its bottom position and try to connect the wind variation to the swing movement. This is physical state resetting, and does not work well if there is a child on the swing.

Kalman Filtering

In adjusting the state variables in a model, we do not need to rely directly on observations of those states. For example, an observation on acceleration would tell us something about position, since acceleration depends on position. So what is the best way of going from a set of observations back to estimates of the underlying state variables? That question was answered for linear systems at almost the same time system dynamics was emerging as a field (Kalman 1960). What has become known as the Kalman filter is a set of equations that track the evolution of both the state and measurement uncertainty and thereby provide a map back from measurements to the state variables. This map is called the Kalman gain, and it is a projection matrix from the set of errors a model makes in computing measurements to the adjustments required in the model's state variables.

To better understand this, consider our pendulum example. For notational clarity, we will present model variables with no accent, and use a breve ($\smile$) to represent measured values, and a hat ($\wedge$) to represent estimates based on a combination of model and measured values. For the pendulum, the model computation is

$$p_t = p_{t-dt} + v_{t-dt}dt ,$$ (2)

where p is position and v velocity. When we applied state resetting, we replaced the model-computed position with the measured position directly, but we could also write this as

$$\hat{p}_t = p_t + 1 \cdot (\breve{p}_t - p_t) ,$$ (3)

where $\hat{p}_t$ is our best estimate of position at time t and $\breve{p}_t$ is the measured value of position at time t. We made no adjustment to the other model state variable (velocity), so the gain we used was effectively the vector $\begin{bmatrix} 1 \\ 0 \end{bmatrix}$. In general, the gain is a matrix with a number of rows equal to the number of model states (2 in our example) and a number of columns equal to the number of measurements (1 in our example). Multiplying the gain times the difference between the measurements and the values computed by the model gives the adjustment that needs to be made to the state variables. The result is our best estimate of the states at time t. The estimate is then used to move forward to the next measurement time.

In the following, we use difference equations for simplicity of presentation, and ignore any time-varying exogenous inputs for the same reason. The extension of this approach to nonlinear systems can be done by successive linearization using what is often called an extended Kalman filter (Anderson and Moore 1979). To clarify the relationship of the simplified equations we use to more general dynamic models, we start with the more general form.

Models used in system dynamics can, for the most part, be represented as lumped systems of ordinary differential equations of the form

$$\dot{x} = g(x, P, e, \mu),$$ (4)

where x is the state variable, P a set of parameters, e a vector of exogenous inputs, μ a vector of random inputs (distinguished from e by their lack of measurability), and g a nonlinear function. As with all functional forms, we can create a linear approximation to equation (4) by differentiating with respect to x, e, and μ. This gives us a linear equation in the form:

$$\dot{x} = A_t x + B_t e + C_t \mu.$$ (5)

Here each of the matrices A, B and C is potentially time-varying, since the linear approximation will change as the system states (and inputs) change. Because our purpose is to work with measured data, and because data are only measured at discrete points in time, we can represent equation (5) as a difference equation in the form:

$$x_t = A^*_{t-dt}x_{t-dt} + B_{t-dt}e_{t-dt} + C_{t-dt}\mu_{t-dt} \text{,}$$ (6)

where $A^*_{t-dt} = I + dtA_{t-dt}$ (a discrete sampling approximation of A). For presentation here, and without any substantive loss of generality, we will simplify this by assuming that $\mathbf{B}$ is zero, that $\mathbf{C}$ is the identity matrix, and that dt is 1 (we will also drop the * on $\mathbf{A}$). This gives us the basic equations representing the dynamics of the underlying model as

$$x_t = Ax_{t-1} + \mu_{t-1} \text{,}$$ (7)

where $\mathbf{x}$ is the state variable vector, $\mathbf{A}$ the dynamics matrix, and μ a noise vector assumed normally distributed with covariance $\mathbf{\Phi}$ and independent over time. Most models include not just states, but also a variety of intermediate variables. The compact notation in equation (7) consolidates these down to use only the state variables. This would correspond to using the smallest number of variables with generally complicated equations. While this is not good modeling practice, it is a convenient and necessary representation for the discussion here. The dynamics matrix $\mathbf{A}$, thus, is both an amalgam and a linearization of the underlying model equations.

The measurements $\mathbf{y}$ are given by the equation[3]

$$y_t = Hx_t + \varepsilon_t \text{,}$$ (8)

where $\mathbf{H}$ maps states to observations and is typically a relatively short, but fat, matrix with a small number of nonzero values (often a single 1) in each row. For the oscillator example, since we only have measurements on position, the H matrix would be $[1 \quad 0]$. Finally, ε is a normally distributed set of error terms with covariance $\mathbf{\Psi}$ that are independent across time.

The Kalman filter is considered to be Bayesian in the sense that it requires a prior distribution in order to be initialized, and at each step computes a prior distribution of values given the model, and a posterior distribution given the data and the model-based prior. Using a Kalman filter, thus, does require that an initial distribution of the state covariance be specified, generally by the user. For the pendulum, which starts at rest, we would expect this to be small, but it must be positive definite for numerical reasons, and we arbitrarily set it to the matrix

$\begin{bmatrix} 1 & 0 \\ 0 & 1 \end{bmatrix}$ in the experiment. Once this is done, the computations proceed in time parallel to those for the underlying model.

The assumptions of linearity and independent stationary noise that is normally distributed are, as in most statistical models, quite strong. But by using them it is possible to create equations for the evolution of the state covariance matrix, and the covariance between the state and measurements that run in parallel to the model state equations. Specifically, with Φ representing the covariance matrix for μ, then

$$\Omega_{t|t-1} = A\Omega_{t-1}A^{\mathrm{T}} + \Phi \tag{9}$$

gives the state covariance Ω at time t based on its value at the previous time. This value can then be combined with the measurement matrix and the covariance of the measurement errors to give an estimate of the measurement covariance

$$\Theta_t = H\Omega_{t|t-1}H^{\mathrm{T}} + \Psi. \tag{10}$$

And that, in turn can be used to compute the Kalman gain as

$$G_t = \Omega_{t|t-1}H^{\mathrm{T}}\Theta_t^{-1}. \tag{11}$$

This gain is what will be applied to update the state estimates using the formula

$$\hat{x}_t = x_t + G_t(\tilde{y}_t - y_t), \tag{12}$$

where $\tilde{y}_t$ is the measured value for y at time t. We now have a map from measurements to states, even though not every state variable is measured, and there are measurement errors. Our best estimate of the states does not rely solely on the model or solely on the data, but instead combines information from both sources.

Finally, before we can proceed to the next step, we need to update our estimate of the state covariance, recognizing that the available measurements have helped to bound this:

$$\Omega_t = (I - G_tH)\Omega_{t|t-1} \tag{13}$$

This allows us to proceed to the next point in time. We iteratively do updates of both the model variables and the statistical character of those variables.

We can apply these equations to the damped pendulum using the built-in computations available in Vensim. To do this, we specify Φ and

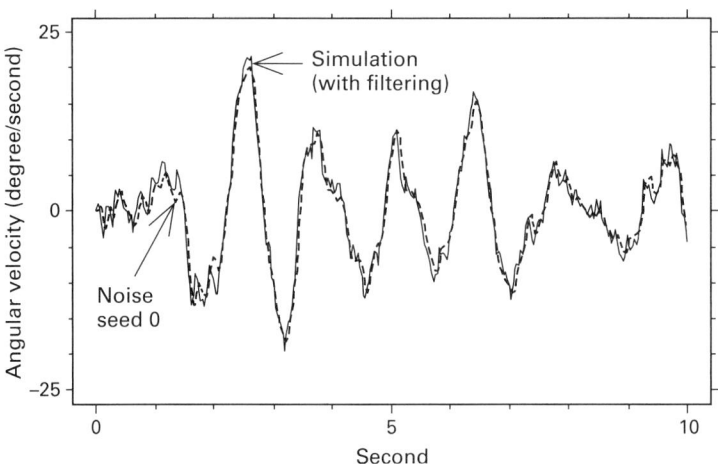

Figure 4.6
Velocity based on Kalman filter applied using measurements of position.

Ψ along with the initial value Ω_0 and allow the software to complete the rest of the computations. We had perfect measurement in our example (so that Ψ is really 0), but we provide a small value for measurement error anyway to allow the computations to proceed.

We can run our model using the correct parameter values of 0.5 for pendulum length and 1.0 for damping with Kalman filtering turned on. Still measuring only position, we get an estimate of velocity that is very close to that from the noise seed 0 run, as shown in figure 4.6.

This is clearly a much better fit than we achieved by just resetting position, and this should come as no surprise. When we reset position, we were effectively making acceleration more like what happened in the experiment. By using a filter we are doing that, but also reaching back and making adjustments to velocity based on the error the model makes in predicting position. Combining the two adjustments gives a better estimate of velocity than just using the one. For this particular example, we could do as well by measuring velocity as change in position from one time to the next, though such a simple computation is not normally available.

In this case, the Kalman gain converges to $\begin{bmatrix} 0.84 \\ 18.8 \end{bmatrix}$ (for linear models, the Kalman gain will normally converge to a constant value, and this model is almost linear). The first element of the gain says that we give an 84% weight to the measurements of position; the second maps position errors to velocity adjustments. The number is quite large, reflecting the generally larger magnitude of velocity relative to position. If the model had been done in minutes instead of seconds, the second element would be much smaller (1/60th the size).

Calibrating with Kalman Filtering

Given our presentation thus far, it may seem natural to combine filtering, a technique for state estimation, with calibration, a technique for parameter estimation. There are rigorous statistical foundations for this approach (Schweppe 1973). Combining filtering and calibration sometimes goes by the name FIMLOF (full information maximum likelihood with optimal filtering; Peterson 1975, 1980). It should be noted, however, that the maximum likelihood part of this moniker depends entirely on the assumption of normality in the driving and measurement noise, as does the expectation that the filtering is optimal. What is true, even when the strong statistical assumptions are not met, is that the approach does try to make use of all available information. This includes both the measured data and the model structure. The approach is also robust to missing data and data series measured with different frequencies, which means more of what is measured can be used.

When we combine Kalman filtering with calibration, the matrix Θ that is computed as part of the filter can be used to weight the errors. In fact, because this matrix varies over time and its values might depend on the parameters being searched over, Vensim computes the full log-likelihood value, including the constant term that is normally left out when calibrating against a weighted squared error. The implication of this is that we can do calibration over both model parameters and noise variance terms and still get meaningful results. This is because while a higher variance will invariably decrease the weighted sum of squares, it will increase the constant term, and those two thus balance one another to give the maximum likelihood estimate.[4]

Performing this calibration with filtering gives parameter estimates of 0.45 (with 95% confidence bounds of 0.39 to 0.54) for length and 1.2 (0.48 to 2.0) for the drag coefficient.

Nonoscillatory Models

The two examples we have shown so far are for models that oscillate, and clearly this type of model is the most likely to follow a track that

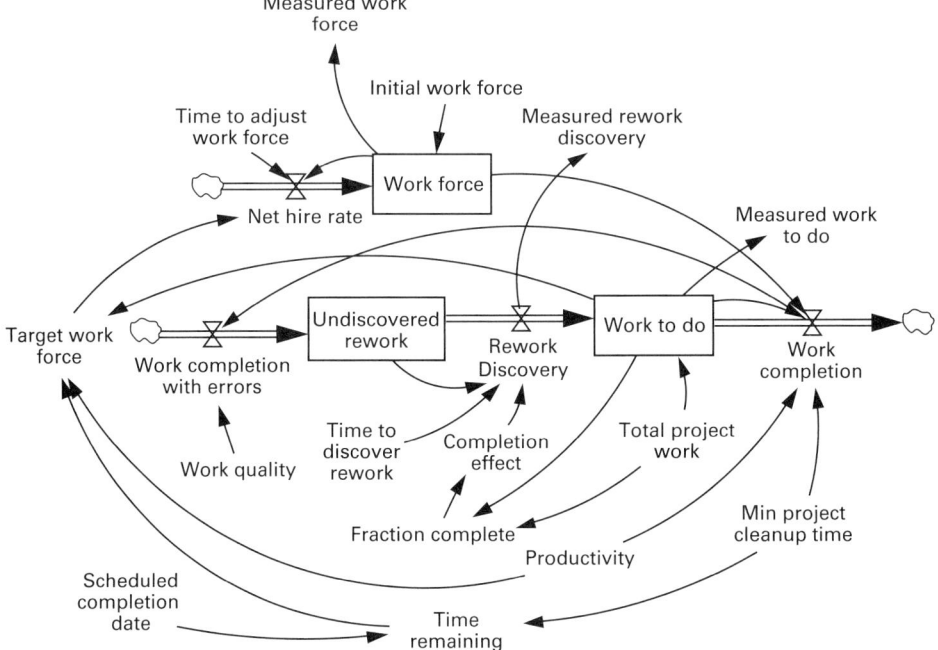

Figure 4.7
A simple project model with rework.

is dramatically different point to point, with only small differences in perturbations. For models that do not oscillate, the implications of trying to track behavior with a deterministic model are more subtle. Consider a simple project model with just three levels: work to do, work force, and undiscovered rework, as shown in figure 4.7.

We do not include work done as a level because it can be determined from initial project specification and work to do. Reducing the number of levels to a minimum is helpful in situations where we might want to reset or adjust their values so that we do not risk inconsistently adjusting two levels that are logically determinable from one another. The gain used in a Kalman filter does not inherently respect such things as conservation of mass, even when the underlying dynamic equations do. In addition, situations where levels are definitionally connected can cause numerical singularities, preventing the computation of the filter equations.

This model is a typical project model (Lyneis and Ford 2007), and if work is of perfect quality (work quality=1), then it shows a quick ramp

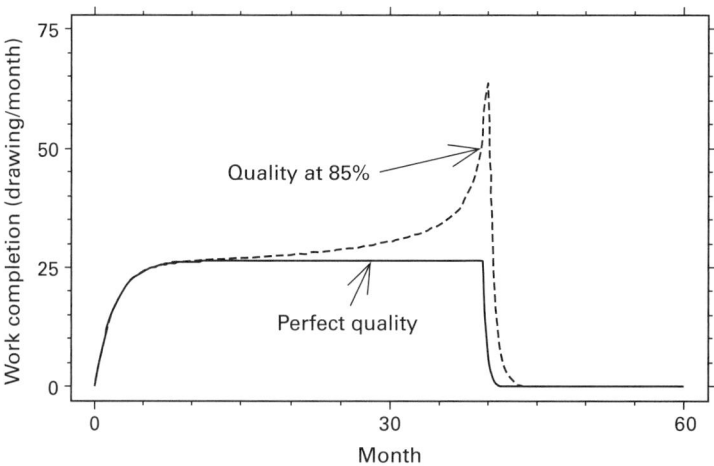

Figure 4.8
Work completion with perfect and less than perfect quality.

of the workforce, steady output, then a quick decline. If quality is lower, however, there is a rise in workforce over the course of the project as work needing correction is discovered (figure 4.8).

To understand the implication of using a filter, we introduce noise into work quality as well as the measurements of work force and work to do. Undiscovered rework, by definition, is not measured, though we will treat rework discovery as measured, again with error. We simulate the model for 30 months with a noise seed of 0, then estimate work quality using simple simulation and a Kalman filter. We can compare these estimates to the parameter values used to generate the results and also look at the implications for the remaining months of the project.

Figure 4.9 shows the behavior of the measured variables over 30 months. Work is measured in "drawing," a common abstraction for research and development projects, but the concepts apply equally to other types of projects such as construction, where "square feet" might be a more appropriate unit of measure. From the behavior observed, we should be able to make some estimate of the underlying quality, especially using the model we have.

Calibrating quality to this data with simple simulation gives us an estimate of 0.69 with 95% confidence bounds of 0.68 to 0.70. Doing the same calibration with Kalman filtering turned on gives us an estimate of 0.68 with 95% confidence bounds of 0.64 to 0.73. The wider

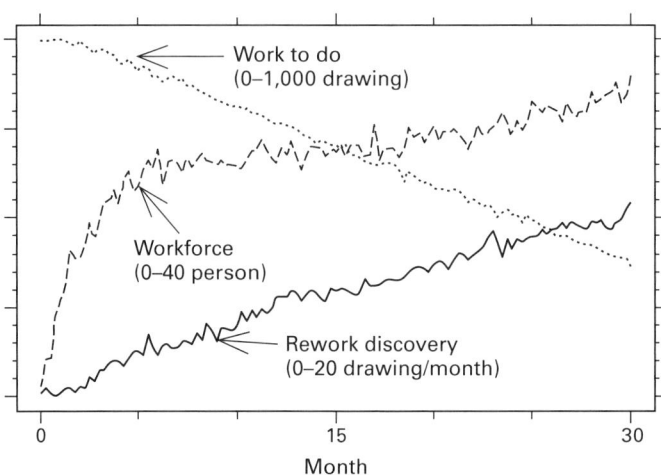

Figure 4.9
Measured variables over the first 30 months.

confidence bounds when using the filter are the result of the more complete specification of the stochastic nature of the model. For the filter, we use the known values for driving and measurement noise (using a small initial variance of 0.1 for each state to reflect the relatively certain nature of the starting point). For the nonfilter optimization, we use the empirically observed errors to compute standard deviations in order to determine weights. These weights actually give artificially narrow confidence bounds. They are based on the assumption that the measurements are exact, which is not the case.

It is worth noting that even for the Kalman filter simulations, the assumptions underlying the statistical model are actually violated. This is because quality is restricted to lie between 0 and 1, and this means that instead of being normally distributed, the driving noise is drawn from a truncated normal distribution with a very strong truncation on the upper bound. The measurement errors are also truncated, though these truncations are further out on the tails and change the distribution only slightly.

In this case, the simple and filter optimizations yield almost the same results for the underlying population parameters. It is interesting, however, to look at the detailed simulation paths of the unobserved level undiscovered rework, as shown in figure 4.10. Here, since our data are drawn from a simulation, we actually know the correct values

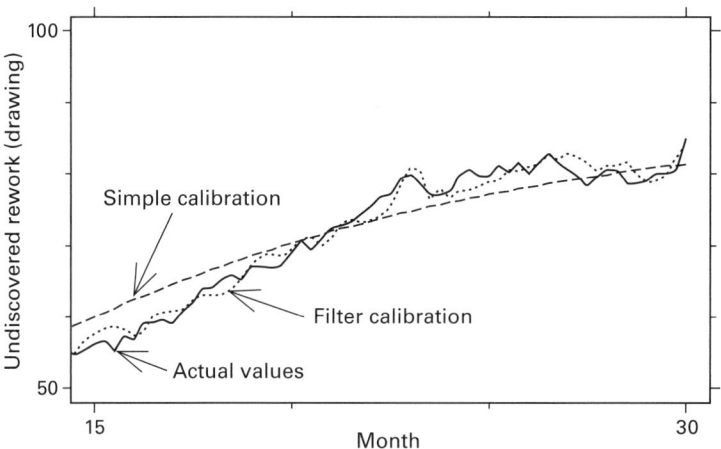

Figure 4.10
Indirectly measured values of undiscovered rework.

(note we zoom in on time 15 to 30 to make it easier to distinguish the runs).

When filtering is turned on, we get a better point-by-point estimate of the unobserved level. This is because the filter is using the model structure, along with the measurements of rework discovery (the outflow) and measured work force (proportional to the inflow) to adjust the trajectory of undiscovered rework.

While it can be useful to get a better idea of what happened in the past, most modeling work is intended to explore what might happen in the future, or as a consequence of different policy choices. To this end, look at the results after 30 months. While we could do this for a single noise realization sequence in the data-generation model, it is more useful to look at the potential spread of values given what has happened to time 30. Again we use a sensitivity simulation on the noise seed in the second half with 2,000 replications. The results of this, along with the trajectories coming from the two calibrations, are shown in figure 4.11.

In this case, the potential range of values spreads relatively quickly after time 30, though it does not spread that broadly. The simple calibration, because it was not tracking the data at time 30, starts outside of the 90% confidence interval, but quickly comes back in range, staying within the 50% band. The filter simulation, on the other hand, always remains in the 50% confidence interval. The two projections converge

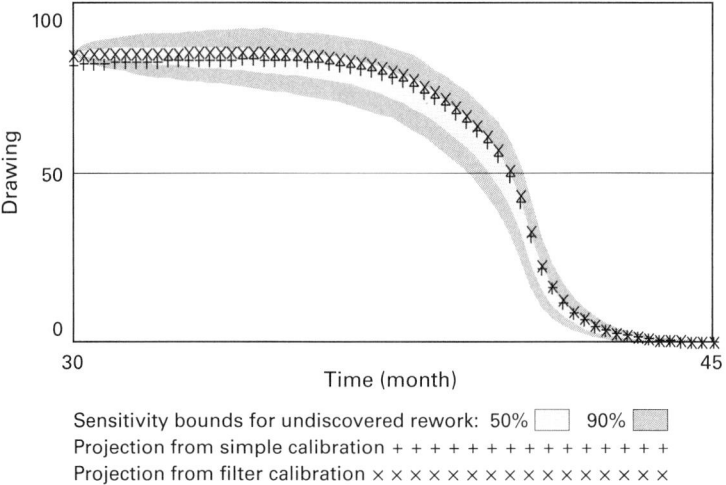

Figure 4.11
Trajectories for undiscovered rework.

at time 40, and the confidence bounds themselves converge by time 45. Eventually, all rework discovery occurs and there is nothing left to be discovered, irrespective of where undiscovered rework started.

In this particular example, using a filter gave us more accurate confidence bounds on our estimate and a better backcast of undiscovered rework. While it would be a mistake to draw a hard conclusion from one example, the results should not be that surprising. Models with clear endpoints, and this includes project models and diffusion models, are not prone to ongoing divergence of trajectories. Eventually, the model reaches a new state in which enough is almost always directly observable (or definitional) to pin down state variables.

Convergent Systems

Convergent systems tend toward some value as time progresses. In feedback terms, they are ultimately dominated by a strong negative feedback loop with little phase shift. In such systems, the value of filtering (or any other adjustment process) is likely to be low. Everything will converge to a clear steady state, and observations are likely available to determine the values of state variables in this condition. The biggest uncertainty about the path rests with the initial conditions, which interim state adjustment can help correct for, but the type of error simple simulation gives is bounded by that initial uncertainty.

Divergent Systems

Divergent systems, typically those displaying exponential growth, can be tricky to compare to data. This is because, as is well recognized in looking at behavior, everything seems to happen right at the end. As Dana Meadows succinctly pointed out (Meadows 1985), if the number of water lilies doubles every day, a pond that will be covered in 30 days is still half open water after 29.

Comparing data and models in such a situation is difficult. If equal weight is given to all observations, then those right near the end matter most. If weights are adjusted based on the state, as is appropriate statistically, the appearance of the trajectory can be very troubling. More importantly, for projection, the most recent observation can be very far off, which means the projection base needs careful scrutiny. In the following discussion, we consider simple calibration using different weighting approaches as well as a state-resetting approach.

For concreteness, consider a simple portfolio return model where the annual return is a normally distributed random variable, with a mean of 10% per year and a standard deviation of 20% per year. Over 100 years, $100 could grow to $1.4 million or, because of the randomness, a great deal more or less. One sample trajectory of the portfolio value is shown in figure 4.12.

This stochastic computation has a number of ups and downs, but ends up at $2 million, higher than the deterministic computation. In retrospectively evaluating that return, we could compute the average of the annual return rates (the arithmetic mean) as 12% per year or the imputed compounded annual return (the geometric mean) as 10.4% per year (for positive returns, the former will always be higher than the latter).

Computing average returns is very sensible when looking at changes in portfolio values, but it is instructive to see what results from calibrating a model against the realized portfolio value trajectory shown above. A naive calibration, one that equally weights all data points, will result in an underlying return of 10.6% with a trajectory as shown in figure 4.13. This is quite close to the geometric mean, and so we end up with a terminal value that is quite similar to the final value in the data.

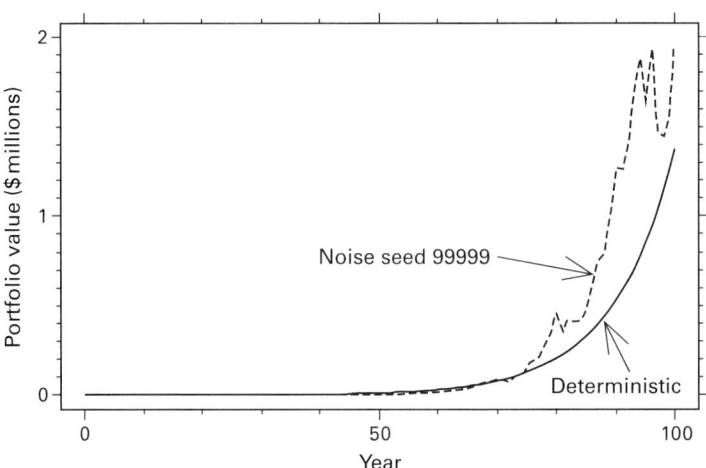

Figure 4.12
Portfolio value with one noise stream draw versus the deterministic result.

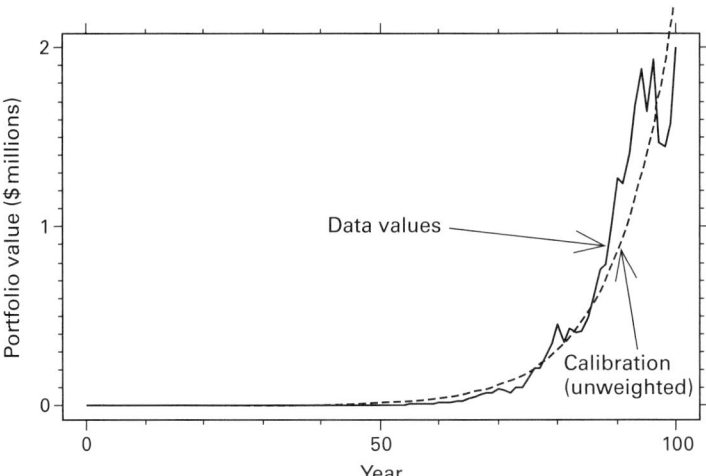

Figure 4.13
Unweighted calibration against portfolio value.

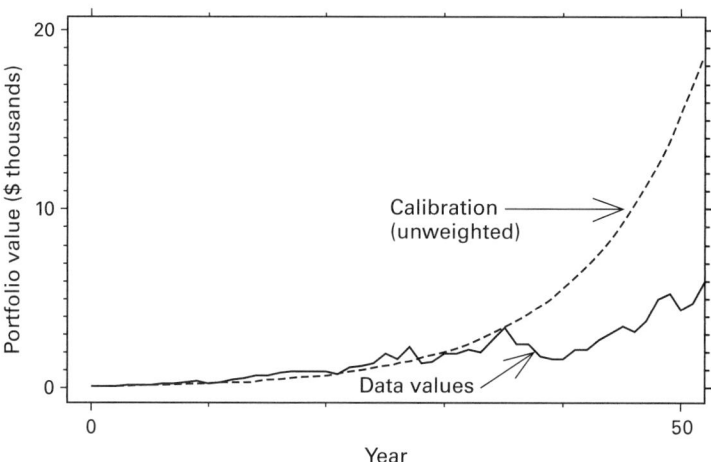

Figure 4.14
Unweighted calibration against portfolio value in the first 50 years.

Though the comparison in figure 4.13 looks like a reasonably good fit, it is hiding completely what happens in the first 50 years, as shown in figure 4.14.

The divergence from the data in the first 50 years is clear, but the numbers are so small they contribute nothing to the numerical fit of the model. When calibrating the model, we are assuming that the error is constant in magnitude, when in fact it is proportional to the portfolio value. There are two solutions to this: the first is to transform the data and look at the log of portfolio value; the second is to use a weight that changes over time. For regressions, using the log is the standard approach, but using a dynamic weight is informative in this case.

We make the weight proportional to 1 over the standard deviation of the expected prediction error point by point. In this case, we have all the information needed to compute that (it is simply $\dfrac{1}{\sigma \cdot V_t \cdot dt}$, where σ is the standard deviation of the average return and V_t the value of the portfolio at time t and dt the solution interval, 1 in this case). Using this weight in the calibration, we get an underlying return of 9% per year. Needless to say, the simulated values to the data provide a very poor fit visually, as the terminal value is only about $500,000.

For this example, we have intentionally selected a random noise sequence that exaggerates the difference between a simple and weighted calibration. This emphasizes the point that the better approach statistically, weighting the errors, yields a simulation that looks nothing like the data. Further, if we intend to look at future values of the portfolio, it would be a huge mistake to use the weighted calibration values. Even the unweighted results are off by 15% at the end of the simulation, which is a poor base for projection.

For this simple model it is, of course, also easy to reset the single-state variable with the realized values. This allows us to derive another estimate of the underlying return. Again, we can do so with and without compensating for the changing magnitude of the expected error. Compensating for the magnitude of the error generates an estimate of 12% per year, exactly matching the arithmetic mean, while treating all errors the same generates an estimate of 5.4% per year. The latter estimate seems almost nonsensical, until we look at the big dip that occurs right toward the end of the simulation. The average return over the final 10 years of the simulation is 5.8% per year. The average with the biggest portfolio value strongly dominates our estimate.

The dramatic difference between these two estimates does not necessarily show up graphically. Figure 4.15 shows the comparison of the weighted and unweighted calibration results with state resetting for the final 15 years. Just looking at this, it is not obvious that one solution is better than the other. With state resetting in place, even poor model choices can generate reasonable-looking comparisons.

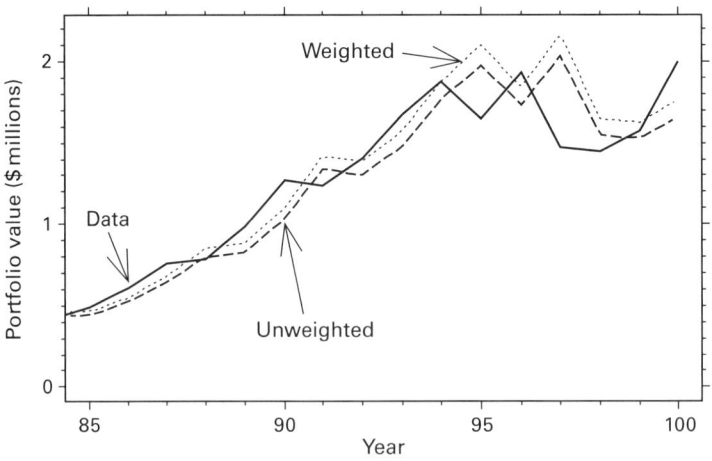

Figure 4.15
Weighted and unweighted calibration with state resetting.

The sensitivity of results to the weights used is also a big issue with state resetting. We saw that using the same weight at all times generated an implausible parameter estimate. Using the wrong weights can do the same thing. For example, basing the weight computation on the end of period value instead of the correct beginning of period value generates an estimate of 4.8% for the underlying return rate. This high sensitivity is related to the particular noise stream realized in our example, but the importance of weighting should not be underestimated when working with divergent systems.

Such high sensitivity to weights makes calibration using a formal Kalman filter difficult for divergent models. Because the driving and measurement variances are changing very rapidly, it is hard to maintain good estimates of the measurement to state covariance. While this may have only a modest effect on the gain computation, it can radically change the likelihood computation. That means estimates based on the Kalman filter can be numerically unstable, and may not have confidence bounds reflecting that instability. Further, in the Kalman filter implementation we have used in the previous examples (Vensim), the driving variance is assumed constant throughout the simulation. Because of this, there is no way to even try running a reasonable Kalman filter in the divergent case.

Population Census Data

Demographic models are a good example of systems for which the most reliable data are collected periodically (typically every 10 years) through a census, and other data on births, deaths, and possibly migration are available with higher frequency (typically annually). To estimate the population between the census periods, the births, deaths, and migration data can be applied starting from the census measurements. If we treat this as a purely mechanical exercise in arithmetic, we could then start again from the new census estimates (or even do a backward calculation from those and take a weighted average of the two computations to get values in between).

Frequently, however, we may want to make small adjustments to the regularly available data to make it more consistent with the census values. For example, such data might be on slightly different age groupings, and require some redistribution as a consequence.

We face precisely this situation with a population data set for Japan we are working with. Every ten years there is a census, with an intercensal estimate between each giving us population measurements every five years. From this, we know with some reliability the number of people in five-year age cohorts from 0 to 100 (with one-year cohorts from 0 to 4), as well as the number of people over 100. Mortality and fertility are based on less complete sampling, but are available annually. Mortality is available by five-year cohorts only till age 80 or, sometimes, 90. To put this all together into a complete model with five-year cohorts to age 100 we add in some adjustment factors:

- Mortality multiplier for ages 80–84, 85–89, and over 90 derived from the infrequently sample mortality number for these age groups relative to the mortality for all people over age 80.
- Mortality multiplier for ages 90–94, 95–99, and over 100, which must be estimated by comparing the generated population to the census values.
- Overall mortality multiplier to correct for possible bias in the mortality numbers that must be estimated.
- Overall fertility correction to correct for possible bias.
- Fraction of male births. This could be computed directly or estimated, and we chose the latter.
- Initial population adjustment—the model does not start at exactly a census point.

All of the above factors except the first are assumed to be time-invariant. Though that is demonstrably not true, it makes it easier to use the model for creating population projections under different assumptions, which was the reason it was developed. It is worth noting that migration is quite small for Japan and is therefore ignored.

Just as in the previous example, there are two ways to approach estimating the above parameters and thus providing a projection of future populations. The first is to run a regular simulation in which the state variables (population by age and gender in this model) simply advance over time based on the rates. The second is to adjust the states based on measured data for population when they are available. In both cases, this model uses a continuous cohort formulation (Eberlein and Thompson 2013) so that each state variable appearing in the model (e.g., population age 90–94) is in fact made up of multiple values (population age 90–90.124, 90.125–90.245 …). Formal filter construction won't work in this situation, but adding in an additional external function to reset the states is fairly straightforward (details on this function, including source code, are included in the electronic materials). The resulting parameter estimates are given in table 4.1.

Table 4.1
Parameter estimates with and without state resetting

Parameter	Simple	Resetting
Fraction births male	0.514	0.514
Fertility mult	1.00	1.00
Initial population adjustment	0.998	0.992
Mortality mult (overall)	0.995	0.973
male (90–94)	1.05	1.05
(95–99)	1.50	1.46
(100+)	2.48	2.50
female (90–94)	0.95	0.98
(95–99)	1.61	1.56
(100+)	2.78	2.81

The results are quite similar, with the exception of the overall mortality multiplier, which is 2% lower in the state-resetting case. The reasons for this warrant further investigation, though that will be left as an exercise for the reader. The resulting projections for population are very similar, though not identical, as can be seen in figure 4.16. Interestingly, in this case, the projections converge toward the year 2050. This is the result of the slightly lower assumed fertility that results from the slightly higher number of fertile females in the resetting simulation during the historical period.

If we look at the population 65 years old and older, on the other hand, there is a more distinct difference, as is shown in figure 4.17. The resetting values are significantly higher and, though there is some convergence later on, remain that way throughout. Interestingly, for the historic portion of the simulation, the values from the two runs are quite similar.

It is likely that the resetting-based projection will turn out to be more accurate, if only because mortality in all age groups continues to decline in Japan, and that will likely make both the projections for older populations underestimates. It is not clear, however, that the resetting approach dominates the simple simulation. While it has some attractive characteristics, it is also harder to explain. In many cases, simplicity is important enough to sacrifice refinements involving state adjustment.

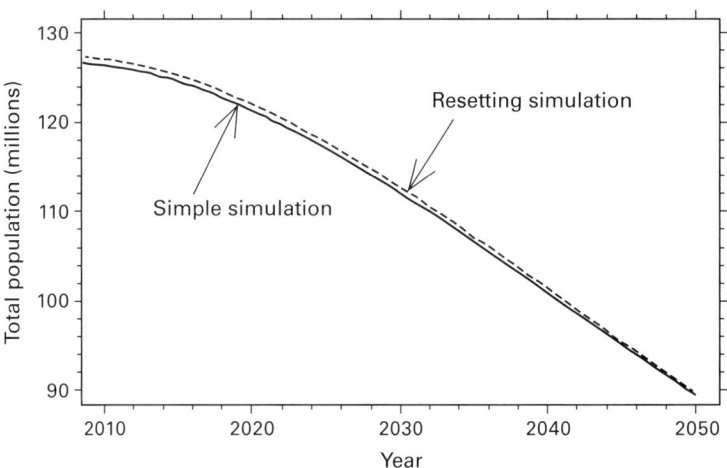

Figure 4.16
Population projections with and without state resetting.

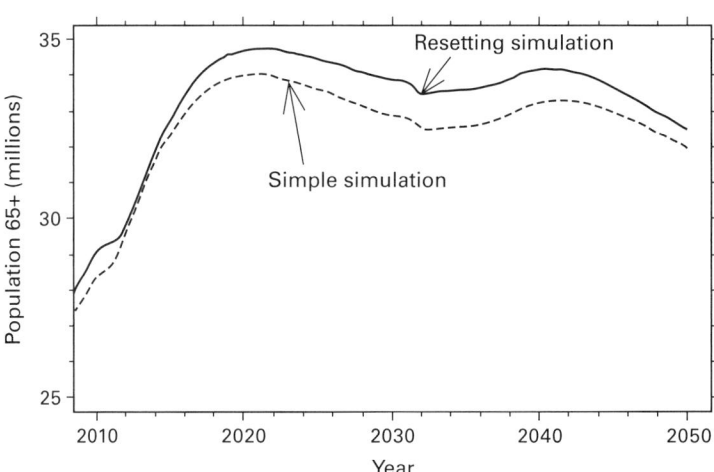

Figure 4.17
Population 65+ projection with and without state resetting.

Challenge 3: Create a model that displays exponential growth for a time and then levels off with a nonlinear capacity effect. Make one version with the capacity effect abrupt: for example, net_births = MIN(capacity-population)/DT,computed_net_births. Make another with a smooth nonlinear effect of capacity on births. Estimate the underlying birth rate for these two models with and without state resetting. How do the results differ?

Exogenous Variables and State Resetting

We have been focusing on models in which the state variables have an equation, but are explicitly moved to make use of available data. Exogenous variables are also moved based on the data, but do not have any model equations for their computation. That is not to say, however, that we cannot compute the same concept in the model.

Consider the model used in the opening example. In this model, incoming requisitions at wholesale are set equal to a delay of requisitions sent from retail. If we had data on incoming requisitions at wholesale, we could break that link and use the data directly. It would still, however, be possible to pass the requests coming from retail through a delay, and those could be compared to the data now driving the wholesale sector. In this way we would decouple the wholesale and retail sectors for the simulation. A similar decoupling could be done for the wholesale and manufacturing sectors. Such decoupling destroys the ability of the model to generate interesting dynamics, but it also turns an oscillatory system into a series of largely convergent subsystems, each of which is much more amenable to ordinary calibration.

This is an approach akin to partial model testing (Homer 2012), in which we are effectively breaking feedback loops in order to get the model (or at least subsets of the model) to replicate observed behavior point by point. Since feedback loops have been broken, the model effectively becomes a collection of smaller models. Thus, using part of a model instead of the whole model in this process is an equally valid approach. Even though system dynamics, by its very nature, calls for synthesis and inclusion, models are built up from components. And those components, especially parameter values, can definitely be developed and refined by analytical and reductionist techniques.

So do not hesitate to dig in using regression or any other appropriate technique. At the same time, don't lose track of the reason you are building the model. The most thorough and rigorous exploration of the specific formulations within a model does not always help answer the question the model is being developed to address.

Conclusion

We can make use of time series data: more than anything else, this point captures the essence of this chapter. Forrester, by constructing an example based on an oscillatory system, showed that it can sometimes be very hard to do this. Kalman found a way to deal with such very hard cases. We have, in a sense, closed the loop.

Practically, the decision to apply filtering or other forms of state resetting in a given situation depends on a number of things. For convergent systems, such as projects and diffusion processes, where everything is tending toward a clear steady state, it is less likely to be helpful. For divergent systems, such as those displaying exponential growth, it can be helpful but it is still challenging to interpret parameter estimates for such systems. For oscillatory systems, filtering or another form of state resetting is absolutely necessary if you want to make any point-by-point comparison to measured data over any significant period of time. One practical way to decide if you need to worry about filtering is to do the experiment that Forrester did in appendix K. Add some noise to your model and simulate it with different noise seeds. Does point-by-point comparison of the different runs still make sense? This test will work even if your model is not simply convergent, divergent, or oscillatory.

If you are using a model that needs filtering to compare it to data, you should seriously consider not making the data comparison. Most system dynamics models are approximations built up using substantial abstractions. While the mathematical form of the resulting models is still quite pure, the relationship of the computational model to the system being represented is inevitably more tenuous. Wrapping a model that is already pretty complicated in another layer of complexity moves it further still from the audience it ultimately needs to serve. Keeping the model simple and being able to articulate why calibrating against historical data is not helping build model confidence can be more effective than trying to prove your model is right by fitting it to data. Still, it is vitally important that you understand what comparing your model to measured data means, and the contents of this chapter should help you build that understanding.

Notes

1. The amount of time required to do the computation would be very large in this case. But, beyond that, numerical random number generators all have limitations when the number of draws made from them is high. Digital computers have limited precision, and this means random values may cycle over long draws and will certainly repeat more than a theoretical random input would.

2. Using the formula T=2*π * SQRT(l/g), we can solve for l as (T/2/ π)^2 * g = (1.5/6.28)^2*9.8 = 0.56.

3. We present only the linear time-invariant form of the measurement equation, but the determination of that from the nonlinear form could be done in the same manner it was for the state equations.

4. Another implication of the use of the full likelihood is that what is reported is the log-likelihood, and not 2x the log-likelihood, as happens when using a weighted sum of squares. Thus, using a chi-square test of significance, we need to multiply by 2 or, put another way, the 95% confidence intervals are found using a change of 2 rather than 4 in the reported payoff measures.

References

Anderson, B. D. E., and J. B. Moore. 1979. *Optimal Filtering*. Englewood Cliffs, NJ: Prentice-Hall.

Barlas, Y. 1989. Tests of model behavior that can detect structural flaws: Demonstration with simulation experiments. In *Computer-based management of complex systems: International system dynamics conference*, ed. P. M. Milling and E. O. K. Zahn. Berlin: Springer-Verlag.

Eberlein, R. L., and J. P. Thompson. 2013. Precise modeling of aging populations *System Dynamics Review* 29 (2): 87–101.

Encyclopedia of Mathematics. 2012. Kolmogorov—Smirnov test. Available at: http://www.encyclopediaofmath.org/index.php?title=Kolmogorov%E2%80%93Smirnov_test&oldid=22659. Accessed February 2, 2013.

Forrester, J. W. 1961. *Industrial dynamics*. Cambridge, MA: MIT Press.

Homer, J. D. 2012. Partial-model testing as a validation tool for system dynamics. *System Dynamics Review* 28 (3): 281–294.

Kalman, R.E. 1960. A new approach to linear filtering and prediction problems, J. Basic Eng., Trans. ASME, Series D. 82:1:35-45.

Lyneis, J. M. and D. N. Ford. 2007. System dynamics applied to project management: A survey, assessment, and directions for future research. *System Dynamics Review* 23 (2/3): 157–189.

Meadows, D. 1985. Nothing is so powerful as an exponential whose time has come. Available at: http://www.donellameadows.org/archives/nothing-is-so-powerful-as-an-exponential-whose-time-has-come/. Accessed January 31, 2015.

Peterson, D. W. 1975. Hypothesis, estimation and validation of dynamics social models. PhD thesis. Massachusetts Institute of Technology.

Peterson, D. W. 1980. Statistical tools for system dynamics. In *Elements of the system dynamics method*, ed. J. Randers, 224–241. Cambridge, MA: The MIT Press.

Schweppe, F. C. 1973. *Uncertain dynamic systems*. Englewood Cliffs, NJ: Prentice-Hall, Inc.

Sterman, J. D. 2000. *Business dynamics: Systems thinking and modeling for a complex world*. Boston: Irwin McGraw-Hill.

5 Combining Markov Chain Monte Carlo Approaches and Dynamic Modeling

Nathaniel D. Osgood and Juxin Liu

It can be very challenging to build a rigorous, empirically grounded dynamic model. Dynamic modelers rarely enjoy access to all desired data when building a model. Lacking sufficient data sources to estimate each model parameter in turn, dynamic modelers commonly seek to leverage empirical data concerning the emergent behavior of the system or its subparts. Model calibration processes can serve as valuable tools for arriving at point estimates—sometimes accompanied by measures of dispersion such as estimated standard deviations—of model parameters based on emergent data, but impose restrictive distributional assumptions, offer limited global insight into the plausibility of other model parameter values, are specific to a particular dynamic model, and cannot be directly translated into understanding of resultant variability in model results or policy tradeoffs. This chapter describes a unified approach for leveraging emergent empirical data that combines, generalizes, and synergizes the benefits associated with traditional calibration and sensitivity analysis while profoundly loosening such distributional assumptions. Specifically, we describe an approach that—given empirical data on emergent behavior of a system over time, one or more simulation models, and a probabilistic model specifying a prior distribution over parameters and likelihood of empirical data given parameter values—permits sampling from diverse posterior distributions: distributions over parameter values, but also over simulation model state, outputs, and intervention results over time. While outside the scope of this introduction, such techniques are also notable for their ability to be applied to inform model selection by deriving posterior probabilities over models.

The structure of the chapter is as follows. In the first section, we briefly note common alternative approaches for exploiting emergent data for insight into parameter values and model results, including the process known variously as "calibration" and "parameter estimation," and filtering, emphasizing some of the constraints associated with such methods. The following section offers an overview of the Bayesian framework and introduces the key notions of prior and posterior distributions. The section "Example and Notation" introduces a chapter-wide example applying the technique in the health science area. The section following discusses the fashion in which posterior sampling is implemented using Markov chains that converge to the posterior distribution, and techniques for assessing and enhancing that convergence. In light of the heavy computational burden exerted by Markov chain Monte Carlo methods, the section "Computational Concerns" discusses computational challenges associated with such approaches and ways of exploiting advances in computer hardware and distributed algorithms to secure higher performance. The subsequent sections discuss the software framework employed here in greater detail, as well as other such frameworks, as well as the steps by which we can perform MCMC for our example, illustrating the results emerging from such analyses. The section "The Impact of Empirical Data Availability on Posteriors" demonstrates the importance of data in limiting the support of the posterior by illustrating how successively larger amounts of data affect the sampled posterior. The penultimate section, "Pragmatics," provides suggestions for dealing with two common challenges confronting those applying MCMC—imposing boundaries on legal parameter values and securing convergence. The final section reviews the chapter and provides some concluding remarks.

Model Calibration

We noted above the traditional use of methods that seek estimate parameter values for a designated dynamic model by leveraging empirical data on emergent behavior of a system. Because of the complex and generally nonanalytic relationship between parameter values and emergent behavior, it is typically not possible to use such emergent behavior to directly identify closed-form estimates of the value of specific parameters. However, frequently such high-level data on emergent behavior—which is often more readily available than data regarding particular model parameters—helps constrain possible interpretations regarding parameter values, and even helps rule out otherwise plausible models.

In different subfields of dynamic modeling, this process is referred to by different names, such as calibration or parameter estimation (Oliva 2003; Schittkowski 2002; Sterman 2000). The most common use of this process is to arrive at "point estimates" concerning model parameters. The process of identifying the best-fit value of parameters is frequently an exercise in maximum likelihood estimation, which in general employs nonlinear optimization algorithms varying in sophistication and complexity. In some cases, such point estimates are complemented by confidence bounds or other interval estimates (Dogan 2007) that help to communicate the precision of those point estimates.

While calibration is a powerful tool for model construction and analysis, it suffers from some significant shortcomings. While point estimates are commonly employed for model parameters, such estimates obscure the degree of confidence that obtains for those calibration results. Even when confidence bounds or other interval estimates are available, major limitations remain. One shortcoming is that such interval estimates are local in nature and provide little sense of the *global* distribution across parameter values. Secondly, the discrepancy (energy, error, penalty, or negative payoff) functions used in calibration typically impose implicit but restrictive distributional assumptions concerning the error term associated with model estimates—for example, the assumption that the model error terms are normally distributed. Many—and likely most—dynamic modelers are unaware that they are imposing such assumptions, and adjusting the error models for different calibration outputs can be challenging. Thirdly, and more importantly, traditionally error bounds offer little insight into the resultant variability to be expected in model outputs or intervention tradeoffs. Achieving such insight requires performing a separate process known as sensitivity analysis, which typically puts aside any appreciation of variability in plausibility of different parameter values arising from the calibration itself, and instead simply examines variability in model output around the point estimate for model parameters arrived at by calibration. Finally, such calibration approaches typically adhere to the assumption that while there is uncertainty[1] with regard to the values of the parameters being calibrated, there is certainty regarding the nature of the model itself—that is, that a particular model is being calibrated. This poses problems for cases in which one seeks to use the empirical observations to rigorously select between multiple models.

Bayesian Framework: Prior and Posterior

While MCMC approaches can be applied in either a Bayesian or non-Bayesian context, the framework presented in this chapter presents a Bayesian approach. At a philosophical level, this means that it treats parameters as random variables, in contrast to frequentist approaches, which treat parameters as having fixed but unknown values. As a consequence of this perspective, the Bayesian treatment of parameters as random variables allows for principled incorporation of information outside of the observed data itself; that is, a prior distribution over parameters[2] (capturing our sense of confidence regarding how the parameters might be distributed prior to observing the new data). On the basis of empirical observations, Bayesian methods derive a posterior distribution that incorporates the "new knowledge" brought by the observations. This posterior distribution takes into account not only our a priori knowledge regarding the parameter values, but also the likelihood of observing the observed data given certain parameter values (and assuming the accuracy of the model).

The method discussed here is distinguished by its ability to sample from a posterior distribution over the parameters—rather than simply point estimates of those parameter values with or without confidence intervals or covariance statistics. While samples from this posterior distribution are in general dependent,[3] the ability to sample offers a significantly more textured picture of the likely parameter space than do the other estimates—thereby allowing recognition of multiple modes (peaks) in the distribution and co-variation across multiple parameters.

Given a simulation and probabilistic model and the ability to sample from a posterior distribution over associated parameters, we can readily compute posterior distributions over emergent simulation model outputs (e.g., a posterior distribution over lifeyears lived across a population, or incremental costs imposed), or over differences in policy outcomes (for example, the posterior distribution of the incremental gain of policy A over policy B with respect to some outcome, or—alternatively—the probability that policy A yields a favorable outcome compared to policy B). The capacity to sample from posteriors over parameter values can also allow for computation of the estimates commonly provided by other methods, such as (joint or marginal) point, interval, mode, percentile, and other estimates for parameters and functions of the parameters or scenario outputs. By contrast, the computation for such estimates can be very difficult when employing traditional calibration tools and frequentist methods.

Example and Notation

Discussion of simulation in the context of MCMC requires terminology and concepts that span two disparate but rich domains. To clarify such discussion, we seek here to introduce consistent notational conventions, and an example model that will be used throughout the chapter.

Simulation Model

Assumptions and Notation   We will treat a simulation model as a vector-based function $\vec{f}(t;\vec{\theta}_s)$ of time, mapping parameters (given by a parameter vector $\vec{\theta}_s$) to trajectories for one or more variables (associated with elements of the vector $\vec{f}(t;\vec{\theta}_s)$) over time. In general, the function $\vec{f}(t;\vec{\theta}_s)$ cannot be expressed in "closed form" (analytically—i.e., cannot be simply specified using a formula). Instead, we generally specify this function implicitly. For example, for a system dynamics model, formulas for flows and initial values of stocks as well as supporting information (e.g., regarding constants, auxiliary variables, etc.) are specified; the function $\vec{f}(t;\vec{\theta}_s)$ emerges from the interaction of such factors. With an agent-based model, the overall system behavior $\vec{f}(t;\vec{\theta}_s)$ would emerge through an interaction of agent behavior using rules, event and message handlers, and so on, as well as the initial populations of agents and such factors. While related (Monte Carlo) approaches have been extended to stochastic differential equations (Mbalawata, Särkkä, and Haario 2012), within this chapter, we confine our discussion to cases where $\vec{f}(t;\vec{\theta})$ is deterministic.

Example   This chapter illustrates the hybrid MCMC-simulation modeling approach by cross-linking a small simulation model and a probabilistic model. The simulation model described here can be downloaded from the section for this chapter on the MIT Press website for this volume. To enhance the transparency of exposition, the example is purposefully very simple and does not demonstrate the full generality of the approach. More sophisticated examples can be found in Dorigatti et al. (2012) and Coelho, Codeço, and Gomes (2011).

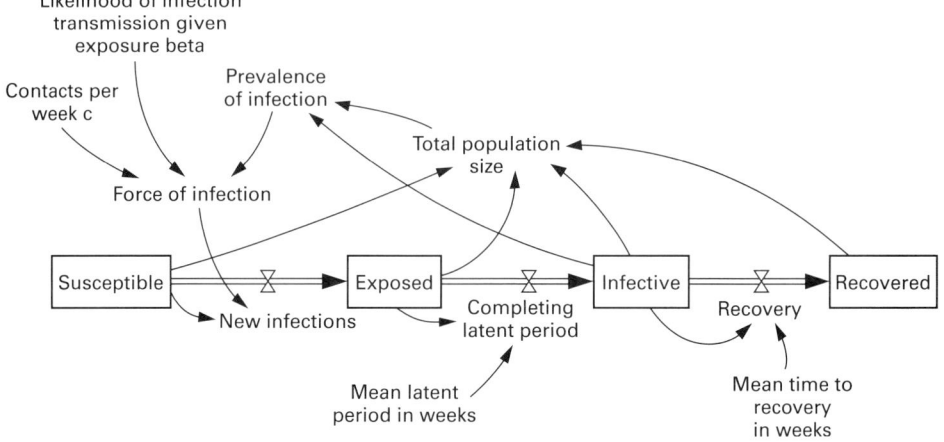

Figure 5.1
System dynamics simulation model for example.

The simulation model employed here is a classic compartmental mathematical epidemiology model characterized using state equations, and rendered here in the popular system dynamics modeling package Vensim (Eberlein and Peterson 1994). The simulation model is depicted in figure 5.1. Additional details concerning model equations and parameter values are included in tables 5.1 and 5.2, respectively. The model characterizes the transmission of an infection in a population. In a division harking back nearly a century to some of the first contributions in infectious disease modeling (Kermack and McKendrick 1927) and whose diverse variants have been extensively applied in mathematical epidemiology (Anderson and May 1991), the population is divided into four categories: susceptible, exposed (latently infected but not yet infectious), infective (with risk of pathogen transmission) and recovered, denoted as S, E, I, R, respectively. Only susceptible individuals suffer risk of infection, and recovery is treated as conferring permanent immunity. The per time unit (here, per week) risk of transmission to a susceptible (i.e., the force of infection) applies the classical principle of mass action, where the risk of transmission to a susceptible conferred by a given contact with another individual is represented as being proportional to the fraction of the entire population that is already infective. Specifically, each susceptible is represented as coming into contact with c individuals per week. The assumption is made that the population is well mixed and, thus, that the fraction of the individuals encountered by a given susceptible who are infected at any given time is equal to the fraction of the entire population that is infected at that time $\left(\dfrac{I}{S+E+I+R}\right)$. Thus, the reasoning follows that a susceptible individual encounters on average $c\dfrac{I}{S+E+I+R}$ infectious individuals per week. Assuming that each such contact between a susceptible and infective confers some small probability β of infection transmission, we follow common convention (Anderson and May 1991) of approximating the infection hazard (i.e., the force of infection—probability per week of infection) faced by each susceptible as the product of β and the number of infective individuals with whom they come into contact per week; that is, $c\dfrac{I}{S+E+I+R}\beta$.

Upon infection, a susceptible individual becomes a latently infected carrier of the pathogen, but one who is not yet capable of communicating it to others. During this time, they are resident in the exposed state (denoted E in the equations). Following a first-order delay associated with a mean incubation period of τ, that individual will proceed on to infectiousness. Similarly, recovery within this model is represented as a simple first-order delay, with mean recovery time of μ.

To aid exposition, the model begins in a state in which there is a single exposed individual embedded within an otherwise entirely susceptible population.

The model time horizon is 32 weeks. To be robust even in the context of parameter assumptions (encountered during the Markov chain exploration) that imply very brief time constants for model processes, the model is Euler integrated with a timestep of 0.0078125 (i.e., 1/128 of a week).

The formulas for this model—which accord with conventions typical in mathematical epidemiology—are given as follows.

In mathematical terms, the system can be written as a set of state equations:

$$\dot{S} = -cS\left(\frac{I}{S+E+I+R}\right)\beta$$

$$\dot{E} = cS\left(\frac{I}{S+E+I+R}\right)\beta - \frac{E}{\tau}$$

$$\dot{I} = \frac{E}{\tau} - \frac{I}{\mu}$$

$$\dot{R} = \frac{I}{\mu}$$

Table 5.1
Parameters for the example dynamic model

Variable Name	Notation	Value	Units
Force of Infection	λ	*Contacts per Week · Prevalence of Infection · Probability of Infection Transmission Given Exposure*	$\dfrac{1}{Week}$
Susceptible	S	INTEG (-*New Infections*,999)	Person
Exposed	E	INTEG (-*New Infections*,1)	Person
Infective	I	INTEG(*Complete Incubation-Recoveries*, 0)	Person
Recovered	R	INTEG (*Recoveries*,0)	Person
Completing Latent Period	i	*Exposed / Mean Latent Period in Weeks*	$\dfrac{Person}{Week}$
New Infections		*Force of Infection · Susceptible*	$\dfrac{Person}{Week}$
Mean Latent Period in Weeks	τ	0.42857 (i.e., 3 days)	Week
Prevalence of Infection		*Infectives / Total Population Size*	Unit
Recoveries		*Infectives / Mean Time to Recovery in Weeks*	$\dfrac{Person}{Week}$
Total Population Size	N	*Susceptible+Exposed+Infective+Recovered*	Person
Probability of Infection Transmission Given Exposure	β	0.005	Unit
Mean Time to Recovery in Weeks	μ	1	$\dfrac{1}{Week}$
Contacts per Week	c	350	$\dfrac{1}{Week}$

Table 5.2
Descriptive statistics for MCMC samples of parameters c and p: mean, standard deviation and standard error (top), quantiles (bottom).

	Mean	SD	Naive Standard Error (Ignores Autocorrelation)	Estimated Time-series SE
c	350.5042	1.15875	2.116e-03	0.0074980
p	0.7539	0.01634	2.984e-05	0.0001386

	2.5%	25%	50%	75%	97.5%
c	348.1210	349.6727	350.6536	351.299	352.8222
p	0.7213	0.7431	0.7541	0.765	0.7855

Because it will serve as the focus of data collection, we denote the rate (in persons per week) of incident cases in the model using the variable i.

Like most simulation models to be employed with MCMC techniques, this simulation model exhibits a number of parameters—only some of which will contribute to the parameter space occupied by the posterior. Specifically, reflecting the fact that contact rates can be challenging to determine for many communicable illnesses (e.g., those spread via airborne or droplet pathways), we will seek a posterior defined over (among other parameters) the final simulation parameter in table 5.1: the number of contacts per week c. It bears remarking that while we define these parameters to be constant in time, we could readily use the approach presented here with models whose parameters vary over time. Given values for parameter c, the model can be simulated. The appendix provides discussion of a variant of this model in which the parameter μ is also unknown. For simplicity, we view the population size $(S+E+I+R)$ and all simulation model parameters other than c (or c and μ for exercise 1) as being fixed and known precisely.

As in most simulation models, there are a variety of outputs from this model, a number of which could in principle be empirically observable. For simplicity, we concentrate on (noisy) observations related to one such variable: The rate *Completing Latent Period*. Operationally, this is very close to the per-week incident case count—for a given time, the count of individuals within the past week who have presented with illness—and we will treat the two as equal. For simplicity in exposition—and to emphasize the capabilities and the limitations of the approach—we use here a synthetic data set, generated using the known parameter values (as specified in the table), but with substantial noise added. The approaches described here can readily be applied to empirical data regarding multiple outputs.

We will then seek to sample from the posterior distribution for parameters. By sampling from the posterior over parameters, we can then sample from the posterior distributions of model output over time for various model variables. Some of these (such as the per-week incident case count) are related to the observables. Others will lack empirical data but will remain of interest, such as the prevalence of infection, or the fraction of individuals who remain susceptible or who are recovered.

Elements of the Probabilistic Model

What must be done to apply the MCMC approach described here for the simulation model above? The key new element that must be brought to the table is a *probabilistic model*. So that the reader can maintain the basic conceptual structure in mind when considering the details of our model, this section is divided into two parts. First, we will briefly outline the essential elements of the probabilistic model, noting the broad ways in which these relate to our example. The next subsection then turns to discussion of the detailed particulars for our example.

Central Probabilistic Model Quantities

Model Parameters   Below, $\vec{\theta}$ denotes a vector of parameters considered in the probabilistic model. $\vec{\theta}$ plays a central role in the technique being presented here, as both the prior $p(\vec{\theta})$ specified by the modeler and the posterior being sampled using MCMC are defined as distributions over $\vec{\theta}$.

Because the probabilistic model is used to evaluate the degree to which points in the simulation model's parameter space yield fits to the observed data, an important subset of the parameters $\vec{\theta}$ will be parameters in the simulation model whose posterior distributions are being sought—for our example, c. In addition, there will commonly be some parameters in $\vec{\theta}$ that are unique to the probabilistic model and that do not appear in the simulation model.[4] In our case, we will consider one additional parameter of the probabilistic model: p, which represents the probability that a given incident case is reported. When generating synthetic data for evaluation of the techniques described here, p holds the value 0.8.

Empirical Data   A second quantity central to the probabilistic model is the empirical data $\vec{y}$. For our example, $\vec{y}$ will consist of (noisy) empirical observations available every week concerning the incident case count; additional types of empirical data could readily be incorporated by extending $\vec{y}$. The empirical data used for the example model is available in the provided Excel spreadsheet (SEIR Historic Data v3 weekly data.xlsx). This file further demonstrates how the sampling of the empirical data is performed to simulate the effects of incomplete reporting.

Prior and Sampling Distributions   The probabilistic model consists of two basic components. The first is a *prior distribution* that encodes our "best guesses" concerning the distribution of parameter values prior to observation. Most of the analysis here will make use of a particularly simple prior distribution (a uniform distribution), but see "The Influence of the Prior" for the impact of an alternative prior distribution.

Secondly, the modeler must specify a sampling distribution[5] $p\left(\vec{y} \middle| \vec{\theta}\right)$ that specifies a likelihood that the empirical (observed) data would be observed given a certain assignment of parameters. For our example, the likelihood function will specify the likelihood of observing empirical data $\vec{y}$ in light of parameters $\vec{\theta} = \begin{bmatrix} c \\ p \end{bmatrix}$. When applying MCMC with simulation models, it will typically be infeasible to provide closed-form expressions to compute the likelihood that the empirical data would be produced from the parameter values. For our example, given assumptions about contact rate c and p alone, it is not possible to specify a closed-form formula to assess the likelihood of a given sequence of observed case counts $\vec{y}$ in the population. This is where the simulation model comes in, allowing us to map parameter values to values for quantities much closer to the observed data—quantities from which the likelihood of yielding the empirical data can easily be computed. For our example, pairing up an assumption regarding contact rate with a particular infection transmission model, one runs a simulation to provide values of the underlying (in persons per week) rate of individuals emerging from latent infection in the population at each point in time at which empirical incident case counts are reported (i). A simple probabilistic model can then be specified to formulate the likelihood that there would be a count of n reported cases in light of there being i actual underlying individuals becoming infectious in the population. In short, the simulation model forms a key link in determining how likely it is the observed data would result from a given set of parameter values by telling us the consequences of parameter assumptions for quantities similar enough to what we empirically observe that the likelihood of the empirical data (given the model and model parameters) can be calculated.

Posterior Distribution   Given the two above components—the prior $p(\vec{\theta})$ and sampling distribution $p(\vec{y}|\vec{\theta})$—we can derive the posterior probability density $p(\vec{\theta}|\vec{y})$ by following Bayes' rule. Bayes' rule relates the likelihood of *a* given *b* to that *of b* given *a*, in light of the probabilities of both *a* and *b*.

Applied in terms of $\vec{\theta}$ and $\vec{y}$, Bayes' rule gives

$$p(\vec{\theta}|\vec{y}) = \frac{p(\vec{y}, \vec{\theta})}{p(\vec{y})} = \frac{p(\vec{y}|\vec{\theta})p(\vec{\theta})}{p(\vec{y})}.$$

The two elements in the numerator of the quotient on the right-hand side are the sampling function and the prior, respectively—the elements of the Bayesian model that we noted above are required. Critically, $p(\vec{y})$ here is simply a constant determined by the empirical data we have at hand—and unaffected by $\vec{\theta}$. We are here seeking to derive a posterior for given specified empirical data and thus have a *fixed* $\vec{y}$ with a *varying* $\vec{\theta}$,

$$p(\vec{\theta}|\vec{y}) \propto p(\vec{y}|\vec{\theta})p(\vec{\theta}).$$

Given observations $\vec{y}$, the ability to compute the posterior value $p(\vec{\theta}|\vec{y})$ allows us to assess the probability density of different possible parameter values in light of the observed data. In the next section, we will see how MCMC algorithms allow us to *sample from* the posterior over not only parameters, but also over model outputs, differences between model scenarios, and other quantities.

Probabilistic Model for Our Example

Having established the basic framework above, we now turn to considering the probabilistic model for our example.

Likelihood Function   Recall from the above that the likelihood function $p(\vec{y}|\vec{\theta})$ in the probabilistic model is a function that—for a specific parameter setting $\vec{\theta}$ (e.g., *c* and *p*)—specifies the probability of observing data $\vec{y}$ (e.g., a sequence of reported cases over time). This function will serve as a key element—and typically the most important element—in computing the posterior probability density (the probability density of the parameters holding particular values in light of the observed data). This subsection examines the formulation of the likelihood function, considering first the likelihood of a single observation at a certain point in time, and then continuing on to consider the likelihood of observing a sequence of observations.

Time-Specific Likelihood Function   To derive the likelihood function for our example, we consider that given a hypothesized true, underlying number of individuals moving from latency to infectiousness within the past week, only a certain fraction of them will in fact be detected and reported. For simplicity, we consider a situation in which there are no false positives—only false negatives—and where the chance that each such individual is detected is a simple constant (p) and independent of the number of other individuals who are infected and reported. Given these simplifying assumptions, the empirical reported count of infected individuals y_t at measurement point t is binomially distributed, with probability p of success ($1-p$ of failure), and with a number of trials equal to the number of individuals i_t in the underlying population who emerged into infectiousness within the preceding week. For simplicity, this count is approximated[6] here as the rate (in persons per week) of individuals i_t emerging into infectiousness at measurement point t, which is itself some emergent function of c—and thus of $\vec{\theta}$. That is, we will treat $y_t \sim binom(i_t, p)$. This binomial relationship in each subsequent week t ($\{t=0,31\}$) will form a key part of the sampling function $p(\vec{y}|\vec{\theta})$. While it is a parameter of the probabilistic—rather than the simulation—model, the parameter p will be a member of $\vec{\theta}$. Thus, we have $\vec{\theta} = \begin{bmatrix} c \\ p \end{bmatrix}$.

Combining Time-Specific Likelihood Functions   Our definition of a likelihood function for a particular week t provides us with a way to calculate the likelihood of a given empirical data point on the basis of a particular output from the model. In general, however, we will have more than one data point used in likelihood functions. For our example, we have empirical observations each week in turn. From the above, we know how to compute the value of the likelihood function for each such week. But how do we combine these components?

A common assumption that is made is one of *conditional independence* of the likelihood functions. Thus, for our example, we have

$$p\left(y_0,y_1,\ldots y_{T-1}\middle|\vec{\theta}\right)=p\left(y_0,y_1,\ldots y_{T-1}\middle|i_0,i_1,\ldots i_{T-1},\vec{\theta}\right)=\prod_{t=0}^{T-1}p\left(y_t\middle|i_t,\vec{\theta}\right).$$

We note that this assumption of conditional independence is far less onerous than assuming independence of the y_i's themselves (which would frequently be highly problematic, due to strong dependence between the observations at different points in time). Instead, the claim is that the dependence between the y_t is accounted for by the $\vec{\theta}$ term.

For our sampling function, as noted above, we treat the observed count of prevalent cases as binomially distributed, with a count of trials of the actual number of underlying incident cases i_t (putting aside the issue as to whether or not they are reported).

Thus the sampling function is

$$p\left(\vec{y}\middle|\vec{\theta}\right)=\prod_{t=0}^{T-1}BinP\left(y_t;i_t,p\right)=\prod_{t=0}^{T-1}\left(\binom{i_t}{y_t}p^{y_t}\left(1-p\right)^{i_t-y_t}\right),$$

where y_t is the empirically observed counts at week t of prevalent cases (rounded to the nearest integer) and $BinP\left(y_t;i_t,p\right)$ is the probability mass function of the binomial distribution, giving the probability of obtaining exactly y_t successes given i_t draws, each of which has a probability p of success.

The Prior   In addition to specifying the sampling function, a modeler combining Bayesian MCMC with simulation modeling must further provide a prior distribution over the parameter space $\vec{\theta}$. By contrast, non-Bayesian MCMC would omit the prior altogether, and would proceed by only sampling from the sampling function itself—rather than from a posterior distribution.

The prior distribution captures in a distributional form our senses of plausible values of parameters before any empirical data is available. The degree of influence of a prior on the posterior distribution sampled with MCMC will typically vary widely depending on the volume of observations on which data is available. When the sample size is very small, the prior may dominate the posterior. For such cases, users need to set up the prior more cautiously, and may wish to examine the impact on the sampled posterior of changed assumptions concerning priors. As examined in the section entitled "The Influence of the Prior," when the sample size is larger (as is the case here), the likelihood dominates the posterior. For such cases, the inference results are not sensitive to the choice of prior, and the user may choose the prior with emphasis on simplicity and convenience.

In most of our examples, we use uniform priors, where the support (range of possible values) for those priors extends from 0 to 1 for p, and from 0 to 500 for c. We will further make the simplifying assumption of independent priors for each of the parameters $\begin{bmatrix} c \\ p \end{bmatrix}$ in turn. Our prior is thus $p(\vec{\theta}) = I_{0 \leq c \leq 500}(c) I_{0 \leq p \leq 1}(p)$, where $I_C(x)$ is an indicator function evaluating to 1 if the criterion C is met by argument x, and 0 otherwise. The section "The Influence of the Prior" briefly examines the impact of using a notably different prior.

The Posterior Distribution   Given a parameter vector $\vec{\theta}$ (here, $\begin{bmatrix} c \\ p \end{bmatrix}$) and observations $\vec{y}$ (here, the sequence of incident cases over time), the value of the posterior distribution for those parameters (up to a constant) can be computed as $p(\vec{\theta}|\vec{y}) \propto p(\vec{y}|\vec{\theta}) p(\vec{\theta})$, where $p(\vec{\theta})$ is the prior distribution, and $p(\vec{y}|\vec{\theta})$ is a likelihood function. Thus, for our example, we have $p(\vec{\theta}|\vec{y}) \propto \prod_{t=0}^{T-1} BinP(y_t; i_t, p) I_{0 \leq c \leq 500}(c) I_{0 \leq p \leq 1}(p)$. In the next section, we will see how this capacity to calculate the posterior probability density of a parameter vector $\vec{\theta}$ can be used to *draw samples* from this posterior distribution—that is, to generate parameter vectors $\vec{\theta}$ such that the probability (density) of drawing a given $\vec{\theta}$ is proportional to its posterior probability density of occurrence.

What Is Markov Chain Monte Carlo? Motivations for Markov Chain–Driven Sampling Approaches

The above described the means by which we could calculate the value of the posterior distribution (up to a constant factor). The ability to evaluate the posterior at a given point in parameter space offers some power in and of itself. For example, we could use it to search for a maximum likelihood estimate, to assess the modality of the space, or even to reconstruct an approximation to the posterior surface.

However, for many purposes, working to simply evaluate the posterior at points in parameter space—such as at gridpoints—is either cumbersome or infeasible. The primary difficulty with such approaches is that the number of gridpoints associated with the space rises geometrically with the number of parameters (dimensions of the space).

While this is not intractable for our simple problem (where $n=2$), even for relatively small n, it can be infeasible to thoroughly explore this space.

A far more efficient strategy is to instead *sample* from posterior space—to draw Monte Carlo[7] samples of $\vec{\theta}$ from that space according to the posterior density. Such sampling eliminates the need to explicitly represent the high-dimensional space. Instead, we can simply compute statistics over the samples (e.g., compute histograms using the samples, or count the fraction with certain characteristics, or compute the mean or extrema of a function of $\vec{\theta}$ using samples), rather than by iterating through an intractably large series of gridpoints. Critically, in contrast to gridding, such sampling will tend to concentrate in higher probability regions of the space, thereby using the limited computational resources to better effect.

MCMC as a Sampling Mechanism

MCMC is a principled way of generating samples from a distribution specified up to some constant. Because (random) samples from the posterior distribution are generated over and over again, the method is termed a "Monte Carlo" method. A "Markov chain" is the basic mechanism used to generate the samples, with the current state of the Markov chain reflecting the most recent sample generated.

Using the samples generated from the posterior, we can readily do many useful things—calculate credibility intervals for $\vec{\theta}$; derive approximations to or sample statistics on marginal distributions for the simulation model outputs; assess the fraction of time for which one intervention yields favorable outcomes compared to another; or sample from the marginal distribution for the difference in gain between two interventions, A and B.

Overview of the MCMC Process

This section describes the rough operation of MCMC simulation. While it is common to rely upon the use of third-party software packages to undertake the core algorithms, it is important that the user of the packages have some sense as to how the algorithms work, so as to be able to detect and address pathological behavior, assess whether the algorithm is "converging" on an appropriate distribution, and so on.

There is a broad family of MCMC algorithms. All members of this family share the basic characteristic of sampling from the posterior for

$\vec{\theta}$ in such a fashion that the stationary distribution of the resulting samples is the posterior. However, there are quite a few variants within the MCMC family. While we present one common and simple algorithm here (termed "random walk Metropolis Hastings"), there are many others, differing across a number of dimensions. We refer the interested reader to Robert and Casella (2010) and Gelman et al. (2004).

Pseudo-code for a single-walker version of the MCMC algorithm used here is shown below. The fundamental operation of the sampling algorithm presented here is to explore parameter space $\vec{\theta}$ (for our example, $\begin{bmatrix} c \\ p \end{bmatrix}$). The algorithm seeds a walker at some location $\vec{\theta}$ in parameter space. Until some convergence criterion is achieved (see below), the algorithm generates candidate samples ($\vec{\theta}^*$) at a randomly chosen local distance from the last sample (with successive such samples leading to "random walks" around this parameter space). For each such candidate value $\vec{\theta}^*$, the algorithm "flips a weighted coin" to determine if that candidate is to be accepted (where the chance of acceptance rises with the ratio $\dfrac{p(\vec{\theta}^* | \vec{y})}{p(\vec{\theta} | \vec{y})}$ of the posterior values between the candidate point $\vec{\theta}^*$ and the current point). The posterior of the candidate is computed by multiplying the prior $p(\vec{\theta}^*)$ and likelihood function $p(\vec{y} | \vec{\theta}^*)$—the latter of which requires running the simulation model.

$i \leftarrow 0$

Pick an initial value $\vec{\theta}_0$ // can draw from crude distribution

Until some convergence criterion is satisfied

   Pick a random value for perturbation $\Delta\vec{\theta}$

   Create a candidate value $\vec{\theta}^* \leftarrow \vec{\theta}_i + \Delta\vec{\theta}$

   if uniform(0,1)$\leq$ $\min\left(1, \dfrac{p(\vec{\theta}^* | \vec{y})}{p(\vec{\theta}_i | \vec{y})}\right)$

       $\vec{\theta}_{i+1} \leftarrow \vec{\theta}^*$ //Accept (transition to candidate location)

   else

       $\vec{\theta}_{i+1} \leftarrow \vec{\theta}_i$ // Reject (remain at current location)

   Emit $\vec{\theta}_{i+1}$ as next sample // emit new sample regardless whether

// it resulted from acceptance or not

   $i \leftarrow i+1$

If the candidate point is accepted, it is emitted. If the candidate point is not accepted, the last sample is re-emitted. The algorithm design allows it to dwell longer at (repeating the samples from) points in $\vec{\theta}$ space that are of high likelihood. At the same time, the fact that there is a nonzero likelihood of the algorithm accepting a candidate sample associated with lower probability density than the current point will help the algorithm avoid being "stuck" in points that are merely locally optimal, and to instead have a chance of wandering through low-likelihood regions to regions of space that are higher likelihood yet.

It turns out that the *acceptance* of a sample is not sufficient for it to be considered a fully legitimate sample from the posterior distribution. There are two further considerations at issue. Firstly, recall that MCMC approaches the posterior distribution asymptotically—that is, given enough samples. To lower the impact of the vagaries of the initial conditions, the algorithm must have completed a minimum number of candidate samples (forming its "burn in period"). Commonly for many MCMC problems, no recording of the accepted samples will take place until thousands—and potentially several tens or hundreds of thousands of samples—have been drawn. For a simulation model, computing the posterior for each such candidate sample is a reasonably expensive operation, and there may not be the luxury of running such a long burn-in period. However, burn-in periods of thousands of samples should be considered; the section "Assessing Efficiency and Convergence" discusses tests that can be run to assess the stationarity of the results. A second consideration—autocorrelation of samples—is discussed in the "Autocorrelation" section.

Within our example, we make use of a pre-existing implementation of the random walk Metropolis-Hastings algorithm given in pseudocode above. Specifically, we use the function MCMCmetrop1R within the MCMCPack package (Martin, Quinn, and Park 2011). Within the implementation included here, we use this package by specifying the (log) posterior function specified above, as well as other parameters specifying the initial point in parameter space, the burn-in period, and the scaling factor for the random perturbation. The log posterior function needs to be called by MCMCmetrop1R with a two-dimensional vector of parameters $\begin{bmatrix} c \\ p \end{bmatrix}$, and to return the appropriate log of the posterior function. However, to do that, the log posterior function requires information on the vectors of empirical data and the times at which that data is measured (and is to be matched). Taking advantage of R's functional programming abilities (Abelson and Sussman 1996), we create a higher-order function generateFnForMC-MCMetrop1R that accepts the vectors (empirical data $\vec{y}$), and which then returns a "log posterior" function that will itself accept a parameter vector and compute $p(\vec{\theta}|\vec{y})$—the log posterior of that parameter vector in light of the (still-remembered) data vectors. Given this function, we can, for example, perform a 100,000 iteration MCMC by calling MCMCmetrop1R in the following way:

```
mcmc100kRuns <- MCMCmetrop1R(generateFnForMCMCMetrop1R(empiricalDataVe
ctor, timesForDataPoints), theta.init=initialPoint, thin=1,mcmc=100000
,burnin=20000,logfun=TRUE,verbose=1,tune=c(0.037,0.037,0.037),
V=diag(4E-4,3));
```

In the treatment above, we have glossed over the question of the location of the starting point in parameter space. Within the R code fragment above, we indicated "initialPoint" as the starting point, but have not specified how a suitable point can be found. For many posterior distributions, it should be suitable to start at any point in parameter space that is associated with a nonzero posterior probability. One way to identify such a point is to draw random vectors in some bounded retanguloid region parameter space until we obtain one with nonzero posterior probability. Within the example code, we have provided the utility function *identifyRandomNonZeroPosteriorTheta* to perform this action. The function can be passed two vectors that delineate the space from which to draw an initial value: a vector of minimum values for each successive parameter in turn, and a vector of maximum values for those parameters. Finally, the function should be passed the function that, when called, will determine the log posterior value.

Assessing Efficiency and Convergence

It can be shown that the algorithm presented in an abstract way above converges to the desired posterior distribution. This guaranteed convergence reflects the dual facts that there is only one stationary distribution of the Markov chain, and that the actual posterior is a stationary distribution for the Markov chain. However, what is not so clear is how quickly or efficiently this convergence will be realized. The MCMC algorithm as a whole involves a set of parameters that can be adjusted—factors such as the thinning rate and perturbation scale—and part of the art of effective use of MCMC methods is to "tune" these factors to achieve the best performance of the algorithm.

In general, to examine this, it is important to observe statistics generated from the MCMC algorithm, such as the acceptance rate, autocorrelation observed between accepted samples, and the within-walker versus between-walker variance. Based on those statistics, we can adjust algorithm parameters such as the size of the perturbations, the duration of the burn-in period, and the thinning rate to apply for accepted samples. While deferring a more extensive treatment to other references (Gelman et al. 2004), we briefly discuss these statistics and relevant adjustments here.

Acceptance Rate The first—and perhaps most basic—criterion for consideration is the acceptance rate: the fraction of candidate samples that are accepted. There is a need to avoid both extremes of the acceptance rate. If the acceptance rate is too low, it suggests that a great deal of computational effort (primarily simulations) are being expended without contributing to the movement of the walker. That is, the walker is spending too much time in low–posterior-density areas of parameter space. Frequently this is caused by spending too much time evaluating candidates in new areas of the space that are not fruitful. One way of lessening the risk of this is to decrease the scale of the perturbation that yields a candidate sample.

By contrast, if the acceptance rate is too high, it suggests that the algorithm is spending too much time around high–posterior-density areas of the parameter space, and that greater exploration may be needed. To enhance exploration of the space, it may be advisable to heighten the scale of exploration, such that bigger perturbation steps are taken.

Based on theoretic analysis that makes use of specific assumptions, past studies of acceptance rates have suggested an acceptance rate of multidimensional spaces of approximately 23% as most appropriate (Gelman et al. 2004), but others recommend considerably higher rates (above 50%, such as 75%). Because the former analysis was based on a context significantly different from that applying here, such an acceptance rate guideline is only approximate.

Initial-State Transients   As noted above, the Markov chain used for a single MCMC walker must begin in an initial state (i.e., with an initial parameter vector), which is typically not particularly privileged and may even be selected randomly. While the walker will tend to diffuse away from this initial state (and toward areas of higher posterior value), it is to be expected that the initial component of the sequence of accepted points in parameter space will be strongly influenced by the imposed initial state. Because the successive perturbation of the walker's position includes an important stochastic component, over time, the walker's location in parameter space will tend to grow less correlated with the initial state. Eventually, the walker can be expected to transition across broad areas of the parameter space—having effectively "forgotten" the initial position.

Recall that the distribution of values emitted by the algorithm will tend to converge to the posterior distribution. If we were to take all accepted values from the start and prematurely terminate the algorithm, the sampled set of values from parameter space would tend to be significantly skewed by the initial location. To minimize the effects of this initial "transient" on the statistics, MCMC methods traditionally establish a fixed-size (e.g., 10,000 accepted points) "burn-in" period, during which the accepted samples are not reported. A common pattern is to discard the first half of the samples. Once the algorithm reaches beyond the burn-in period, the Markov chain is assumed to have diffused sufficiently around the space to be fairly independent of the initial state.

An additional technique that helps lower the degree to which the vagaries of initial state selection bias the research findings is to make use of several chains. That is, rather than "putting all eggs into one basket" and starting a single walker (at a single initial point), we deploy a set of such walkers, each beginning at different points. There are two notable derivative advantages of this approach. Firstly, the progress of the different walkers is essentially decoupled, and can thus proceed in parallel. This "embarrassingly parallel" problem can then be naturally mapped onto multiple computing units, such as cores of a given processor, multiple processors on a single computer, or different computers. A second major advantage of employing multiple walkers (quite independent as to whether the walkers are executed concurrently or sequentially) is that statistics can be compared between the walkers, thus giving a sense as to the degree to which walkers starting at different initial points are yielding comparable results.

Autocorrelation   Recall that the goal of MCMC is to serve as a means of drawing samples from a distribution (here, from the posterior distribution for the parameters). By default, such samples typically exhibit substantial autocorrelation—that is, a given sample will often be dependent upon previous samples. Such autocorrelation arises naturally from use of random walkers. While many uses of such samples—such as those presented here—are robust in the context of such samples, some uses benefit from the ability to assume that the samples are statistically independent. (An example would be applications where sample means of the distribution need to be taken and a calculation is required to determine how many such samples to take.) Because each successively accepted sample within many MCMC algorithms is based upon a local perturbation of the immediately previous accepted sample, significant correlation—and, thus, statistical dependence—between such samples is to be expected, and should be considered in decisions involving sample size. Such statistical dependence can be readily identified using autocorrelation plots.

In order to reduce the correlation between the generated samples, MCMC implementations commonly include a "thinning factor" k. An integer value of k greater than 0 indicates that only one out of every k accepted samples is to be reported to the user, with the remainder discarded following acceptance. However, because thinning can adversely affect estimate efficiency, the reader experiencing high autocorrelation is encouraged to rely on longer chains (i.e., a larger number of Monte Carlo iterations) in place of thinning.

Assessing Convergence   There are several approaches to assess convergence of the MCMC algorithm for a given walker. One option is to use the Heidelberger and Welch stationarity and interval half-width tests. Given a sequence of samples from an MCMC algorithm, the coda tools allow for examining this statistic using the command heidel.diag(mc). Another option is to make use of the multiple walker strategy advanced above. Specifically, we continue the evolution of that walker until the variability observed within the walker is comparable to that observed between walkers. This variability—and, thus, convergence—can be measured for one or more scalar estimands (typically parameters, but potentially other quantities as well, such as outputs from the simulation models). Within the coda tools, if one has samples produced by multiple walkers (mcA, mcB), one can combine them using the function call:

```
multiWalkers <- mcmc.list(mcA, mcB).
```

Computing the within- and between-walker variance allows for determination of both a lower bound on the variance of the posterior given the empirical data (the within-sequence variance itself) and an upper bound on that variance (defined by a linear combination of the between- and within-walker variance). These, in turn, allow for computation of two useful quantities—scale reduction factor R for each estimand whose variance is being taken, and the estimate of the effective number of independent simulation samples. Space constraints prevent a discussion of these quantities here; interested readers are referred to Gelman et al. (2004).

Computational Concerns

The technique presented in this chapter is well suited to a broad range of modeling situations in which data arrives relatively infrequently (e.g., with daily or lower frequency). Compared to traditional approaches such as maximum likelihood, the current method offers greater generality and flexibility. However, this power comes at considerable computational cost. It is important that modelers planning to employ MCMC methods anticipate such cost in advance and in their scheduling.

Running the MCMC algorithm requires high computational effort. Traditional MCMC using parametric probabilistic models (rather than simulations) are routinely run for "chain lengths" of tens of thousands or hundreds of thousands of accepted samples. Such experiments commonly require hours or days to run, reflecting a need for a burn-in period, the large fraction of posterior evaluation that is typically for rejected samples (computation that has not directly contributed to exploration of the space), and—sometimes—the need for thinning.

When combining MCMC with dynamic models, the computational cost with posterior evaluation is considerably higher, because a simulation is required to evaluate the sampling function. Such simulations may themselves require anything from a fraction of a second (such as in our example) to hours. Making efforts to secure economies in the MCMC evaluation is thus very important.

One important technique to reduce evaluation times is to take advantage of the embarrassingly parallel character of the multiple walkers and to evaluate different walkers on different computational units—cores, processors, or distributed machines. While there is some coordination overhead required for such an arrangement—configuring the system for a particular set of resources (e.g., cluster of machines), achieving appropriate load-balancing of the different computational units, retrieving and integrating results, and so on—the benefits are likely to repay this investment. While the simple system presented here does not yet support this option, we hope to evolve it in that direction.

Even given concurrent execution of this sort, evaluation of longer-running models (e.g., agent-based models with large populations) is likely to remain problematic. Such simulations frequently require minutes of simulation per realization, meaning that a particular walker may require many months of computational effort. Scale modeling (Osgood 2007) may be a viable option for reducing the computational time involved for individual-based models, but could still entail many days' worth of effort.

Software Framework

The code presented here makes use of a simple software framework designed to ease the process of performing MCMC together with a popular class of simulation models. Specifically, this framework—which is described further in the appendix—is designed to leverage the statistical software package R—which is widely used for MCMC—to help perform MCMC analyses together with models built in the system dynamics simulation package Vensim. This pairing of software packages allows for a natural division of responsibilities. Specifically, R can draw on its third-party MCMC package support, extensive library of statistical features, rich graphical display capabilities, support for a variety of export formats, and sophisticated and concise language to specify the probabilistic model and report results. When R computes the sampling function, it calls Vensim to perform the simulation.

While support for use of dynamic models with MCMC is not nearly as extensive as support for statistical models—or of dynamic models by themselves—several additional options do exist for those seeking to pursue this approach. The most recent versions of Vensim themselves provide direct support for MCMC. While much of the work to diagnosis convergence behavior must be conducted outside of Vensim, use of this functionality offers virtues in terms of both performance and convenience. The popular WinBugs framework (Ntzoufras 2011; Lunn et al. 2012) now provides direct support with dynamics described by classes of ordinary differential equations.

Running the MCMC Analysis

The MCMC analysis can be run using the *MCMCmetrop1R* function of the *MCMCpack* package. Here, we run the *MCMCmetrop1R* algorithm for 300,000 ($3*10^5$) iterations, preceded by a burn-in period of 20,000 iterations, with a single walker and no thinning.

First, we define a posterior, to be used both in identifying a suitable initial point and in computing the chain itself:

```
fnPosteriorAllDataPoints2Parms <- generateFnForMCMCMetrop1R2Parms(vecE
mpiricalData,vecTimesForDataPoints, logPriorUniformFn2Parms)
```

We then issue the following command:

```
mcmcBoundedRuns2ParmsLatencyPeriod300kTune1Pt01   <-   MCMCmetrop1R(fnP-
osteriorAllDataPoints2Parms,identifyInitialTheta(lowerSearchBounds2Par
ms,upperSearchBounds2Parms,fnPosteriorAllDataPoints2Parms), thin=1,mcmc
=300000,burnin=20000,logfun=TRUE,verbose=1,tune=c(1,0.01), V=diag(1,2))
```

Loading the "coda" library, we plot the results:

```
library(coda)
plot(mcmcBoundedRuns2ParmsLatencyPeriod300kTune1Pt01)
```

The traceplots and marginal density estimates are given in figure 5.2, with c occupying the top row and p the bottom. Traceplots show the value of a specific parameter (vertical axis) by MCMC iteration (horizontal axis). The density estimates provide a smoothed approximation to the histogram of values received in order to approximate the underlying density. Further information can be found in the documentation for coda's *plot* function.

By invoking

```
heidel.diag(mcmcBoundedRuns2ParmsLatencyPeriod300kTune1Pt01)
```

a)   b)

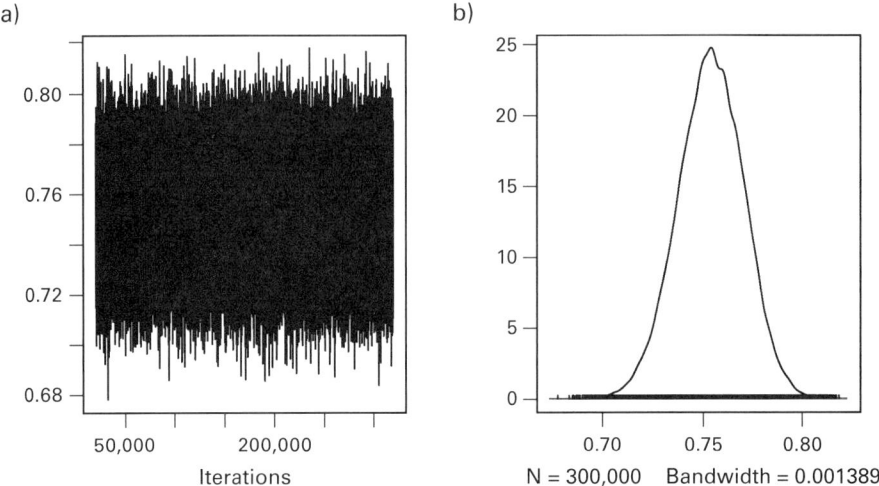

Figure 5.2
Traceplots and marginal density estimates for the MCMC results. Traceplots are in the
left column, density estimates on the right. Results for c are on the top row, those for p
on the bottom.

we can use the Heidelberger and Welch diagnostics to assess convergence of the Markov chain. The Heidelberger and Welch method consists of two parts—the stationarity test and the confidence interval
generation and evaluation procedure—which test if the half-width of
the CI generated is less than some threshold ε times the sample mean
of the retained iterations. A satisfactory test result should indicate that
both parts succeeded, as is the case in this situation (results not shown).

While not covered here, the interested reader may wish to consult
the review paper by Cowles and Carlin (1996) for description of additional diagnosis methods.

With confidence concerning convergence of the chain, we then issue
the command

```
summary(mcmcBoundedRuns2ParmsLatencyPeriod300kTune1Pt01)
```

to obtain basic descriptive statistics on the chain, including means
(350.504 for c, 0.754 for p) as well as standard deviations and quantiles
for the marginal distributions. It also includes the standard errors—
both naïve (a form that ignores autocorrelation) and a variant that
estimates the time series standard error based on the spectral density.
These results are shown in table 5.2.

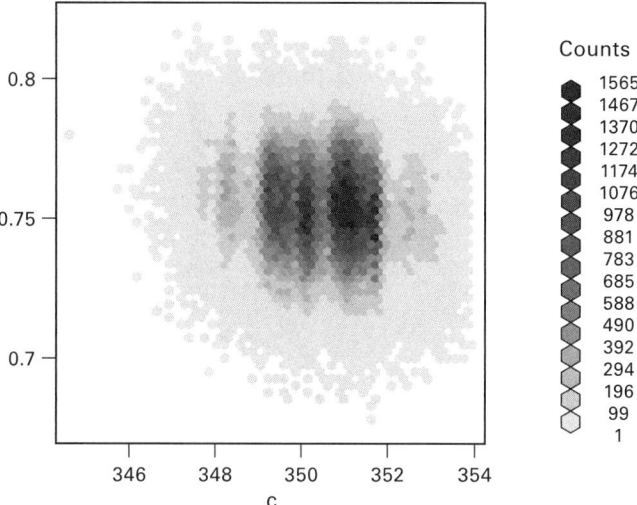

Figure 5.3
Sampled posterior distribution for parameters c (x axis) and p (y axis) for all data points.

Running the Metropolis algorithm on the given example samples from the distribution. The posterior is a scalar field in two dimensions (with one dimension for each of c and p), and is depicted in the graph below (with the horizontal axis corresponding to c, and the vertical to p). The graph is generated using the command:

```
plot(hexbin(mcmcBoundedRuns2ParmsLatencyPeriod300kTune1Pt01[,1],
mcmcBoundedRuns2ParmsLatencyPeriod300kTune1Pt01[,2], xbins=50))
```

Several features are immediately evident from figure 5.3 and the figure for the marginal above. First, it can be recognized that the distribution is multimodal, with several "ridges" associated with local maxima in the region from c=349.5 to 352, but with broad extent in terms of p.

The samples for the multivariate distribution lead to implied distributions for other model variables at each point in time. For example, figure 5.4 shows two-dimensional histograms for infective (top) and exposed (bottom) individuals over time for the baseline scenario.

Similarly, we can simulate the implied distributions when an intervention is undertaken, as shown in figure 5.5. The figure depicts successive histograms of samples for counts of infective (top) and exposed

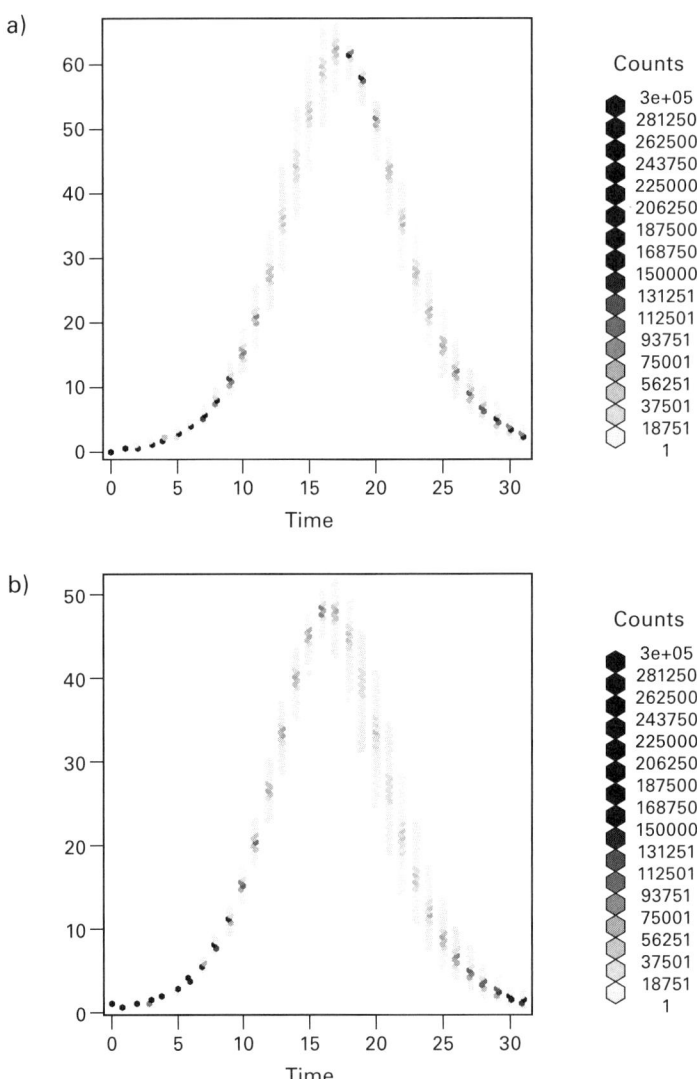

Figure 5.4
Successive histograms of samples for count of infective (top) and exposed (bottom) individuals in the baseline.

(bottom) individuals in the presence of a hygienic intervention that lowers the value of transmission constant beta by 25%.

Likewise, as depicted in figure 5.6, we can compare the distributions associated with the difference between the output variables in the context of the intervention and baseline. The figure shows successive histograms of samples for the difference in counts of infective (top) and exposed (bottom) individuals between the intervention and baseline scenarios.

Exercise 1: Try deriving the posterior distribution for three parameters. In particular, try additionally considering the mean latent period (μ) as a parameter, beyond c and p originally considered. (Code to perform this analysis and results may be found in the supplemental materials.)

The Impact of Empirical Data Availability on Posteriors

It is quite common that new data becomes available for a modeling project over time. There are a variety of approaches that seek to incorporate such new datapoints into model parameter and state estimates. These span a range from the Kalman filter (Gelb 1974)—on the high-efficiency, more restrictive assumption side—to sequential Monte Carlo methods, which involve heavier computational burden, but offer greater generality. Both such filtering approaches offer particularly key advantages when the dynamic models involved are stochastic, as it is not feasible to use MCMC methods alone to rigorously sample from the space of stochastic perturbations involved. While particle filtering (Doucet, De Freitas, and Gordon 2001; Doucet and Johansen 2009; Kantas et al. 2009; Murphy 2012) offers an exceptionally powerful and general method for estimating the latent state of such a stochastic process (e.g., values of stocks and flows), it is not in general feasible to use particle filtering as a method for simultaneously estimating the values of static parameters. Instead, more sophisticated and computationally burdensome methods such as particle MCMC (Andrieu, Doucet and Holenstein 2010) must be employed to estimate both latent state and parameter values in stochastic systems.

For deterministic systems, we can pursue incremental incorporation of new data with Markov chain Monte Carlo (so-called "batch" Monte Carlo methods) approaches. An important difference from the particle filter is that batch Monte Carlo methods are not defined recursively—each new data point leads not to an update of a previous site

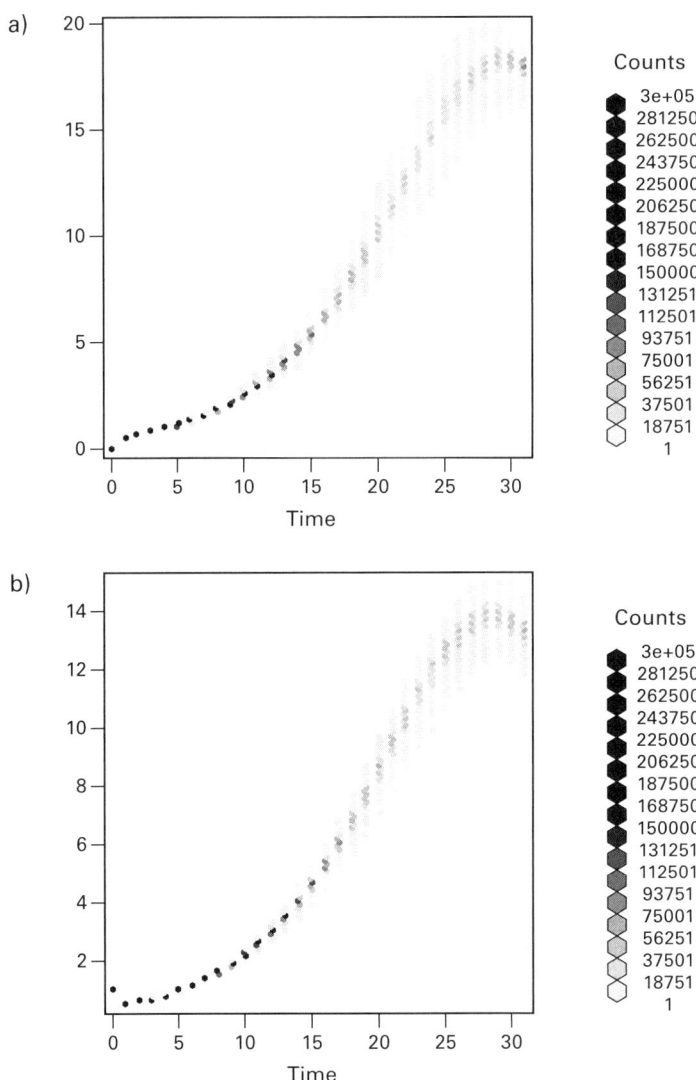

Figure 5.5
Successive histograms of infective (top) and exposed (bottom) case counts under intervention.

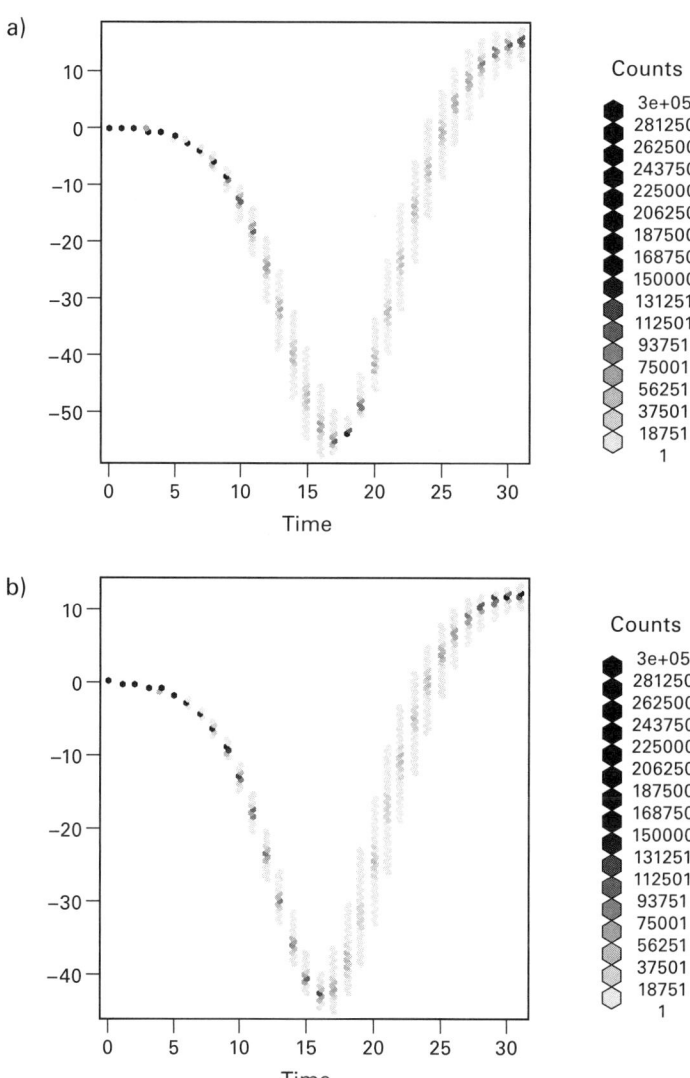

Figure 5.6
Histograms of intervention-induced change in counts of infective (top) and exposed (bottom) individuals.

of estimates, but instead to a repeat of the entire MCMC process considering the new data point. While this requires a greater amount of computation, it does have the virtue of not privileging the importance of early data, with an accompanying sidestepping of the problem of "lock-in" effects.

We illustrate this process here, with a particular eye toward appreciating the way in which the availability of empirical data affects the posterior. Recall that we are dealing with 32 data points for the two-dimensional histogram of MCMC samples depicted in figure 5.3. For comparison, figure 5.7 depicts successive projections of the samples for c and p for the case where we consider only the first 32, 16, or 8 parameters, respectively.

Figures 5.8 and 5.9 depict corresponding differences in the count of infective and exposed individuals between the intervention and baseline scenarios for the first 16 (figure 5.8) and 8 (figure 5.9) data points.

It can be recognized that the support of the posterior resulting from considering just subsets of the observations is far broader than that resulting from all of the observations. In short, the degree of uncertainty concerning each parameter is much larger. Successive observations in this case do reduce the uncertainty, although the large degree of dependence between the successive observations greatly reduces the amount of information each confers to the posterior.

Exercise 2: Modify the code to perform the examples above to consider only the first four data points. Specifically, please examine the impact on the posterior parameter distribution, the distribution for the count of infective and exposed individuals under the baseline and intervention cases, and the incremental differences between these cases.

Pragmatics

This section discusses two important pragmatic considerations that frequently arise when using the MCMC methods: enforcing boundaries on the possible domain of posteriors, and steps that can encourage convergence of the MCMC sampling process.

Boundary Violations and Reparameterization

In computing the posterior (up to a constant), there is a common desire to limit consideration to certain ranges of parameter values. These bounds derive from multiple sources. In some cases, these limits may be imposed by the nature of the parameters. In some cases (such as for

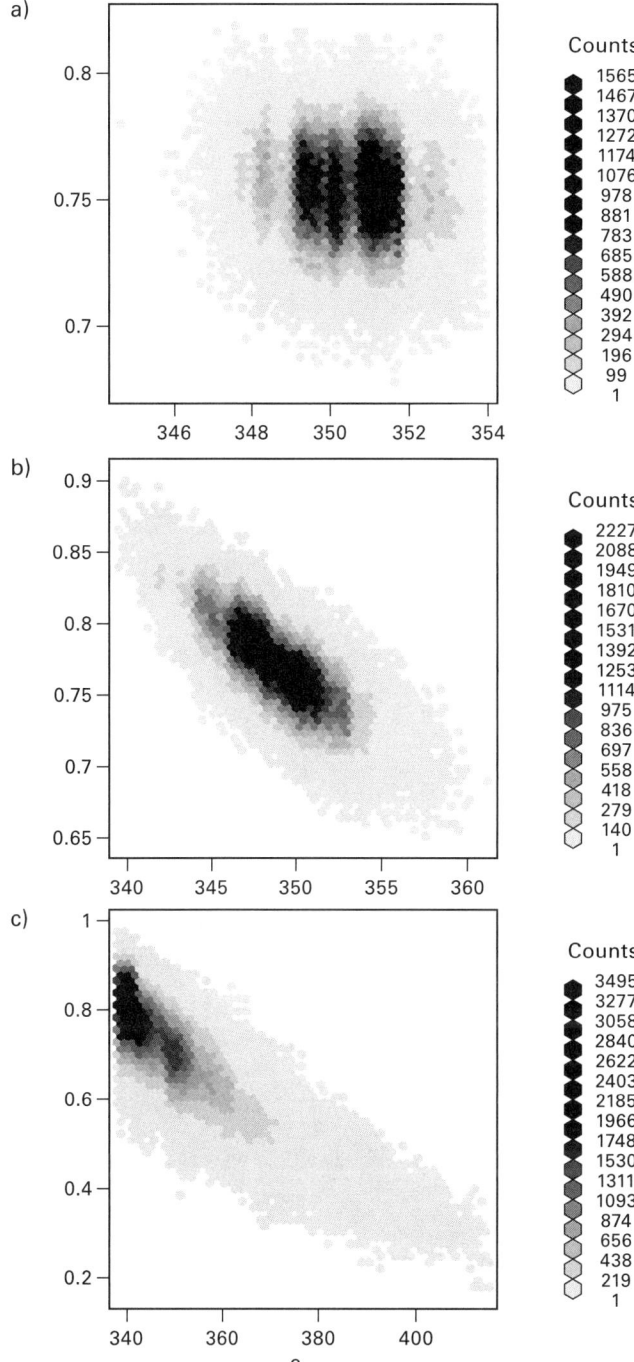

Figure 5.7
Sampled posterior distribution for parameters c (horizontal axis) and p (vertical axis) for first (from top to bottom) 32, 16, and 8 data points.

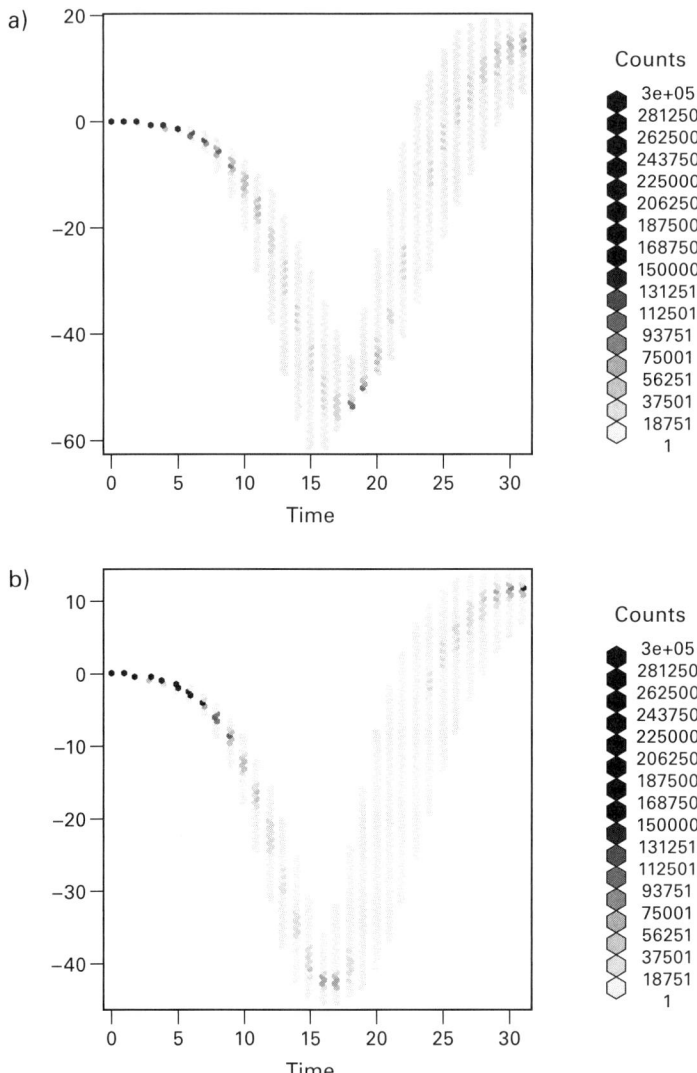

Figure 5.8
First 16 data points: posterior distribution samples for intervention-induced change in infective (top) and exposed (bottom) case count.

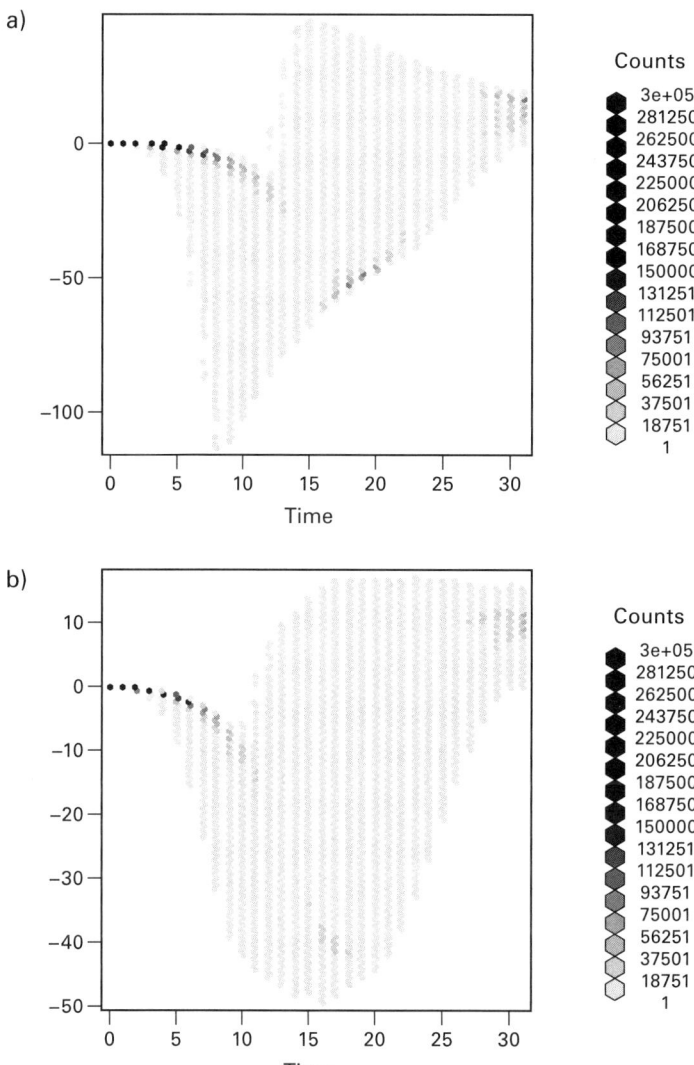

Figure 5.9
First 8 data points: posterior distribution samples for intervention-induced change in infective (top) and exposed (bottom) case count.

parameter p of the probabilistic model), the parameter will be a probability, whose value by definition must lie in the range from 0 to 1. In other cases, parameters represent physical quantities where negative values would be unphysical. This is, for example, true for mortality and natality rates, and for our parameter c (as "negative contact" is not meaningful). In a different subclass of cases, the bounds represent not logical or physical limits but the limits within which we are certain from evidence—formal or informal—that the value lies.

For some MCMC algorithms, we can design jump functions that directly impose such limits on the range of candidate values that are considered. However, for other algorithms—such as random walk Metropolis-Hastings—there is normally no mechanism explicitly requiring user specification where these bounds on candidate samples can be enforced. How are we to constrain candidate samples to lie within certain bounds before they are considered for acceptance? Two common approaches are *resampling* and *reparameterization*. We briefly discuss each here. For simplicity of the code, the example discussed here makes use of resampling rather than reparameterization to represent the model.

Resampling   Resampling-based approaches to boundary violations impose a "candidate proposal" stage that considers suitability of *proposed* candidates before they are promoted to *full* candidate status. Within this candidate proposal stage, a proposed candidate is repeatedly drawn from the normal procedure (without changing the location within the Markov chain—that is, based on the last *accepted* state of the Markov chain) until the proposed candidate lies within the acceptable bounds. At that point, the proposed candidate is promoted to full candidate status, and the algorithm continues as normal (for example, by determining whether to accept or reject the given sample). We suggest that the acceptance rate still only be calculated with respect to *full* candidate samples (i.e., that it does not consider the count of proposed candidates rejected within the denominator).

The resampling method offers many virtues to recommend it. In addition to being intuitive, it is simple to implement, even with arbitrarily geometrically complex boundary conditions. As a result, it imposes minimal risk of programming error or analysis. The major disadvantage of resampling lies in the loss in efficiency: given the "blind" character of the resampling, much time may be spent drawing proposed candidates before suitable candidates are found. This problem can be particularly acute as the number of parameters (number of dimensions of the parameter space) grows, since the fraction of the entire area that lies very close to the boundaries grows accordingly.

Reparameterization   A second approach that helps address boundary constraints—"reparameterization," or transformation of variables— also confers many other benefits as well. The availability of this approach reflects the fact that in representing a parameter space, we have the option of choosing our coordinate system. Given this flexibility, it sometimes behooves us to make use of a coordinate system that is maximally convenient for our purposes. Two notable ways in which a coordinate system can help the MCMC process are by facilitating exploration over a wide dynamic range of parameter values and by eliminating the need for bounds-checking by mapping boundaries (e.g., 0 or 1) to limits that can never be reached (such as negative and positive infinity, respectively, such as via logit transformation). Compared with resampling, reparameterization requires somewhat greater care to implement, and greater analytic consideration on the part of the MCMC user. Bearing in mind that readers who find the mathematics too daunting may wish to opt for the resampling option as an alternative for enforcing parameter bounds, we discuss the impact of reparameterization on the prior in the appendix.

Securing Convergence

While the MCMC process is well-defined, conceptually clear, and rigorous, the actual execution of that process is best pursued with careful observation, iteration, and learning. A central question is whether the Markov chain has successfully converged to its stationary distribution. A number of techniques aim to help understand patterns observed within the algorithm output and speed convergence of the algorithm. We highlight some of these techniques here.

The Influence of the Prior   To better understand the effect of the prior distribution on results—for example, if such higher-density regions are being dictated by the assumed prior distribution—we may also wish to evaluate both the sampling function $p(\vec{y}|\vec{\theta})$, and sample from a posterior distribution that assumes a different prior. To illustrate the relative impact of the prior and sampling distribution on the results shown here, we examine the impact of assuming a highly biased prior on the posterior. Specifically, in contrast to the bounded uniform prior assumed for most of this chapter, we examine in this section the effects of alternatively assuming a prior distribution of the form:

$$p(\vec{\theta}) = p\left(\begin{bmatrix} c \\ p \end{bmatrix}\right) \propto lognormal\left(c; \mu = \log\left(\frac{350}{2}\right), \sigma = 0.5\right),$$
$$\cdot\, triangle\,(p; a = 0, b = 1, c = 0.5)$$

where "triangle" denotes a triangularly shaped probability density function (piecewise linear with two pieces) whose values are 0 at $p=0$ and $p=1$, and which reaches a maximum value at $p=0.5$. The observant reader will note that this function is seriously biased on the lower side in terms of both c and p (E[c]=224.7; E[p]=0.5 compared with the true values of c=350, p=0.8). Figure 5.10 depicts the results of using MCMC sampling to depict the structure of the prior and emphasizes the biased nature of the joint distribution. To investigate the impact of the biased prior on the posterior, we use MCMC to sample from a posterior that uses this biased prior. Figure 5.11 shows the joint empirical density resulting from this sampling. The similarity between figure 5.11 and figure 5.3 demonstrates how—even with few data points—the shape of the posterior can overwhelm the influence of the prior.

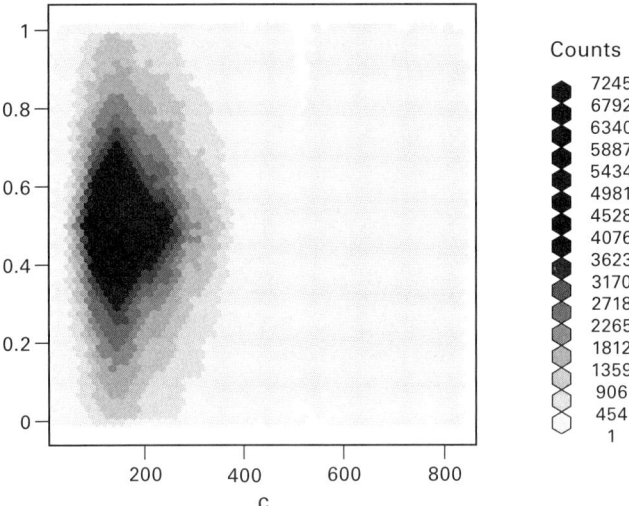

Figure 5.10
MCMC sampled (c, p) values for prior. Blue lines show the location of true values of c and p.

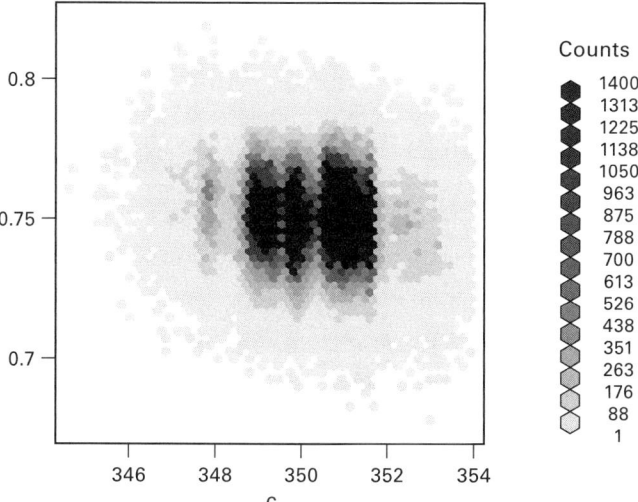

Figure 5.11
Samples from posterior distribution with a highly biased prior; note limited distortion
that results.

**Detecting and Dealing with Implicit Inter-Parameter Dependencies
in the Posterior**   Within a simulation model, it is typical for many
parameters to contribute to governing evolution of a particular model
variable. In some cases, changes in these parameters will contribute in
ways that are sometimes similar, and sometimes work in opposition.
Given these influences, when we are considering—as we do here—the
posterior imposed by matching some data using model output using
a probabilistic model, it is natural that the posterior will exhibit dependencies between the influence of different parameters. That is, to put
it another way, it would be surprising indeed if each different parameter exhibited a statistically independent effect on the (joint) posterior
distribution.

The dependency of the posterior distribution on parameters can take
many forms. Sometimes (as in figure 5.10), the dependencies between
parameters will be nearly linear. In other cases (such as for the solution
to the three-parameter exercise depicted in the supplemental material),
this dependency will be highly nonlinear; for example, laying on a thin
curve. In any event, when there are high degrees of dependency exhibited between two parameters (e.g., if the posterior density is high only
when the value of both of the parameters is high, or when one is high

and the other is low), exploration of the posterior via uncorrelated random perturbations in the parameter space can be rather inefficient. In essence, we end up moving via random perturbations across a higher dimensional space (the nominal dimension of the parameter space) when exploring a structure that exhibits lower intrinsic dimensionality.

For such cases, it can be helpful to directly make use of the knowledge of the structure of the space when exploring it. We note here two ways of doing this:

- **Analytically reparameterize the model.** In some cases, we may have the flexibility to choose to perform MCMC with respect to a different set of parameter values—a set of parameters that exhibit less direct (especially less linear) correlation.
- **Data-driven reparameterization.** If early modeling results demonstrate significant dependencies between model variables that can be characterized statistically (e.g., using regression), one can often formulate one parameter (call it p_0) as a function of one or more other parameters ($f(p_1 \ldots p_k)$). Given such an expression, one can then replace p_0 in MCMC with the residual $p_0 - f(p_1 \ldots p_k)$. While p_0 may exhibit pronounced dependency on the parameters $p_1 \ldots p_k$, the residual is unlikely to exhibit such dependency.

Tests with Synthetic Data   In cases where a modeler experiences difficulty securing convergence, it is often helpful to attempt to identify the issue underlying the problem. Some of these difficulties arise from the fundamental challenge of exploring a high-dimensional parameter space via random walk. In such cases, approaches such as Hamiltonian Monte Carlo (Neal 2011) and other techniques may be of help in focusing the effort on more promising components of the space. In other cases, the difficulties reflect latent inconsistencies between the data (on the one hand) and the simulation or probabilistic model (on the other). In yet others, the difficulties may reflect identifiability constraints—the fact that there is too little empirical data to constrain the posterior. To recognize such cases, it is often helpful to examine whether convergence difficulties are experienced when operating with data of known pedigree. This could involve data known to be consistent with the probabilistic and dynamic model, data with especially high sample size, or both. To study this, the modeler can frequently test convergence with a deliberately chosen sample size and a synthetic data set with known properties and adherence to the assumptions of the probabilistic and simulation models. In many cases, creation of such a data set is straightforward and can additionally aid in testing and evaluation of the MCMC approach.

Conclusion

This chapter has provided an overview of a technique for cross-leveraging the principled, general, and conceptually straightforward MCMC approach together with dynamic simulation models. This combination provides a rigorous way to derive posterior distributions for model parameters. The approach provides insights traditionally drawn in a fragmented way from separate steps of sensitivity analysis and calibration, can be used for a wide range of intervention- and outcome-oriented analysis, and avoids the need to rely upon point estimates alone or in combination with confidence intervals commonly emerging from calibration. MCMC also provides an excellent fit for the needs of dynamic recalibration of models when new data arrives only on the order of days or more. This reflects its formulation around a well-defined but general probabilistic model that can typically be reused for newly arrived data points, and its ability to sidestep many of the restrictive and sometimes compromising assumptions associated with other techniques, such as the extended Kalman filter (Gelb 1974). While relatively computationally intensive, MCMC occupies a point on a spectrum balanced between analysis duration (on the one hand) and restrictiveness of assumptions that is well-suited to many applications, and which can exploit the growing availability of computational resources for parallel and distributed analysis.

Within the conceptual framework articulated here, the simulation model is responsible for evaluating the implications for model outcomes or interventions of candidate samples from the posterior distribution, and a well-defined, explicit, and general Bayesian probabilistic model is used to assess the likelihood that a given set of such outcomes would yield the observed empirical data. In addition to priors over the parameter values (whose impact is frequently dominated by the information associated with empirical data), this Bayesian model includes—most notably—a sampling function that uses parametric or nonparametric techniques to specify the likelihood that the empirically observed data would result from (1) the parameter value at this point in space, and—critically—(2) from the simulation model results given those parameter values.

The implementation of this approach is centered around the need to draw samples from a posterior distribution that lacks closed-form specification (most notably because of the dependence of that posterior upon simulation results that are themselves typically not a closed-form function of the parameter values). In the MCMC method, a Markov chain each of whose states is associated with a particular point in parameter space is used to transition across that space, stochastically drawing values of parameters from the distribution (hence the Monte Carlo designation). Reflecting the distribution of the posterior from which it is sampling, the Markov chain is designed to spend more time sampling from regions with higher posterior density.

The approach is very powerful and general, but is associated with a number of diagnostic outputs that should be inspected to help ensure efficient and correct operation. This chapter presented a number of diagnostic and supportive tools to help guard against overdependence on the vagaries of initial point selection for exploration, to reduce the covariance between samples, and to ensure that large enough sets of samples are drawn, and presented methods to help increase the efficiency of the approach and speed convergence.

In addition to characterizing the conceptual approach, we have illustrated here use of a practical software framework that allows for relatively straightforward mechanics for interfacing third-party MCMC packages in the popular statistical software R with the popular simulation software Vensim for the purposes of MCMC analysis. The multipiece library—described in further detail in the appendix—is designed to ease the practical hurdles and distractions associated with interfacing software packages to support MCMC analysis, and to allow the modeler to make use of this powerful technique in a way that supports focus on performing high-quality and efficient MCMC analyses for modeling insight.

Notes

1. Within this document we will use the term "uncertainty" in a colloquial sense, referring to situations where the exact value of a parameter is unknown to us.

2. While not formally examined in this document, it bears noting that this can include consideration of alternative models.

3. As will be seen below, ensuring independent samples requires additional work, such as "thinning" these samples—that is, reducing autocorrelation between samples by only making use of one out of every k nominal samples.

4. Despite the fact that some parameters in $\vec{\theta}$ are not required by the simulation model, for uniformity in notation, the balance of this document expresses the simulation model output $\vec{f}(t;\vec{\theta}_s)$ as $\vec{f}(t;\vec{\theta})$. That is, we will conceptually consider the simulation model as being parameterized by the full set of parameters used for the probabilistic model (where some of those parameters simply have no use within $\vec{f}(t;\vec{\theta})$; this requirement imposes no need to change the simulation model).

5. A note on terminology, which can be confusing to those not from statistical backgrounds: The same mathematical expression— $p(\vec{y}|\vec{\theta})$ —goes by two different names, depending on whether it is viewed as a function (probability density over) of $\vec{y}$ given a fixed parameter assignment $\vec{\theta}$, or as a function of $\vec{\theta}$ for a fixed observation (vector) $\vec{y}$. In the former case, consider the likelihood of obtaining different observations $\vec{y}$ for a particular assumption of parameters $\vec{\theta}$. Because the expression denotes a probability density and integrates to 1 over all possible values of those observations $\vec{y}$, it is termed the *sampling distribution* (or *data distribution*). For the latter case—which views this as a function of $\vec{\theta}$ for a fixed observation (vector) $\vec{y}$, it is *not* a distribution—that is, it does not integrate to 1 over all values of $\vec{\theta}$, and is thus called simply a *likelihood function*.

6. We recognize and accept the fact that this approximation neglects to take into consideration a delay.

7. While it is possible to formulate general-purpose samplers constructed around cumulative distribution functions (CDF), constructing a formal representation of these functions also falls prey to the "curse of dimensionality." Monte Carlo techniques provide a general-purpose technique for sampling from an arbitrary distribution while avoiding the curse of dimensionality.

References

Abelson, H., and G. J. Sussman. 1996. *Structure and interpretation of computer programs*. 2d ed. Cambridge, MA: MIT Press.

Anderson, R. M., and R. M. May. 1991. *Infectious diseases of humans: Dynamics and control*. Oxford: Oxford University Press.

Andrieu, C., Doucet, A., and R. Holenstein. 2010. Particle Markov chain Monte Carlo methods. *Journal of the Royal Statistical Society* B 72 (3): 269–342.

Coelho, F. C., C. T. Codeço, and M. G. M. Gomes. 2011. A Bayesian framework for parameter estimation in dynamical models. *PloS One* 6 (5): e19616.

Cowles, M. K., and B. P. Carlin. 1996. Markov chain Monte Carlo convergence diagnostics: A comparative review. *Journal of the American Statistical Association* 91 (434): 883–904.

Dogan, G. 2007. Bootstrapping for confidence interval estimation and hypothesis testing for parameters of system dynamics models. *System Dynamics Review* 23 (4): 415–436.

Dorigatti, I., Cauchemez, S., Pugliese, A., & Ferguson, N. M. 2012. A new approach to characterising infectious disease transmission dynamics from sentinel surveillance: Application to the Italian 2009–2010 A/H1N1 influenza pandemic. *Epidemics* 4 (1): 9–21.

Doucet, A., N. De Freitas, and N. Gordon. 2001. *Sequential Monte Carlo methods in practice*, vol. 1. New York: Springer.

Doucet, A., and A. M. Johansen. 2009. A tutorial on particle filtering and smoothing: Fifteen years later. In *Handbook of nonlinear filtering*, 12, 656–704.

Eberlein, R., and D. Peterson. 1994. Understanding models with Vensim. In *Modeling for learning organizations*, vol. xxii, ed. J. D. W. Morecroft and J. D. Sterman. Portland, OR: Productivity Press.

Gelb, A., ed. 1974. *Applied optimal estimation*. Cambridge, MA: MIT Press.

Gelman, A., J. Carlin, H. Stern, and D. Rubin. 2004. *Bayesian data analysis*. 2nd ed. Boca Raton, Fla.: Chapman & Hall/CRC.

Kantas, N., A. Doucet, S. S. Singh, and J. M. Maciejowski. 2009. An overview of sequential Monte Carlo methods for parameter estimation in general state-space models. In *Proceedings of the 15th IFAC Symposium on System Identification*. Vol. 15.

Kermack, W. O., and A. G. McKendrick. 1927. A contribution to the mathematical theory of epidemics. *Proceedings of the Royal Society of London Series A* 115 (772): 700–721.

Lunn, D., C. Jackson, N. Best, A. Thomas, and D. J. Spiegelhalter. 2012. *The BUGS book: A practical introduction to Bayesian analysis*, vol. 98. Boca Raton: CRC Press.

Martin, A. D., K. M. Quinn, and J. H. Park. 2011. MCMCpack: Markov Chain Monte Carlo in R. *Journal of Statistical Software* 42 (9): 1–21.

Mbalawata, I. S., S. Särkkä, and H. Haario. 2013. Parameter estimation in stochastic differential equations with Markov chain Monte Carlo and non-linear Kalman filtering. *Computational Statistics* 28 (3): 1195–1223.

Murphy, K. P. 2012. *Machine learning: A probabilistic perspective*. Cambridge, MA: MIT Press.

Neal, R. 2011. MCMC for using Hamiltonian dynamics. In *Handbook of Markov Chain Monte Carlo*, ed. S. Brooks, A. Gelman, G. Jones, X.-L. Meng, 113–162. Boca Raton: CRC Press.

Ntzoufras, I. 2011. *Bayesian modeling using WinBUGS*, vol. 698. Wiley.

Oliva, R. 2003. Model calibration as a testing strategy for system dynamics models. *European Journal of Operational Research* 151 (3): 552–568.

Osgood, N. 2007. Lightening the performance burden of individual-based models through dimensional analysis and scale modeling. *System Dynamics Review* 25 (2): 101–134.

Robert, C., and G. Casella. 2010. *Introducing Monte Carlo Methods with R*. New York: Springer.

Schittkowski, Klaus. 2002. *Numerical data fitting in dynamical systems: A practical introduction with applications and software*, vol. 77. New York: Springer.

Sterman, J. 2000. *Business dynamics: Systems thinking and modeling for a complex world*. Boston: Irwin/McGraw-Hill.

II Model Analysis

6 Pattern Recognition for Model Testing, Calibration, and Behavior Analysis

Gönenç Yücel and Yaman Barlas

In different stages of modeling and model analysis, one faces the task of evaluating model output for different purposes. For instance, we compare historical time-series data to the model's output behavior in parameter estimation (model calibration) and validation tasks. In order to test the credibility of the model through extreme condition tests, we check the fit between the model output patterns and the expected "real" behavior patterns under the same extreme conditions. In order to explore the behavior space of the model, we try to identify and categorize the types of behaviors the model generates with different parameter sets. Finally, in scenario and policy analysis, we try to explore the degree of change in the model behavior as a result of alternative policies or under different scenarios.

When we specifically focus on policy-oriented dynamic modeling, the long-term development trajectories of the key variables are of primary importance. As a natural consequence, output evaluations are mostly done in a pattern-oriented manner; that is, it is the qualitative pattern features of the output that are of primary relevance, rather than point-by-point numeric features. The exact values of the model output at certain time points are of little relevance, and the main concern is the overall dynamic pattern characteristics of the model behavior—for example, whether the model generates exponential growth, oscillations, or S-shaped growth behavior.

This pattern orientation brings about a burden with respect to the model output evaluation tasks. Evaluation of a dynamic behavior in terms of its pattern features is a qualitative task by its nature and renders standard statistical tests and measures mostly inapplicable. This situation forces the modelers to conduct such evaluations by visual inspection, which has major problems. First of all, the evaluation process becomes subjective and informal. Secondly, visual, nonautomated inspection of the model outputs makes the aforementioned testing processes extremely time-consuming. In the end, most modelers are forced to be content with a very limited number of evaluations (much less than the ideal) due to time constraints. Modelers can afford to conduct only a small number of extreme condition and sensitivity runs, and they can explore only a very limited portion of the behavior space of the model visually or manually, due to the combinatorial nature of all possible runs. Thus, the subjective nature of the output evaluations and the limitations of manual simulation experimentation stand as serious drawbacks.

Hence, overcoming the problems mentioned above calls for automating the output pattern-evaluation task. This is where pattern recognition, a computational approach for identifying a set of given entities based on their special features, becomes highly relevant and potentially fruitful for dynamic modelers. A pattern-recognition algorithm that is custom designed for the task at hand can perform automated identification and classification of model outputs based on their dynamic behavior features. By coupling a pattern-recognition algorithm with existing dynamic modeling platforms, both the subjectivity and the time constraint issues can be overcome, since an exhaustive number of tests can be conducted in a computerized manner. The modeler/analyst will be able to conduct a large number of behavior evaluations, in a well-defined, objective and consistent manner, in different stages of a modeling study.

As a generalization, the aforementioned output evaluations are of the following two types:

- Similarity evaluation of two given dynamic behaviors: Does the model output pattern resemble the historical behavior pattern of the real system? Does policy X lead to a different behavior than policy Y?
- Class identification of a given dynamic behavior: What type of behavior pattern does the model generate under some condition C?

Considering these two different types of output evaluation, it is possible to automate both similarity evaluation and class identification tasks with the use of pattern-recognition algorithms. Considering these capabilities, the potential contribution of pattern-recognition methods to the dynamic modeling field can be used at different stages of the modeling process.

Model Calibration and Parameterization   In the parameter estimation (calibration) stage, a pattern-recognition algorithm can conduct the evaluation of how similar a model output with certain parameter values is to the desired behavior. This evaluation helps to identify the parameter values that yield the best fit to the historical behavior of the real system being modeled.

Model Testing   Extreme condition tests are a very important part of the model verification and validation process (Barlas 1996). A pattern-recognition algorithm can be used to check whether the model output is similar to the hypothesized behavior under certain extreme conditions. Considering the uncertainties in model parameters, testing the robustness of the model behavior through sensitivity analysis is an important task. A pattern-recognition algorithm can be used to automate the task of finding the parameter ranges within which the model behavior is robust, and it can also report the parameter values for which the pattern of the model output changes.

Behavior Analysis   An extensive exploration of the behavior space of a model is a difficult and rare practice in modeling fields. Exploring the behavior space of even a simple model through visual inspection is a very demanding task. By replacing the visual inspection, a pattern-recognition algorithm makes an extensive and efficient behavior-space exploration feasible.

Policy Design and Analysis   A pattern-recognition algorithm can also be used for identifying robust policies that lead to desired behavior modes, or that alleviate the undesired ones. Once the analyst specifies the classes of the desired and undesired behavior modes, a pattern-recognition algorithm may be used to conduct a search on the policy parameters in order to identify the set of values that satisfy or optimize the user objectives.

In summary, pattern recognition offers opportunities for strengthening the toolbox of dynamic modelers and has the potential to contribute to the efficiency and effectiveness of the modeling process at various stages. However, such a contribution will not come directly from pattern-recognition algorithms. The potential benefit comes from the fact that pattern recognition provides the missing link that is required to couple modeling practice with advanced computational techniques and tools; that is, automation, or computerization of the manual process of output evaluation, identification, and classification. Such a contribution widens the scope of methods that can be used and tools that can be designed to support the modeling process.

In the following sections, we discuss the current state of the art of pattern recognition in the context of dynamic modeling and provide detailed examples on how existing pattern-recognition tools can be used by modelers. Section 2 provides a brief introduction to the basics of pattern recognition in general, and recognition of dynamic behavior patterns in particular. After this general introduction, the third section is devoted to summarizing the current state of the art in pattern-recognition applications in the dynamic modeling context. After presenting a set of tools and algorithms relevant for dynamic modeling, we provide a set of examples on how these tools can be implemented and used in different stages of the modeling process. The chapter concludes with a brief discussion of the road that lies ahead for widespread utilization of pattern-recognition tools and algorithms in the dynamic modeling fields.

Pattern Recognition

Pattern recognition is one of the most important subfields of machine learning, which involves the design and development of algorithms that allow computers to evolve skills and behaviors based on available data and past experience (Alpaydin 2010). Simply put, pattern recognition deals with the assignment of labels to a set of given objects by computer algorithms. These input objects, commonly referred to as data instances, can be anything from customer profiles (i.e. a collection of demographical and historical transactions data) to digital images (i.e., a collection of light and color data for each pixel). Given a large collection of such instances, pattern recognition is mainly about the automatic discovery of regularities in this data set through the use of computer algorithms and the use of these regularities to take actions, such as classifying the instances into different categories (Bishop 2006; Duda, Hart, and Stork 2001; Theodoridis and Koutroumbas 2003). Pattern recognition is a wide research field with applications to many different domains, which differ mainly based on the type of objects to be recognized. Among those, pattern-recognition research that focuses on temporal data objects (temporal data mining or dynamic pattern recognition) has the utmost relevance for dynamic modelers (Mitsa 2010).

Pattern-recognition approaches can be considered as falling into two main categories. The first category is *supervised pattern recognition*, in which a computer algorithm is supervised by an expert prior to the recognition task. In the supervised case, a set of training data, in which all instances are already labeled with their ideal classes, is used to train the pattern-recognition algorithm. The algorithm learns the distinctive characteristics of each class by iterating through these labeled instances. For example, experts manually analyze and label a certain number of customers as profitable and some others as unprofitable in a customer profiling application. Then the profiles of these customers with their labels (i.e. profitable or unprofitable) are fed to the algorithm, which eventually learns to distinguish these two types of customers. This type of pattern recognition is commonly referred to as *classification*, since a set of classes are determined beforehand, and the computer algorithm is asked to label a given instance with one of these prespecified, finite classes. A set of different techniques, which include distance-based classifiers, Bayes classifiers, decision trees, support vector machines, and neural networks, can be employed for the classification task (Wu et al. 2007; Mitsa 2010). Among those, distance-based classifiers, and especially k-nearest neighbors (k-NN) classifiers, are the most commonly used classification techniques in analyzing temporal data objects (Chen, Ye, and Hu 2007). In k-NN, the object being classified is assigned to the class most common among its k nearest neighbors from the labeled training set.[1] In the case of k=1 (1-NN), the object is simply assigned to the class of the nearest neighbor.

In the unsupervised case, as the name implies, there is no expert supervision for the computer algorithm; neither a set of predefined class labels nor a set of prelabeled training data exists. In this case, the main task of the pattern-recognition process is to identify natural groups of similar examples in the given set of objects. This type of pattern recognition is called clustering. The algorithm has a built-in method to evaluate the similarity of two objects. Based on this similarity measure, the algorithm tries to split the input data into clusters. The overall goal of the clustering algorithm is to organize the given objects into groups where the within-group similarity is minimized and the between-group dissimilarity is maximized (Liao 2005; Zhang et al. 2011). The eventual number of clusters is not known beforehand, and it is dependent on the distribution of instances within the given input set and the similarity measure used. As in the case of classification, a variety of techniques can be utilized to cluster a given set of objects.

These include partitioning techniques such as K-means and fuzzy c-means, hierarchical clustering, density-based clustering, self-organizing maps, and relocation clustering (Liao 2005; Fu 2011; Mitsa 2010). In the context of temporal data mining, the most commonly used clustering approaches are agglomerative hierarchical clustering and k-means clustering (Gullo et al. 2009; Zhang et al. 2011). In hierarchical clustering, a hierarchy is constructed by progressively merging clusters. Initially, each individual object is treated as a cluster by itself. Then, the closest two clusters are merged at every iteration. Each agglomeration occurs at a greater distance between clusters than the previous agglomeration, and one can decide to stop clustering either when the clusters are too far apart to be merged (distance criterion) or when there is a sufficiently small number of clusters (number criterion). In k-means clustering, the given set of objects is clustered into k clusters in which each object belongs to the cluster with the nearest mean, serving as a prototype of the cluster.

The pattern-recognition field mainly deals with methods and approaches that are developed for static objects. In other words, recognition is based on a set of features derived from an input object's state at a single point in time. However, it is quite likely that the data objects being analyzed are dynamic—their states, their features, changing over time—and the goal is to label them based on temporal dynamics of their features. In this case, the data to be fed into the pattern-recognition algorithm is related to the change in an object's features over time. Then we speak about temporal data mining or dynamic pattern recognition, which is the subject of this chapter. The most common practice in temporal data mining is to treat the whole trajectory of a dynamic pattern as a single, static object (Angstenberger 2001; Fu 2011). However, this is not a straightforward task. It generally demands transforming the raw patterns into simpler representations and specifying a way to evaluate similarity between patterns. Once they are resolved, dynamic patterns to be analyzed can be fed into arbitrary classification and clustering algorithms, such as an unsupervised neural network algorithm, self-organizing map, or hierarchical clustering algorithm (Wang, Smith, and Hyndman 2006).

Whether it is classification or clustering, object representation and similarity or distance measurement are very important in pattern recognition. This is especially valid in the case of recognition of dynamic patterns. Each dynamic pattern of time-series data is a high-dimensional object in its original form, since each data point corresponds to a single dimension. This causes serious efficiency problems for pattern-recognition algorithms. Therefore, dimensionality reduction and time-series representation are crucial issues in time-series classification and clustering (Fu 2011; Zhang et al. 2011; Li, Guo, and Qiu 2011; Liao 2005). In a good transformation, objects to be classified or clustered are represented with a smaller number of dimensions, and the extent of similarity between objects is preserved.

Depending on what is considered important, a dynamic pattern can be represented in many different ways. Representing time-series data with global properties such as mean, variance, trend, and periodicity is one of the simplest representations that can be used to reduce the dimensionality of the object. Alternative representations can be obtained using discrete Fourier transform (DFT; Bagnall and Janacek 2005) or discrete wavelet transform (DWT; Wu, Agrawal, and El Abbadi 2000). Another common dynamic pattern-representation approach converts the time series to symbolic form. First, the pattern is discretized into segments, and then each segment is converted into a symbol (Fu 2011). Model-based representations are also frequently used in pattern classification and clustering. In this approach, a model, be it a simple statistical model or a hidden Markov model, is fitted to the time-series data. The coefficient of the used model serves as a compact representation of the original dynamic pattern (Corduas and Piccolo 2008). As Gullo et al. (2009) indicate in their review of dimensionality-reduction approaches, there is no absolute winner among these methods in every domain. Depending on the task at hand and the nature of the time series being analyzed, the best-performing transformation may differ. An extensive survey on different approaches to time-series representation can be found in Fu (2011).

As mentioned above, the second important issue in time-series recognition is measuring similarity between time series, which is generally regarded as one of the biggest problems (Cuberos et al. 2004; Zhang et al. 2011; Li, Guo, and Qiu 2011). The similarity measurement can be performed either based on the raw time-series data or based on their transformed versions. In both cases, the most popular approach is to use the Euclidean distance (Fu 2011). If the raw time series are being used, the Euclidean distance is measured based on the differences of the time series at each time point. If the similarity measurement is based on the transformed representations (feature vectors), the distance is based on the differences at each dimension of these feature vectors. However, being similar with respect to the Euclidean distance is one of many similarity alternatives. Bagnall et al. (2006) discuss three fundamentally different classes of similarities. According to this classification, two time series that behave similarly at the exact same time point are said to be *similar in time.* A correlation-based or Euclidean distance metric performs well in measuring this kind of similarity. Alternatively, having common shape features independent of where they appear over time corresponds to *similarity in shape.* Dynamic time warping (DTW) is proposed as the most successful way to measure this type of similarity (Ratanamahatana and Keogh 2005). Finally, if two time series have similar autocorrelation structures, then these time series are said to be *similar in change.* The general tendency in measuring similarity in change is to fit an ARMA model to a time series, then measure similarity based on the coefficients of this model (Kalpakis, Gada, and Puttagunta 2001; Zhang et al. 2011). As can be seen here, two time series can be evaluated as very similar or dissimilar depending on the overall objectives of the pattern-recognition task at hand. Due to the purpose-dependent nature of the similarity evaluation, a variety of similarity measurement approaches can be found in the literature. A wide selection of different distance functions that can be used to compare raw or transformed time series is presented in a recent survey by Liao (2005).

In conclusion, it is nearly impossible to have a single well-performing pattern-recognition algorithm that fits all purposes, considering both transformation and similarity measurement issues. Such algorithms need to be selected and custom-tailored for the problem at hand. Therefore, we look into relevant cases in the next section.

Readers interested in learning more about the topic may refer to Alpaydin (2010), Duda, Hart, and Stork (2001), and Theodoridis and Koutroumbas (2003) for a comprehensive introduction to pattern-recognition and machine-learning methods in general. Alternatively, a course offered by Adrew Ng from Stanford University through Coursera[2] is a notable resource for the interested readers. Apart from these general resources, Mitsa (2010) is an important textbook that focuses particularly on the use of pattern recognition for temporal data analysis. Finally, the surveys by Liao (2005) and Fu (2011) provide an extensive snapshot of the state of the art for those who wish to review diverse implementations of pattern-recognition methods for time-series analysis.

Dynamic Modeling and Pattern Recognition

Pattern-recognition tools and methods have been widely used in time-series classification and clustering in various domains as diverse as economics, finance, medicine, linguistics, marketing, biology and astronomy (Corduas and Piccolo 2008; Antunes and Oliveira 2001; Angstenberger 2001). The increased availability of empirical time-series data as well as methodological developments enabled the recent increase in the number of such applications.

In this section, we will discuss two main contributions that aim to enable using dynamic pattern-recognition tools in the modeling context. The first contribution is a supervised pattern-recognition (classification) algorithm that is designed to recognize most fundamental behavior trajectories encountered in the dynamic modeling practice. The second study is about the use of unsupervised pattern-recognition (clustering) approaches in dynamic modeling. After a detailed introduction to these contributions in this section, we will provide examples of how they can be used at different stages of a modeling process.

A Classifier for Dynamic Behavior Patterns

This hidden Markov model–type of classifier (Theodoridis and Koutroumbas 2003; Therrien 1989; Duda, Hart, and Stork 2001), was developed by Barlas and Kanar (1999) to recognize a set of basic behavior patterns that are frequently encountered in theory and practice, as shown in figure 6.1. The algorithm analyzes a behavior pattern by splitting it into a fixed number of segments (six segments in particular). Each segment is characterized with three features: mean, slope (first derivative), and curvature (second derivative). In other words, during the feature extraction stage, a feature vector of 18 elements is constructed for each given behavior pattern. The algorithm builds upon the premise that the likelihood of a segment type being followed by another particular segment distinguishes the basic behavior patterns. For example, in an exponential growth pattern, a segment with positive slope and positive curvature is most likely followed by another segment of the same character (i.e., positive slope, positive curvature). However, for an S-shaped growth pattern, the likelihood of successive segments to have a positive slope but negative curvature is significantly higher. Based on this premise, each basic behavior pattern is represented with a different Markov chain of segment types. This Markov model incorporates the probabilities of segment types being followed by other

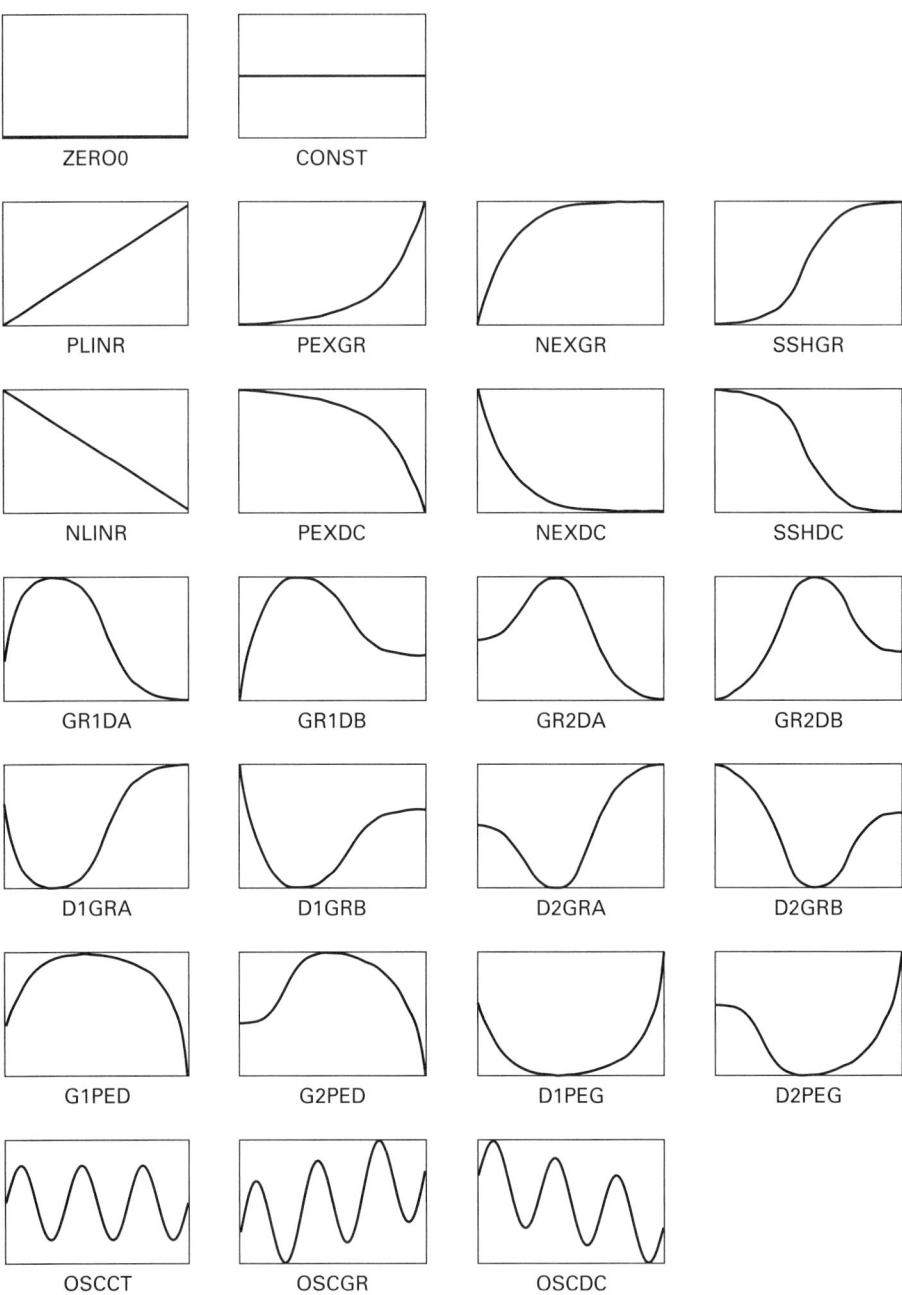

Figure 6.1
Behavior patterns that can be recognized by the algorithm of Barlas and Kanar (1999).

segment types (i.e., state transition probabilities). These probabilities are estimated for each basic behavior pattern via a separate "training" process.

This algorithm is trained using a set of prelabeled behavior instances; it is *supervised*. During the "training" process, for each pattern class, a collection of behavior patterns (approximately 100) that belong to that particular class is used as the training set. The patterns in these training sets are variants of the basic pattern class with some noise added. Using the training set, transition probabilities that best represent the basic pattern are estimated and learned by the algorithm. The procedure is repeated for all basic behavior patterns to parameterize their hidden Markov models.

Once the algorithm is "trained," it can recognize to which basic class a given behavior pattern belongs. In the pattern-recognition phase, the algorithm takes a behavior pattern (an output from a model) as an input, and first normalizes it in order to filter out the impact of the y axis scale. Then the pattern is split into segments, and the three features (mean, slope, curvature) of each segment are extracted. Once the feature vectors are extracted, the algorithm analyzes the sequence of these segments and calculates the likelihoods that this particular sequence belongs to basic pattern classes. The Markov models embedded in the algorithm are used to calculate this likelihood. As a quantitative measure of this likelihood, the algorithm relies on the normalized state-optimized likelihood (Barlas and Kanar 1999). This is done for all basic pattern types, and the algorithm indicates the most likely pattern class to which the given behavior belongs. A pseudo-code that summarizes the way the algorithm works is given in box 6.1. More extensive discussions on both training and classification processes related to the pattern-recognition algorithm are given in Barlas and Kanar (1999) and Kanar (1999).

A Method to Measure Similarity of Dynamic Behavior Patterns

As discussed earlier, similarity measurement is a purpose-dependent and challenging aspect of dynamic pattern recognition. Yücel (2012) proposes a method for measuring the similarity of two dynamic model behaviors based on their qualitative features. The proposed approach starts from the premise that the characteristic of a dynamic behavior lies in the sequence of atomic behavior modes that it consists of. An atomic behavior mode can be identified with a constant sign of slope and curvature. Nine such atomic behavior modes exist, which are given

---

Box 6.1

Pseudo-code of the classifier by Kanar (1999)

Receive the dynamic behavior, X, to be classified

Normalize X, and obtain X_{norm}

Split X_{norm} into equal-length segments, and obtain $S_1, S_2, ...,S_n$

Iterate over all segments {

Pick a segment S_i

Calculate mean for the segment, i.e., a_i

Calculate slope for the segment, i.e., b_i

Calculate curvature for the segment, i.e., c_i

}

Iterate over all defined behavior classes {

Pick a behavior class, B_j

Calculate the state-optimized likelihood of X_{norm} belonging to the B_j class,

i.e., p_j

}

Calculate the maximum likelihood, $p_{max}=max(p_j)$

Report the behavior class j for which the $p_j=p_{max}$

---

in figure 6.2. The overall pattern of a model behavior can be represented as a concatenation of these atomic pieces,[3] and it is the sequence and the number of elements in this sequence that characterize a model behavior in terms of its dynamic pattern characteristics. For example, an S-shaped growth can be characterized with the following sequence using the atomic pieces given in figure 6.2[4]: (0,0), (+,+), (+,0), (+,-), (0,0). This sequence is purely related to the dynamic pattern of the behavior. Independent of its scale or other numeric features, all S-shaped growth behaviors are identical in terms of this sequence. Therefore, if we aim to measure the difference in terms of pattern characteristics, we can base the measurement on each behavior's atomic behavior sequence. This is the main idea behind the proposed measurement algorithm, which consists of two major stages.

The first stage is converting a given behavior into a sequence of atomic behavior modes (transformation). The second stage is related to measuring the distance between two sequences. These stages are elaborated in the following subsections.

Conversion to a Sequence of Atomic Behavior Modes   The conversion process will be explained using sample time-series data. The data are plotted in figure 6.3, and the individual points that constitute the series are given in the second column of table 6.1.

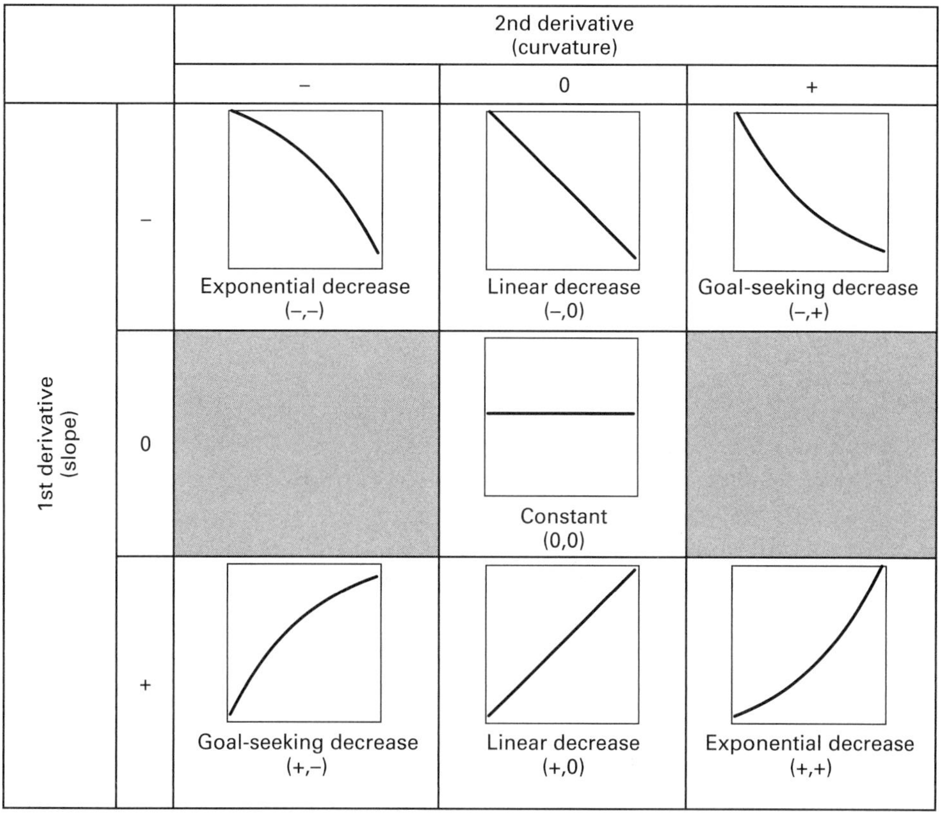

		2nd derivative (curvature)		
		−	0	+
1st derivative (slope)	−	Exponential decrease (−,−)	Linear decrease (−,0)	Goal-seeking decrease (−,+)
	0		Constant (0,0)	
	+	Goal-seeking decrease (+,−)	Linear decrease (+,0)	Exponential decrease (+,+)

Figure 6.2
The atomic behavior modes.

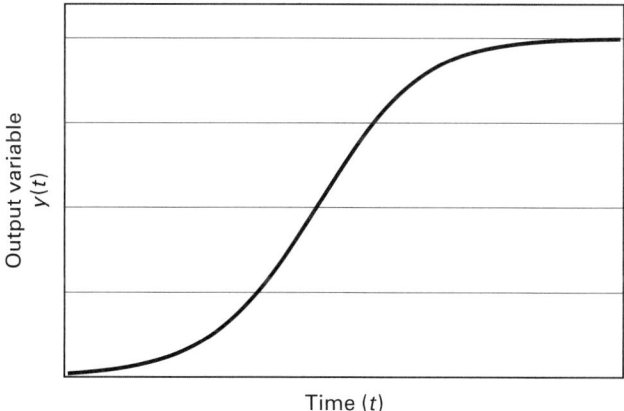

Figure 6.3
Sample dynamic behavior.

Given a set of time-series data, the slope and the curvature values are estimated at each data point. As an approximation, the following equations are used in these calculations:

$$slope\,(i) = \frac{x(i+1) - x(i)}{d(i)}$$

$$curvature\,(i) = \frac{slope\,(i+1) - slope(i)}{(\frac{d(i+1) + d(i)}{2})}\,,$$

where $d(i)$ is the time interval between the ith and $(i+1)^{th}$ data points .

The slope and curvature values for the example are given in the third and fourth columns of the table, respectively.

Then these slope and curvature values are converted to (+), (0), or (-) based on the sign of the slope/curvature (see columns 5 and 6). For each point, the combination of signs gives the atomic behavior mode that represents the behavior phase to which that point belongs (column 7). As a final step, the points that belong to the same phase are grouped together to identify different phases of the overall time-series data (column 8). This last step produces a sequence of atomic-behavior modes that represent the given time-series data. For the example used

Table 6.1
Behavior conversion example

t	X(t)	Slope	Curvature	Sign Slope	Sign Curvature	Atomic	Phases
0	5	0	0	0	0	(0,0)	(0,0)
1	5	0	0	0	0	(0,0)	
2	5	0	0	0	0	(0,0)	
3	5	2.28	0.93	+	+	(+,+)	(+,+)
5	10.48	4.39	1.41	+	+	(+,+)	
7	20.67	7.32	1.39	+	+	(+,+)	
9	36.70	9.68	0.28	+	+	(+,+)	
10	46.38	9.96	-0.49	+	-	(+,-)	(+,-)
15	86.47	4.04	-1.11	+	-	(+,-)	
20	97.93	0.67	-0.22	+	-	(+,-)	
25	99.71	0.09	-0.03	+	-	(+,-)	
28	99.91	0.02	-0.01	+	-	(+,-)	

here, the given time-series data can be represented with the following sequence of three phases: (0,0), (+,+), (+,-).

As the reader must have noticed, this conversion process can be seen as a qualitative abstraction of the numeric data. Once it is converted, neither the length of the data series (i.e., number of data points), nor the features related to the numerical scale (mean, max, min, etc.) has any relevance. The sequence only contains information about the pattern features of the time series. As a result of this abstraction, this approach is invariant to scale differences and translation (e.g., horizontal or vertical shifts in characteristic points such as inflection points), which is a key feature expected from structural similarity measures (Angstenberger 2001; Duda, Hart, and Stork 2001).

Measuring the Similarity Between Two Behavior Patterns   Once the given behaviors are converted into sequences of atomic patterns (i.e., feature vectors), the next task is to quantify the similarity of these sequences. In the proposed approach, this similarity is estimated by a distance measure, as follows.

Let x and y be two data series to be compared.

For now, assume that their feature sequences are composed of p phases.

$$d_{x,y} = \frac{\sum_{i=1}^{p}\left((x_{1,i} - y_{1,i})^2 + (x_{2,i} - y_{2,i})^2\right)}{p} \qquad x_{1,i}, y_{1,i} : \textit{The slope term}(1 \textit{ for }(+),$$

0 for (0), -1 for (-))

$x_{2,i}, y_{2,i} : \textit{The curvature term}(1 \textit{ for }(+), 0 \textit{ for }(0), \text{-}1 \textit{ for } (\text{-}))$

This measure gives a sort of average error per phase between behaviors. The value of d lies in between 0 and 8, 0 meaning qualitatively perfect match, and 8 meaning two totally different patterns.[5]

The measurement approach is quite easy and straightforward as long as two behaviors' feature vectors are of the same length (i.e., $p_x = p_y$). This, however, is not a very common case. For example, the previously given S-shaped behavior and a goal-seeking growth behavior; the feature vector of the latter is composed of two sections (i.e., (+,-), (0,0)). In other words, $p_x = 3$ and $p_y = 2$. This creates a serious problem in measuring the similarity between these two, since it becomes impossible to calculate $d_{x,y}$ according to the given equation. As a solution, a subroutine (i.e., sister creation) is proposed. A sister of a feature vector is identical to the original one in terms of the sequence of atomic behaviors, but it is longer in length. During sister creation, sister vectors of the short feature vector that are of the length of the long vector are created so that the problem reduces to comparison of two equal-length vectors. The following example clarifies the concept of sister and the operation of sister-creation.

Short feature vector: (+,-), (0,0)
Long feature vector: (0,0), (+,+), (+,-)
Sample sisters of the short vector:
a. (+,-), **(+,-)**, (0,0)
b. (+,-), **(0,0)**, (0,0)

As can be seen from this example, both sisters are identical to the time series with the shorter feature vector in terms of pattern characteristics (they all correspond to perfect goal-seeking growth). The only difference from the original feature vector is an extra section that is identical to one of the existing sections (see the bold sections in a and b above). In this extended representation, a certain phase of the pattern is represented by two entries in the feature vector, rather than one. In sister a, it is the growth phase that is represented by two sections, and in the case of sister b, it is the stabilization phase. This way, the short feature vector is extended without distorting the behavior-pattern information it carries. Since the sisters are both of length 3, now it is possible to use the previously given equation to compare them to the long feature vector. In this case, the similarity is calculated as follows,

$$d'_{x,y} = \min_{s \in S(y)} d_{x,s},$$ where S(y) is the set of sisters of y.

For the case given above, the short vector has only two sisters. Depending on the lengths of the two vectors being compared, the number of possible sisters may be quite large. The proposed approach does a random sampling among all possible sisters with a sample size of 20, which is experimentally evaluated to be satisfactory.

Example Applications

To prove the potential benefits of pattern classification and similarity clustering in dynamic modeling, we introduce a set of demonstrative examples that rely on the introduced classifier and the similarity measurement approach in the previous section. In all of these examples, we use a simple test model, which is introduced in the following subsection.

Test Model

The test model that is used in the rest of this chapter is a simple second-order linear system. The equation set of the model is given below:

$$\frac{\partial x(t)}{\partial t} = \frac{\hat{x} - y(t)}{\propto}$$

$$\frac{\partial y(t)}{\partial t} = \frac{x(t) - y(t)}{\beta}.$$

For the sake of illustration, this system can be conceptualized as a temperature adjustment system, where x(t) corresponds to the actual temperature and y(t) to the measured temperature. The model constants $\hat{x}$, α, and β are target temperature, temperature adjustment time, and measurement delay, respectively. This system can also be represented by the stock-flow diagram given in figure 6.4.

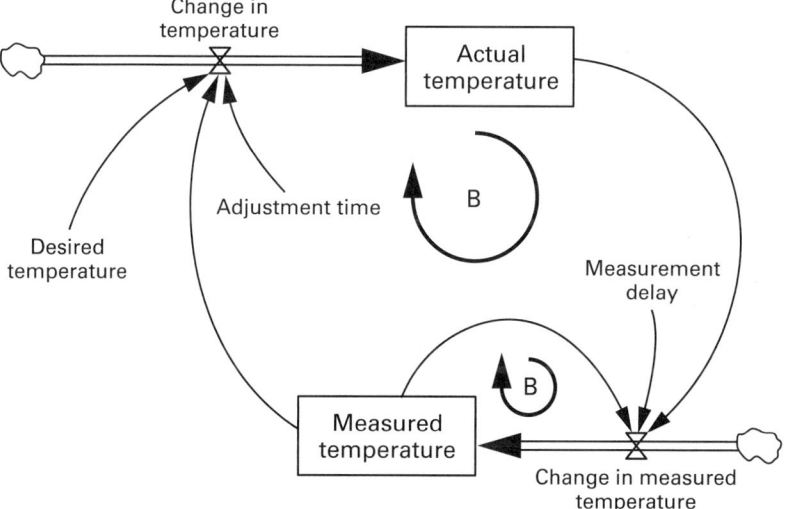

Figure 6.4
Stock-flow diagram of the test model.

In the following subsections, we will be using pattern-recognition tools for conducting automated testing, parameter calibration, and output clustering on this test model.

Model Testing with the Dynamic Behavior Classifier

The idea of using a pattern classifier to automate the model-testing process was initially demonstrated by Barlas and Kanar (1999) as a proof-of-concept application. In a follow-up project, Boğ and Barlas (2005) developed a standalone software named SiS, which has the aforementioned behavior classifier embedded into it. The software is Java-based, and it offers full integration with models that are developed with the Vensim® software. The self-executable version of SiS for Windows operating systems, which does not require any installation, is included in the e-companion of this chapter.

The software offers a set of functions, but this example will focus on the "automated validity testing" feature of it and conduct a couple of extreme condition tests on the model.

Step 1: Choosing the Function Mode and Loading the Model   When the SiS application is opened, the user is expected to specify the usage mode. In order to conduct validity tests, the "validity testing by setting parameters" option must be chosen. As mentioned earlier, SiS can communicate with Vensim models. However, these model files should be in the binary format (.vmf files). By default, Vensim models have the .mdl extension. In order to be able to use SiS, we need to save our test model in the binary format.

Once we have the binary format Vensim® model (TestModel.vmf in this case), we select and load that model to the SiS application. After successful loading, SiS requires the model to be simulated once in the beginning. After clicking on the "Simulate" command, click on the "Get Variable List" button, and this will populate the drop-down list with the variables of the selected model. In this example, we select the "Actual Temperature" variable to inspect. Then, click on the "Graph Results" button in order to plot the behavior of the selected variable with the default parameter values of the model (figure 6.5).

Step 2: Specifying the Indirect Structure Test   An indirect structure test requires a hypothesis about the expected model behavior in a specific condition (i.e., with a specific parameter value set). We consider the following hypothesis to be tested on our test model:

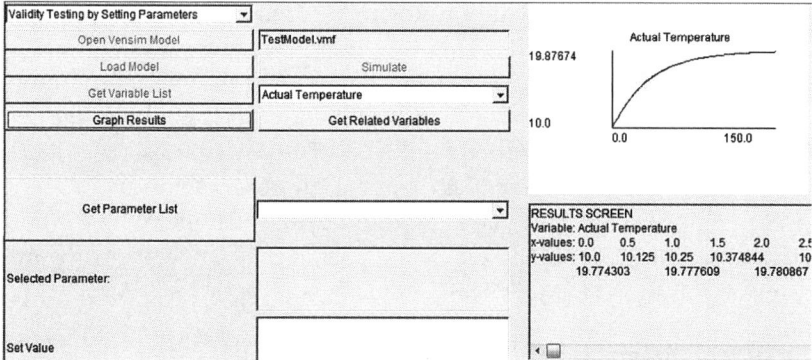

Figure 6.5
Default behavior of the model in SiS interface.

Hypothesis 1: When desired temperature is lower than the initial value of the actual temperature variable, the model is expected to demonstrate a negative exponential decline (converging to the desired temperature) behavior.

In order to specify this hypothesis in SiS, we first click on the "Get Parameter List" command, which populates the list of parameters. Then, we choose the parameter that we wish to specify, desired temperature, and specify its test value to 0. When we click on "Confirm Value," SiS sets the parameter to this test value and displays a confirmation message in its dialogue box. As the final step, we have to choose the expected behavior mode. SiS offers a pop-up list that contains the abbreviations of major behavior types[6] (figure 6.1). From this list, we choose NEXDC, which stands for negative exponential decline. After selection, SiS will display a sample behavior that corresponds to the selected behavior type. As an optional feature, we can also specify whether we expect the variable to converge to zero or to a nonzero value. At this stage, we finished specifying the above hypothesis in a form that SiS can evaluate.

Step 3: Running the Test and Interpreting the Results   When we click on the "Run the test" command on the SiS interface, what happens in the background can be summarized as follows.

- SiS communicates with Vensim and sets the user-specified parameter values that define the test condition. Then, SiS sends a

command to the simulation software to run the model under the specified parameter values.
- Once the model is run, SiS gets the behavior of the variable of interest from the simulation software and passes this behavior to the pattern classifier embedded within SiS. The pattern classifier identifies and returns the most likely class of the observed behavior.
- SiS compares the class that is returned by the classifier with the behavior pattern hypothesized by the user. If they agree, the software evaluates the test as a "PASS." Otherwise, it reports a "FAIL."

When we run our example test on SiS, we receive the output screen given in figure 6.6. Next to the initial behavior of the "Actual Temperature" variable, SiS displays the new behavior that is obtained under the test conditions. Below these plots, it is possible to see the brief conclusion of the test. SiS indicates that the observed dynamic behavior fits to the class of NEXDC (i.e., negative exponential decrease) the best. Since this is exactly the same behavior mode we hypothesized in our test, SiS evaluates this structure-oriented behavior test successfully, indicating that the model passed the test. Below the dialog box where these conclusions are displayed, SiS also provides more detailed information about the pattern classification. This table displays the degree of fit of the model behavior to all behavior classes that are defined in SiS. By definition, these numbers are all negative, and a smaller absolute value means a better fit.

As can be seen from the example, SiS is able to compare a qualitatively stated behavior mode with the actual model output in terms of their match to basic behavior classes with the help of a pattern-classification algorithm embedded to it. As a software with such pattern-recognition capabilities, SiS should be considered as an important first step toward a comprehensive automated model-testing tool. In its current form, it lacks some features, such as conducting tests in batches and user-friendly reporting of the results. Despite such shortcomings, it stands as a good demonstration of how a pattern classifier can be turned into a modeling support tool.

Model Calibration with the Dynamic Behavior Classifier

A pattern classifier enables an algorithm to judge whether a model generates a prespecified behavior type. Utilizing this ability, Boğ and Barlas (2005) discuss that their aforementioned software SiS can also be used for model-calibration purposes. Given a feasible range for a

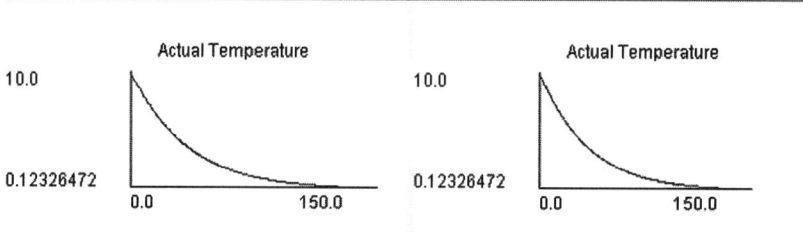

RESULTS SCREEN
    Outcome of the Hypothesis:
    PASSED   [Best fit class is NEXDC]

PATTERN RECOGNITION RESULTS

ZERO0	-10.0	GR1DA	-7.624572...	SSHDC	-4.597822...	G2PED	-18.50226...
CONST	-10.0	GR1DB	-6.643798...	PEXDC	-30.44091...	OSCCT	-21.68090...
PLINR	-10.0	GR2DA	-4.554411...	D1GRA	-13.85986...	OSCGR	-10.0
NLINR	-10.0	GR2DB	-5.300322...	D1GRB	-8.838284...	OSCDC	-21.68090...
NEXGR	-18.28290...	D1PEG	-6.133559...	D2GRA	-4.571861...		
SSHGR	-25.01014...	D2PEG	-5.484089...	D2GRB	-1.875817...		
PEXGR	-33.36098...	NEXDC	-0.261016...	G1PED	-14.20368...		

Figure 6.6
Results of the first hypothesis test in SiS interface.

model parameter, SiS runs the model with different values of this
parameter and evaluates the resulting behaviors. If a resulting behavior
belongs to the class of the targeted model behavior, the software reports
the parameter values that lead to this behavior. During this search
process, the software scans the feasible space of the parameter by incre-
menting its value by a fixed amount. In other words, it conducts a
uniform scan of the feasible range. This is an easy-to-implement type
of search, but it is not very efficient when feasible ranges are wide and
the number of parameters to be specified is large.

Building on the idea of using a pattern classifier for model calibration, Yücel and Barlas (2011) proposed another approach that incorporates a more advanced and efficient search algorithm compared to SiS.

The pattern-oriented parameter specifier (POPS) is a combination of an optimization heuristic and the pattern-classification algorithm of Barlas and Kanar (1999). Figure 6.7 provides an overview of the structure of POPS and how it conducts a parameter search. As seen in the figure, POPS has three key components. The *simulator* generates dynamic behavior of a model using particular parameter values. The *evaluator*, which is the dynamic behavior classifier introduced in the third section of this chapter, inspects the generated behavior with respect to the proximity of its pattern characteristics to the ones of the desired pattern. The last component, the *generator*, is responsible for searching new parameter values to be tested. This last component benefits from an optimization heuristic that uses the pattern proximity scores of the previously generated parameter values returned by the evaluator component.

POPS conducts a search based on three basic user inputs:

• model structure in the form of a differential/difference equation set
• parameter set and the feasible ranges of these parameters
• desired behavior pattern (to be selected from the basic patterns repertoire recognized by the classifier; see figure 6.1).

Briefly, the generator generates candidate parameter vectors (a scalar vector with each element corresponding to a model parameter being specified). These vectors are evaluated based on the degree of pattern-wise fit between the desired behavior and the behavior obtained by using the parameter values specified by this vector. In order to do so, the simulator simulates the model with the parameter values from the candidate parameter vector, and the evaluator evaluates the degree of fit of the resultant behavior to the desired one. As mentioned earlier in the chapter, the classifier calculates an error related to the likelihood that a given dynamic behavior belongs to a desired behavior class. This likelihood value is used as the degree of fit between the desired and the model-generated behavior. The generator uses the results of this evaluation in order to generate new candidate parameter vectors, as the evaluation indicates whether the tested parameter values are good (yielding a behavior that is close to the desired one). The purpose of the generator is to search the feasible parameter space to find "good" parameter values. This is done with the means of a genetic algorithm (GA), which is an optimization heuristic that relies on evolutionary principles (Goldberg 1989; Holland 1995). In an earlier study on automated model calibration, Eksin (2008) also utilizes GA for parameter

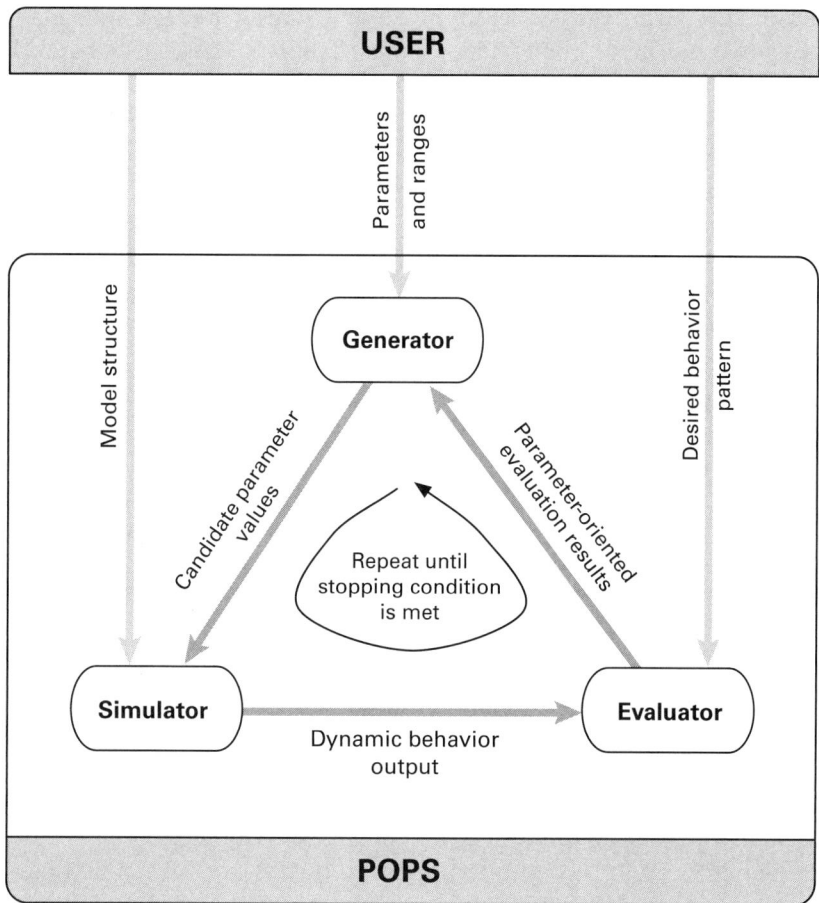

Figure 6.7
An overview of the main components of POPS (Yücel and Barlas 2011).

search. Briefly, an initial set of feasible solutions goes through a synthetic evolution process in GA. Solutions yielding better results have a higher probability of staying in the set (i.e., survival) and being used in creation of new feasible solutions (i.e., mating). In this way, the algorithm aims to explore the feasible solution space. This search process continues until a stopping condition is met (e.g., time constraint or lack of improvement in the last n steps). The technical details of the GA implementation in POPS can be found in Yücel and Barlas (2011).

As an example, we use POPS to identify parameter values for our test model that generate a specified behavior type.

Step 1: Preparing the Model for POPS and Setting the Desired Behavior   In its current form, POPS is a constellation of MATLAB functions. Therefore, the differential equations that correspond to the stock variables have to be entered into MATLAB software[7] in order to use POPS. In other words, the model needs to be translated into the MATLAB environment. The MATLAB function that corresponds to the model (i.e., model function) should take a parameter vector as an input. Each entry in this vector corresponds to a model parameter being searched. This function should conduct a simulation with these parameter values when it is called. In our example, we will search values for measurement delay and adjustment time parameters. Therefore, the model takes a two-dimensional vector as an input.

As mentioned above, the model function conducts a simulation when it is called. Additionally, this function will also evaluate the resulting output behavior. The last line of the model function is related to the evaluation of model output. We insert a command that passes the model output and the desired behavior to the main function of the pattern classifier (i.e., testhy). The command is as follows:

```
[a,b,fitness2] = testhy(Output_1, id('gr2db'), 0)
```

In this command, we specify the desired behavior mode in this command using the abbreviation convention of SiS (e.g., nexgr for negative exponential growth, gr2db for growth and decline; see figure 6.1). The model function returns the error in the likelihood of model behavior belonging to the specified target behavior class.

The MATLAB file that corresponds to the test model can be found in the electronic supplement of this chapter.

Step 2: Specifying the Feasible Ranges for the Parameters   Before starting a parameter search, we specify the range for the search. This has to be done in the optimizer module of POPS (i.e., optimizer.m file). We specify the lower and upper bounds for both parameters as 0 and 20, respectively.

Step 3: Running the Parameter Search   In order to run POPS, we have to call the controller function of POPS, which initiates the parameter search algorithm and returns the results. Since the search algorithm is probabilistic in nature, it is recommended to repeat the process while seeking parameters values that would yield a certain model behavior. The controller function also takes care of this repetition task. In our example, we will conduct the search with 30 repetitions. For this purpose, we enter the following command in the command window of MATLAB:

```
controller(30)
```

Once the search process is completed, the controller will create a data file (.mat file) that stores the parameter values found in these 30 replications.

Using this procedure, we have conducted a value search for adjustment time and measurement delay parameters with different desired behavior patterns. The desired patterns included negative exponential growth (nexgr), s-shaped growth (sshgr), growth-and-decline (gr2db), and oscillations (osscs). For each of these searches, POPS returned 30 parameter value tuples that yield a model behavior to fit the class of the desired behavior. The values returned by POPS for adjustment time (alpha) and measurement delay (beta) are plotted in figure 6.8. The model behaviors that are obtained using these parameter values can be seen in figure 6.9.

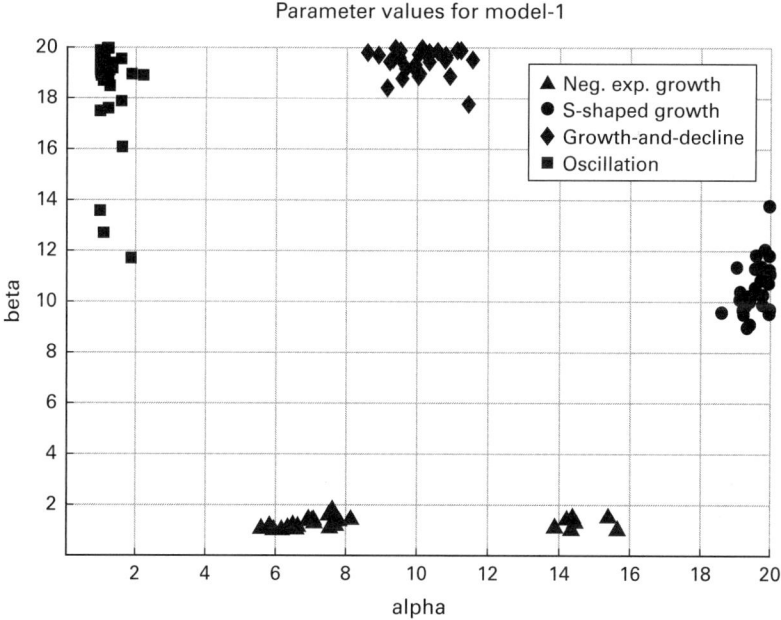

Figure 6.8
Parameter values identified with POPS for different target behavior modes.

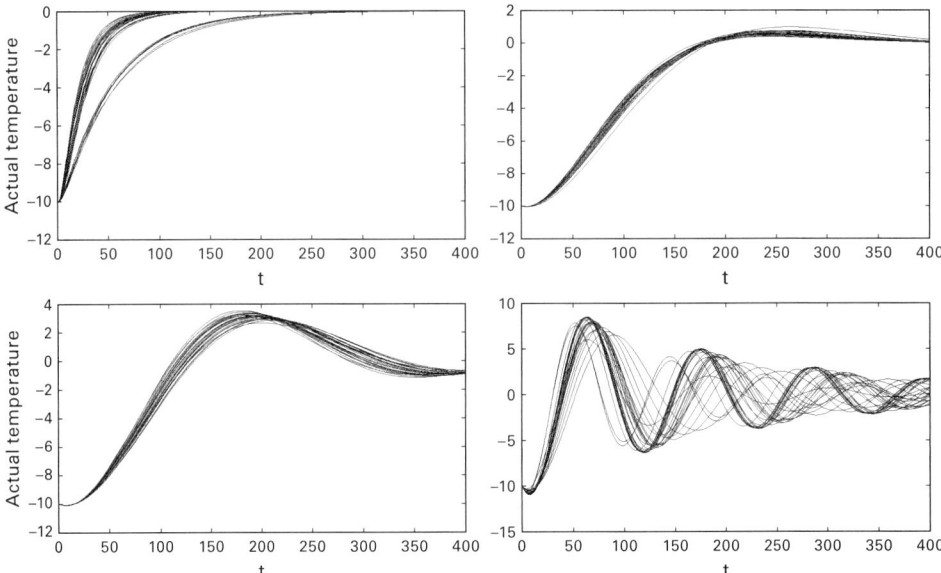

Figure 6.9
Model output with the parameter values identified with POPS.

The example given above involves simultaneous fine-tuning of only two parameters; hence the task can be assessed as an easy one. However, when we consider a simultaneous search process with a large number of parameters for a nonlinear model, the potential assistance of a tool like POPS becomes apparent. POPS is an important proof-of-concept application that shows how computational optimization approaches such as genetic algorithms can be of service to modelers when they are coupled with a pattern classifier.

Behavior Analysis with Dynamic Behavior Clustering

As a third application, we demonstrate how the similarity measurement approach discussed in the third section can be used for pattern recognition in model behavior analysis.

Given a large set of dynamic behavior patterns, it is possible to calculate their similarity in a systematic and consistent manner using the aforementioned similarity measurement approach proposed by Yücel (2012). Once this measurement is done, the problem reduces to static data clustering. The calculated similarities (or pattern-wise distances, as they may be called) can in turn be used as input to a regular

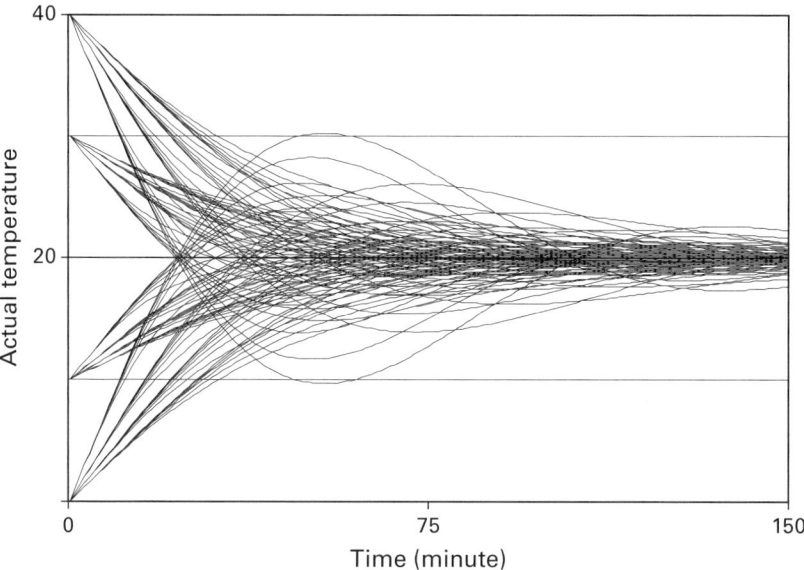

Figure 6.10
Resulting behavior patterns in the multivariate experiment with the test model.

clustering algorithm, and a large number of model outputs can automatically be grouped according to their dynamic behavior characteristics.

We will again use the test model to demonstrate the benefits of such an automated clustering algorithm. Although our test model is a fairly simple one, it is capable of generating various behavior patterns. In order to explore the model's behavior, we conduct a multivariate experiment where we simulate the model with different adjustment time, measurement delay, and initial actual temperature values.[8] The resulting set of model outputs is given in figure 6.10. Looking at the aggregate plot, it is observed that the model generates different types of dynamic output. However, filtering out different behavior types and linking them with specific parameter configurations is not a straightforward task even for this small set of experiments. It requires going over each run individually in order to link the parameter values and the resultant pattern, which can be a very time-consuming and ineffective approach. In this third example, we will use pattern recognition for automatically clustering these model behaviors and identifying which parameter values lead to which class of behavior.

We will utilize a Python-based implementation of a hierarchical clustering algorithm that uses the previously introduced similarity measurement approach. This algorithm is implemented as a part of the exploratory modeling and analysis (EMA) Workbench (Delft 2014). A compact version of the EMA Workbench that includes only the clustering-related components (i.e., behavior pattern clusterer, or BPC) can be found in the e-companion of this chapter. Technical instructions for setting up BPC are given in the appendix.

BPC has two main components. The first component takes an Excel (.xlsx) file that contains time-series data to be clustered as the input and returns a vector of pattern-wise distances between these. The second component is a conventional hierarchical clustering algorithm that uses these distances to identify pattern clusters in the given set of time series.

As a demonstrative example, we will use BPC to identify behavior clusters in the output of our multivariate experiment with the test model.

Step 1: Preparing the Model Output for the Clustering Process As mentioned above, the BPC algorithm uses an Excel (.xlsx) file as the input. Therefore, the model output has to be exported to such a file. In our case, we use the "Export Dataset" feature of Vensim® to export the results of the experiments first to a tab-delimited file, then save that file as an Excel workbook. Details about the format of this input file are described in the appendices of this chapter.

Step 2: Setting Up and Running BPC Before running the clustering algorithm, the name of the input file has to be specified in the main procedure of the clusterer.py file (see appendix B for locating the file and the procedure). In our case, the name of the file is TestModel_Demo.xlsx. Therefore, we set the inputFileName variable to "TestModel_Demo."

The last line of code in the main procedure calls the cluster procedure. This procedure has a set of arguments that can be fine-tuned by the user. A detailed and up-to-date description of these input arguments can be found at the web portal of the EMA Workbench.[9] One of these arguments is cValue, which stands for the cut-off value of the hierarchical clustering stage. Since the best cut-off value differs between different data sets, we recommend experimenting with this value a few times until the most informative clustering result is identified. In our

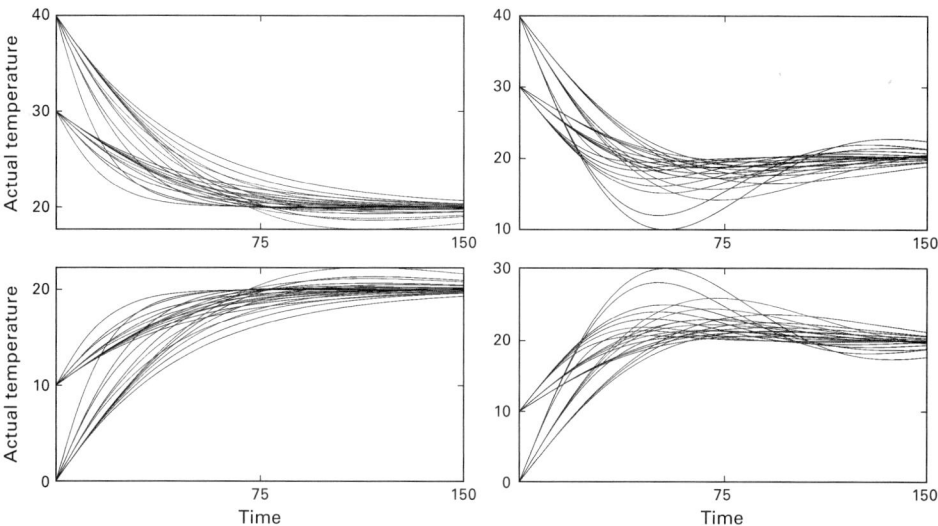

Figure 6.11
Pattern clusters identified by the clustering algorithm.

example, a cValue of 4 yielded only two clusters, which are not homogenous enough (i.e., different types of behavior are still in the same cluster). When we try the algorithm with a cValue of 0.5, we obtain 13 clusters. After a couple more trials, we obtained the most satisfactory result with a cValue of 1.4. The behavior clusters identified by BPC with a cValue of 1.4 are given in figure 6.11.[10]

As can be seen above, we are able to group similar dynamic behaviors together and label them automatically using BPC.[11] As BPC assigns cluster indices to each individual model output, we do not need to go over each run in order to identify parameter values that led to growth-and-decline behavior, for example; it suffices to filter the model runs that are labeled as belonging to cluster 3. The most important point that needs extra emphasis here is the fact that this grouping is done purely by a series of computer algorithms in an automated manner. The potential contribution of such a tool can be appreciated by imagining a large set of model outputs with very different pattern characteristics.

Although the implementation discussed above may appear to require too much technical work and hassle just for grouping model outputs, it is described simply to demonstrate that in principle it is possible to design an algorithm that can distinguish model outputs based on their qualitative dynamic pattern features. Widespread use of such an algorithm depends on its being part of a user-friendly software package that can link with existing modeling software, so that it requires little programming experience and knowledge.

Discussion

This chapter provides a brief introduction to the field of pattern recognition and its current state in analyzing dynamic patterns. As discussed in the opening section, pattern-recognition tools and methods have great potential for automating important stages of a modeling process via computerizing the output pattern-evaluation task. Although it is a fairly simple task for the modeler to visually evaluate a given pattern, the added value of automating this task is in the opportunity to use advanced computational tools for exhaustive parameter search and output analysis purposes. Such tools enable the modeler to conduct an in-depth parameter search, model testing for sensitivity and robustness, policy optimization, and behavior-space exploration—tasks that would be impossible by visual inspection, due to their combinatorial complexity.

We provide three practical examples on using three pattern-recognition–supported tools designed to assist different stages of the modeling process. These examples are fairly simple and intended to serve as tutorial exercises. Despite being simple, we believe that they help to grasp the working principles and to picture the potential added value of using pattern-recognition tools in much more complicated cases. These examples also stand as proof-of-concept applications on what can be done by coupling pattern-recognition algorithms with other computational tools such as optimization heuristics.

The advantages offered by pattern-recognition tools naturally come with a cost. Apart from specialized standalone software such as SiS, most of these tools operate on a programming (e.g., Java, Python) or scientific computing environment (e.g., MATLAB, R). In order to use such a pattern recognition tool with a simulation model, one needs to bridge the model and the tool. Alternative ways to establish the connection include formulating the model in the native environment of the pattern-recognition tool (e.g., the case of POPS), exporting the model output to a format that can be read by the tool (e.g., the case of the clustering algorithm), or developing a program that will establish the connection between the model and the tool. All of these options require at least a basic knowledge of the environment in which the pattern-recognition tool functions, as well as putting some effort into establishing the link between the model and the tool. This stands as the major cost of benefiting from pattern-recognition approaches until more standalone software with pattern-recognition capabilities that are tailored for model output analysis, like SiS, are available for modelers.

Pattern recognition is a very large field of research with diverse techniques and application domains. Covering the breadth of this field, with all its technical and mathematical background, in this chapter would naturally be an impossible task. Therefore, we have provided a brief introduction to the key concepts of the field and introduced three tools that are developed specifically for dealing with output patterns generated by dynamic models. In that respect, the chapter focuses mainly on using already developed pattern-recognition tools rather than designing and developing a new pattern-recognition algorithm from scratch. The latter demands a deeper understanding about the computational and mathematical background, as well as an expertise in key aspects such as representation, feature selection, and similarity measurement.

Exercise

Using the SIS software, conduct the following extreme-condition test for the test model. The hypothesis to be tested is as follows.

Hypothesis: When desired temperature is higher than the initial value of the actual temperature variable, and the measurement delay is equal to adjustment time, the model is expected to demonstrate a negative exponential growth (converging to the desired temperature) behavior.

The test model, SIS software, as well as the solution to this exercise can be found in the electronic supplement of this chapter.

Notes

1. In this context, nearness refers to the degree of similarity between objects.

2. https://www.coursera.org/course/ml; accessed January 11, 2013.

3. In some cases (e.g., the cases with stochasticity), the raw model output may not be represented as such. The overall pattern needs to be extracted from the raw data, for example by smoothing, noise filtering methods, or both.

4. Each atomic pattern is represented by the sign combinations in the parentheses in order to make it easier to refer to the atomic behaviors throughout the paper.

5. The worst match between two phases occurs when both slope and curvature signs are different. Since slope and curvature can only take the values of -1, 0, or 1, the total error in such a case will be:

$(slope_1-slope_2)^2+(curvature_1-curvature_2)^2 = (2)^2+(2)^2 = 8$.

Therefore, maximum value of average error per phase is 8.

6. See the appendix for a full list of abbreviations that are used in SiS.

7. The MATLAB® code that corresponds to this model can be found in the e-companion.

8. The parameter values used during these experiments are as follows:

Adjustment time: 20, 30, 40 and 50
Measurement delay: 5, 10, 15, 20, 25 and 30
Initial actual temperature: 0, 10, 20, 30 and 40

We run the model with all 120 combinations of these parameter values.

9. http://simulation.tbm.tudelft.nl/ema-workbench

10. Constant cluster is omitted.

11. BPC labels clusters with numbers (e.g., cluster 1, 2, etc.). Clusters are named (e.g., growth-and-decline cluster) by the user based on subjective assessment.

References

Alpaydin, E. 2010. *Introduction to machine learning*. Cambridge, MA: MIT Press.

Angstenberger, L. 2001. *Dynamic fuzzy pattern recognition with applications to finance and engineering*. Dordrecht, The Netherlands: Kluwer Academic.

Antunes, C., and A. Oliveira. 2001. Temporal data mining: An overview. Paper presented at the Workshop on Temporal Data Mining (KDD 2001), San Francisco, CA.

Bagnall, A., and G. Janacek. 2005. Clustering time series with clipped data. *Machine Learning* 58 (2–3): 151–178.

Bagnall, A., E. Keogh, C. A. Ratanamahatana, S. Lonardi, and G. Janacek. 2006. A bit level representation for time series data mining with shape based similarity. *Data Mining and Knowledge Discovery* 13 (1): 11–40.

Barlas, Y. 1996. Formal aspects of model validity and validation in system dynamics. *System Dynamics Review* 12 (3): 183–210.

Barlas, Y., and K. Kanar. 1999. A dynamic pattern-oriented test for model validation. Paper presented at the 4th System Science European Congress. Valencia, Spain.

Bishop, C. M. 2006. *Pattern recognition and machine learning*. New York: Springer.

Boğ, S., and Y. Barlas. 2005. Automated dynamic pattern testing, parameter calibration and policy improvement. Paper presented at the 23rd International Conference of the System Dynamics Society, Boston, MA.

Chen, X., D.-Y. Ye, and X.-L. Hu. 2007. Entropy-based symbolic representation for time series classification. Paper presented at the Fourth International Conference on Fuzzy Systems and Knowledge Discovery, Haikou, Hainan, China.

Corduas, M., and D. Piccolo. 2008. Time series clustering and classification by the autoregressive metric. *Computational Statistics & Data Analysis* 52 (4): 1860–1872.

Cuberos, F. J., J. A. Ortega, L. Gonzalez, and F. Velasco. 2004. A methodology for qualitative learning in time series. Paper presented at the 18th International Workshop on Qualitative Reasoning. Northwestern University, Evanston, Illinois.

Delft, T. U. 2014. Exploratory modelling and analysis (EMA) workbench. Available at http://simulation.tbm.tudelft.nl/ema-workbench/contents.html. Accessed July 8.

Duda, R., P. Hart, and D. Stork. 2001. *Pattern classification.* New York: Wiley-Interscience.

Eksin, C. 2008. Genetic algorithms for multi-objective optimization in dynamic systems. Paper presented at the 26th International Conference of the System Dynamics Society, Athens, Greece.

Fu, T.-C. 2011. A review on time series data mining. *Engineering Applications of Artificial Intelligence* 24 (1): 164–181.

Goldberg, D. E. 1989. *Genetic algorithms in search, optimization and machine learning.* Reading, UK: Addison-Wesley.

Gullo, F., G. Ponti, A. Tagarelli, and S. Greco. 2009. A time series representation model for accurate and fast similarity detection. *Pattern Recognition* 42 (11): 2998–3014.

Holland, J. 1995. *Hidden order: How adaptation builds complexity.* Reading, UK: Addison-Wesley.

Kalpakis, K., D. Gada, and V. Puttagunta. 2001. Distance measures for effective clustering of ARIMA time-series. Paper presented at the IEEE International Conference on Data Mining, San Jose, CA.

Kanar, K. 1999. Structure-oriented behavior tests in model validation. MSc thesis, Boğaziçi University, Istanbul, Turkey.

Li, H., C. Guo, and W. Qiu. 2011. Similarity measure based on piecewise linear approximation and derivative dynamic time warping for time series mining. *Expert Systems with Applications* 38 (12): 14732–14743.

Liao, W. T. 2005. Clustering of time series data—a survey. *Pattern Recognition* 38 (11): 1857–1874.

Mitsa, T. 2010. *Temporal data mining.* Boca Raton: CRC Press.

Ratanamahatana, C. A., and E. Keogh. 2005. Three myths about dynamic time warping data mining. Paper presented at the Fifth SIAM International Conference on Data Mining, Newport Beach, CA.

Theodoridis, S., and K. Koutroumbas. 2003. *Pattern recognition,* 2nd ed. San Diego: Academic Press.

Therrien, C. W. 1989. *Decision estimation and classification.* New York: Wiley & Sons.

Wang, X., K. Smith, and R. Hyndman. 2006. Characteristic-based clustering for time series data. *Data Mining and Knowledge Discovery* 13 (3): 335–364.

Wu, X., V. Kumar, R. J. Quinlan, J. Ghosh, Q. Yang, H. Motoda, G. J. McLachlan, A. Ng, B. Liu, P. Yu, Z.-H. Zhou, M. Steinbach, D. J. Hand, and D. Steinberg. 2007. Top 10 algorithms in data mining. *Knowledge and Information Systems* 14 (1): 1–37.

Wu, Y.-L., D. Agrawal, and A. El Abbadi. 2000. A comparison of DFT and DWT based similarity search in time-series databases. Paper presented at the 9th International Conference on Information and Knowledge Management, ACM. McLean, VA.

Yücel, G. 2012. A novel way to measure (dis)similarity between model behaviors based on dynamic pattern features. Paper presented at the 30th International Conference of the System Dynamics Society, St. Gallen, Switzerland.

Yücel, G, and Y. Barlas. 2011. Automated parameter specification in dynamic feedback models based on behavior pattern features. *System Dynamics Review* 27 (2): 195–215.

Zhang, X., J. Liu, Y. Du, and T. Lv. 2011. A novel clustering method on time series data. *Expert Systems with Applications* 38 (9): 11891–11900.

7 Linking Structure to Behavior Using Eigenvalue Elasticity Analysis

Rogelio Oliva

The link between system structure and dynamic behavior is one of the defining elements of dynamic modeling. In a sense, a simulation model can be viewed as an explicit and consistent theory of the behavior it exhibits. Although this point of view has certain merits, not least the fact that it lifts the discussion from outcomes to causes of these outcomes and from events to underlying structure (Forrester 1961; Sterman 2000), we are concerned here with a more compact explanation of the system's behavior. In fact, most dynamic modeling projects report their results in terms of simpler explanations of the observed results, typically in terms of dominant feedback loops and, occasionally, external driving forces to the system that produce the salient features of the behavior.

For simple systems with relatively few variables, it is usually easy to use intuition and trial-and-error simulation experiments to explain the dynamic behavior as resulting from particular feedback loops. In larger systems, this method becomes increasingly difficult, and the risk of incorrect explanations rises accordingly. There is a need, therefore, for analytical methods that add consistency and rigor to this process.

These analytical tools are important to the practitioner because the structure-behavior link is the key to finding leverage points for policy initiatives. And they are important to the theorist because a system dynamics theory of a particular phenomenon is an account of how certain feedback loops cause certain dynamic patterns of behavior to appear. The qualitative understanding of the model behavior is often at least as important as the particular numerical predictions obtained, even in applied studies. Yet the rigor of such an account depends directly on the rigor with which a structure-behavior link can be established in a given model.

Eigenvalue elasticity analysis (EEA) is a set of methods to assess the effect of structure on behavior in dynamic models. It works by considering observed model behavior as a combination of characteristic behavior modes and by assessing the relative importance of particular elements of system structure in influencing these behavior modes. Elements of the model structure that have a large influence on particular behaviors can provide important clues to the modeler to identify areas for further testing and study, and for policy analysis. The method represents a high degree of mathematical rigor compared to the traditional experimental simulation methods normally used in the field. The method uses linear systems theory to (1) decompose the observed behavior into its constituent *behavior modes*, such as oscillation, growth, and exponential adjustment, and (2) outline how a particular behavior mode, and its appearance in a given system variable, depends upon particular parameters and structural elements (links and loops) in the system. In this manner, the method provides a precise account of the relationship between structure and behavior.

The EEA method enables large-scale models to be analyzed systematically in a manner that is not possible or practical using trial-and-error simulation. Given the rigor and relative sophistication of the method, it may also provide legitimacy to dynamic model analysis in fields that are traditionally dominated by analytical mathematics, such as economics and econometrics.

The purpose of the elasticity analysis is to analyze the relative importance of structural elements, not to estimate the strength of system elements or values of system parameters. Therefore, the method works very much from a given model structure and parameter set and then tells you something about what would happen if you modify the structure or parameters. It thus fits mostly in the interpretation and policy analysis stages of model building. It may also prove useful in the model building and testing stage, to the extent that it can help identify structures that produce unwanted or puzzling behavior.

The EEA methods are similar in aims and scope to the *pathway participation method* (PPM; Mojtahedzadeh, Andersen, and Richardson 2004). The main difference between the two approaches is that while PPM emphasizes identifying a single "dominant" structure that drives the behavior of a particular variable, and does so relying primarily on partial system structures, the EEA approach provides an overview of the relative influence that different pathways simultaneously have on a variable, and does so considering global system properties (see

Mojtahedzadeh 2008 for a comparison of the two methods). Duggan and Oliva (2013) summarize other methods that rely on iterative and sensitivity-based approaches to explore dominant structure.

A word of caution is in order: like any other quantitative method, there is always an unavoidable element of judgment and interpretation when employing the method in practice. Moreover, the results of the EEA may require some work to interpret: since the method involves a translation from patterns of behavior over time to complex number eigenvalues, the results can appear highly abstract. New measures are being tested to facilitate more direct interpretation, but these are still in the developmental stage.

Background and Formulation

The method of using eigenvalue elasticities is based on the tools from modern linear systems theory (Chen 1970; Luenberger 1979), applied to a linearized model. The method was first introduced in system dynamics with Nathan Forrester's doctoral dissertation (1982). He used the method in the context of a macroeconomic model to explore various stabilization policies. However, the method was only peripheral to the dissertation, with most of the emphasis being traditional simulation experiments. Some attempts were made using eigenvalue analysis in the National Model project at MIT, but the limited availability of software and difficulty in interpreting the results prevented the method from gaining extensive use.

In 1996, Kampmann (2012) reintroduced the method by combining it with network and graph theory to reveal some fundamental relationships between feedback loops and eigenvalue elasticities. In particular, he highlighted that there is typically a very large number of alternative loop descriptions of a system and introduced the notion of an independent loop set from which all other feedback loops can be said to be derived. He further demonstrated how eigenvalues, or behavior modes,[1] are in a sense determined only by the *loop gains* in the independent set, while the appearance of these behavior modes in the behavior of individual variables is a function of the *link gains* in the system.

The traceability of eigenvalue elasticity to specific feedback loops, together with the availability of software to support numerical and algebraic analysis (e.g., Mathematica, Maple, MATLAB) and the advances of the PPM, triggered a stream of research to test and expand the usefulness and applicability of EEA.[2] Kampmann and Oliva (2006) automated some of the computational requirements to perform EEA and tested the method across three types of models. They found that the utility of the method depended on the model structure and that it was most useful for large-scale quasi-linear models. Güneralp (2006) developed a new measure that takes all model modes into account and proposed a normalization approach for elasticity values. Gonçalves (2009) and Saleh et al. (2010) extended the eigenvalue approach to focus on the overall trajectory of a state variable, and in particular the contribution of the eigenvector, which allows for the analysis of both short- and long-term impact on changes in link and loop gains. These papers, however, have focused on explaining how the method works and guiding the interpretation of results. As such, the authors have chosen simple and well-behaved models in which it is relatively easy to map the method's outcomes with the observed behavior and structure. To date, there is no documented case of the benefits of the application of the EEA methods to a realistic model, where structural dominance analysis is hypothesized to be most effective.

The following subsections provide an analytical description of the EEA method.

Characterizing Linear and Nonlinear Systems

A dynamic model can be represented mathematically as a set of ordinary differential equations,

$$\frac{d\mathbf{x}(t)}{dt} = \dot{\mathbf{x}}(t) = \mathbf{f}(\mathbf{x}(t), \mathbf{u}(t)), \tag{1}$$

where $\mathbf{x}(t)$ is a column vector of n states variables (levels), $\mathbf{u}(t)$ is a column vector of p exogenous variables, $\mathbf{f}$ is a corresponding vector function, and t is simulated time. The system is said to be linear (nonlinear) if $\mathbf{f}$ is a linear (nonlinear) function of its arguments. Given the model structure in equation (1), knowledge of the initial conditions $\mathbf{x}_0$, and the path of the input variables $\mathbf{u}(t)$, the behavior of the model is completely determined. In this sense, the model structure described in equation (1) constitutes a "theory" of the behavior $x(t)$.

The approaches considered in this chapter are based on tools from linear systems theory (Chen 1970), and they approximate the nonlinear model in equation (1) with a linearized version, using the first-order Taylor expansion around some operating point $\mathbf{x}_0$, $\mathbf{u}_0$; that is,

$$\dot{\mathbf{x}}(t) \approx \mathbf{f}(\mathbf{x}_0, \mathbf{u}_0) + \frac{d\mathbf{f}}{d\mathbf{x}}(\mathbf{x} - \mathbf{x}_0) + \frac{d\mathbf{f}}{d\mathbf{u}}(\mathbf{u} - \mathbf{u}_0),$$

or, by redefinition of the variables $\mathbf{x} \rightarrow \mathbf{x} - \mathbf{x}_0 - \mathbf{f}(\mathbf{x}_0, \mathbf{u}_0)(t - t_0)$ and $\mathbf{u} \rightarrow \mathbf{u} - \mathbf{u}_0$,

$$\dot{\mathbf{x}}(t) \approx \mathbf{A}\mathbf{x}(t) + \mathbf{B}\mathbf{u}(t), \tag{2}$$

where $\mathbf{A}$ is a constant $n \times n$ matrix of partial derivatives $\partial f_i / \partial x_j$ and $\mathbf{B}$ is a constant $n \times p$ matrix of partial derivatives $\partial f_i / \partial u_j$, and all partial derivatives are evaluated at the operating point. These matrices of first-order partial derivatives are know as *Jacobian* matrices. Both the eigenvalue and eigenvector elasticity analysis are based upon this approximated system.

Initially, one may primarily be concerned with the endogenous response of the system, in which case one can set the exogenous or control variables to a constant.[3] In the absence of changes in exogenous inputs, the resulting behavior of any given state variable $x(t)$ can be written as a weighted sum of a set of behavior modes,

$$x_i(t) = w_{i,0} + w_{i,1}e^{\lambda_1 t} + \ldots + w_{i,n}e^{\lambda_n t}, \tag{3}$$

where the λ's are the eigenvalues of the system's Jacobian matrix A and the weights w are constants that depend upon the eigenvectors and the initial conditions of the system (see Saleh et al. 2010 for derivation).

Equation (3) yields three important insights. First, each of the system eigenvalues represents a behavior mode. For real eigenvalues, the behavior mode $e^{\lambda t}$ amounts to an exponential growth ($\lambda > 0$) or adjustment ($\lambda < 0$). Complex eigenvalues appear in conjugate pairs $\delta \pm i\omega$, which give rise to oscillations $e^{\delta t}\sin(\omega t + \theta)$ of frequency ω that are either expanding (if $\delta > 0$) or damped oscillations (if $\delta < 0$). The absolute value of λ is known as the *natural frequency* $f_n = |\lambda| = \sqrt{\delta^2 + \omega^2}$, while the imaginary part of λ is known as *damped frequency* $f_d = \omega$ (see figure 7.1).

Second, the behavior of every state variable in the system is a *constant* weighted sum of the system behavior modes. That is, the behavior of every state variable in the system is the result of how each of the behavior modes λ is projected into that state variable x_i. Finally, the system core behavior modes are structurally determined, as they are derived from the eigenvalues λ of the system matrix A.

Different "flavors" of EEA emphasize each of the three insights in different ways (Kampmann and Oliva 2008, 2009). For example, EEA can be used to develop "structural explanations of behavior," since it can pinpoint which system elements are responsible for generating a

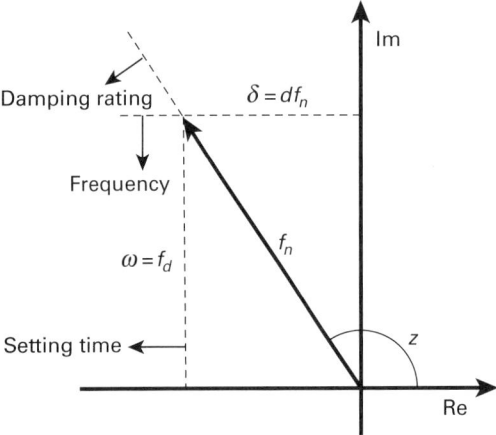

Figure 7.1
Characterization of eigenvalues in the complex plane.

particular behavior mode λ. Alternatively, the tools can be used to derive effective policy recommendations by isolating the system elements that affect the projection of a particular reference mode in a stock (w), or altogether change the system reference modes λ. Before describing in detail the tools to perform these analyses, the next subsection presents an example of these computations with a simple model.

Simple Example   Consider the Lotka-Volterra model in figure 7.2, defined by the following equations

$$\frac{dx}{dt} = Bx - Dx, \quad \frac{dy}{dt} = By - Dy,$$

$$Bx = ax, \quad Dx = bxy, \quad By = cxy, \quad Dy = dy,$$

where x represents the prey population; y the predator population; and the parameters, $a, b, c,$ and d, determine, respectively, the natural growth rate of the prey population in the absence of predators, the efficiency of predation, the predator reproduction rate (per available prey), and the natural death rate of predators in the absence of prey. With appropriate initial conditions and parameter values, the model reaches a limit cycle, with the two populations rising and falling alternatively (see figure 7.2).

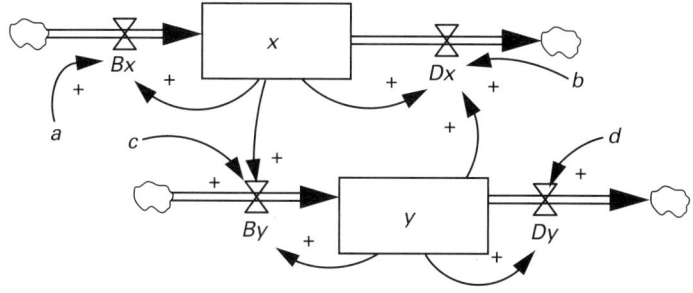

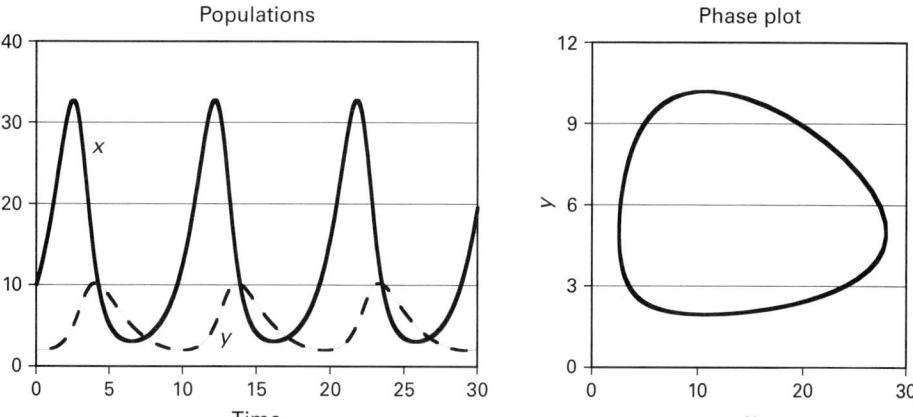

Figure 7.2
Structure, trace, and phase plot of Lotka-Volterra model. Parameter values: $a=1$, $b=0.2$, $c=0.04$, and $d=0.5$; initial conditions, $x_0=10$ and $y_0=2$.

Since the model has no exogenous inputs, the matrix representation of the system can be derived directly:

$$\begin{pmatrix} \dot{x} \\ \dot{y} \end{pmatrix} = \begin{pmatrix} a-by & -bx \\ cy & cx-d \end{pmatrix} \begin{pmatrix} x \\ y \end{pmatrix}.$$

The eigenvalues of the *Jacobian* or system matrix **A** are the conjugated pair

$$\left(a-d+cx-by \pm \sqrt{(a+d-cx)^2 - 2by(a+d+cx)+b^2y^2} \right)/2.$$

The eigenvalues are a function of x and y, so their value will change throughout the simulation. The eigenvalues at time zero are 0.25±0.194i, and at that point in time the behavior of the two stocks can be represented by the following equations:

$$x = 20 - 45.018\, e^{0.25t} sin\,(0.224 - 0.194t)$$

$$y = 8 - 9.004\, e^{0.25t} sin\,(0.729 - 0.194t),$$

where the first term represents a scaling parameter, the coefficient of the second term the weight w of the behavior mode on that stock (in this case the two eigenvalues collapsed into a single oscillatory behavior mode), and the first term inside the sin function represents the phase lag of this projection. Although these trajectory equations are only valid around the operating point x=10, y=2 (i.e., the trajectory equations will change with the eigenvalues), they give a clear indication of the tendency of the system, and it is possible to assess the impact of the system structure on the behavior at that point in time. To fully understand the behavior of the system, it would be necessary to replicate the analysis at different operating points. Before leaving this example, however, it should be noted that for most stock values within the limit cycle, the eigenvalues take complex values, indicating an oscillatory behavior mode.

The main strategies for exploiting the information available in this description of the system are described in the following subsections.

Eigenvalue Elasticity and Influence

The EEA is concerned with assessing how system structure affects the behavior modes (λ) as well as the projections of those behavior modes in a particular stock (w). A measure of the impact on an eigenvalue λ when one changes individual elements a of the system matrix is the eigenvalue elasticity,

$$\varepsilon = \frac{\partial \lambda}{\partial a} \frac{a}{\lambda}.$$

The most granular element of system structure is a link between two variables, and the strength of that relationship is measured through what is call the link gain—that is, the ratio of the output to the input. For example, in the model above, the gain of the link between x and By is cy. Clearly, all elements a of the system matrix $\mathbf{A}$ are combinations of these individual link gains, and thus it is possible to make assessments of eigenvalue elasticity to each link gain and model parameter.

For a complex-valued eigenvalue, the elasticity measure will also be a complex number. One may define the elasticities of the real and imaginary parts separately, that is, as the real numbers

$$\varepsilon_\delta = \frac{\partial \delta}{\partial a}\frac{a}{\delta}, \ \varepsilon_\omega = \frac{\partial \omega}{\partial a}\frac{ga}{\omega},$$

respectively. Note that it is not the case that $\mathrm{Re}\{\varepsilon\} = \varepsilon_\delta$ or $\mathrm{Im}\{\varepsilon\} = \varepsilon_\omega$, since

$$\mathrm{Re}\{\varepsilon\} = \frac{\varepsilon_\delta \delta^2 + \varepsilon_\omega \omega^2}{\delta^2 + \omega^2}, \ \mathit{Im}\{\varepsilon\} = \frac{(\varepsilon_\omega - \varepsilon_\delta)\delta\omega}{\delta^2 + \omega^2}.$$

Kampmann and Oliva (2006), however, found that it is often easier to work with the so-called *influence measure* instead, defined as

$$\mu = \frac{\partial \lambda}{\partial a}a, \ \mu_\delta = \frac{\partial \delta}{\partial a}a, \ \mu_\omega = \frac{\partial \omega}{\partial a}a. \tag{4}$$

For the influence measures, it is indeed the case that $\mathrm{Re}[\mu] = \mu_\delta$ and $\mathrm{Im}[\mu] = \mu_\omega$. In addition to simplifying interpretation, the influence measures also remove technical difficulties involved when eigenvalues are close to zero.[4]

Loop Eigenvalue Elasticity Analysis (LEEA)
Kampmann (2012) showed that it was possible to express the characteristic polynomial[5] of the system matrix **A** (that is, the polynomial whose zeros are the eigenvalues of **A**) in terms of the gains of the loops in what he termed an *independent loop set* (ILS). The loop gain is defined as the product of the gains of its constituent links; for example, in the model above, the loop gain of the loop formed by {x, y, x} is (-bcxy). An ILS is a maximal set of loops whose gains can be determined or changed independently of each other through an appropriate assignment or change in the link gains of the system. The gain of any loop outside this set is then dependent upon the loop gains in the ILS. Put differently, the ILS is a *complete description* of the feedback structure of the system, where the many additional feedback loops are redundant.

Once an ILS has been identified (for procedures, see Oliva 2004 and Kampmann 2012), it is straightforward to calculate the gain g of each loop in the set and then use those gains as the basis for exploration of the behavior of the eigenvalues. Specifically, the loop eigenvalue elasticity and the loop influence metrics are defined as

$$\varepsilon = \frac{\partial \lambda}{\partial g} \frac{g}{\lambda} \text{, and } \mu = \frac{\partial \lambda}{\partial g} g \text{ .} \tag{5}$$

While the ILS is not unique in a model, this decomposition focuses the analysis in a relevant subset of loops. In particular, changes in relationships in the model that are not part of a feedback loop will have no effect upon the system eigenvalues. Thus, one can interpret the elasticities or influence measures in terms of how they affect the gains of a set of (independent) feedback loops in the system. Alternatively, one can assess the relative importance of particular feedback loops in generating a particular mode of behavior, where loops with large elasticities (or influence) are considered important for the behavior mode in question.

Dynamic Decomposition Weight Analysis (DDWA)

The eigenvector or dynamic decomposition weight analysis, introduced by Saleh et al. (2010), is concerned with what happens to the weights w in equation (3) when changes are made to the system elements (parameters and link gains). As with the LEEA, one may express the relationship either as influence measures or elasticities. Specifically,

$$\varepsilon_w = \frac{\partial w}{\partial a} \frac{a}{w} \text{, } \mu_w = \frac{\partial w}{\partial a} a \text{ .}$$

Unlike in LEEA, however, where only those links in the model that are part of feedback loops will have any significance, all the links in the model are potentially relevant in the determination of the dynamic decomposition weights. Figure 7.3 presents the main analyses covered by these methods, as well as the inputs and outputs required by each.

In order to conduct a meaningful policy analysis, it is necessary to specify a set of criteria for what constitutes a successful policy change. Forrester (1982) discusses different measures of stabilizing policies and their possible tradeoffs. This issue, however, is difficult to treat in general, since the policy criteria are linked to the purpose of the model and the problem definition, which may involve transient behaviors like overshoot and collapse, for example, in the world model (Forrester 1971), or the settlement in the system to undesirable end states, for example, in the urban dynamics model (Forrester 1969). In this paper, I focus on policies that reduce the oscillatory tendencies of the system, since the model presented is designed to address this issue, and since,

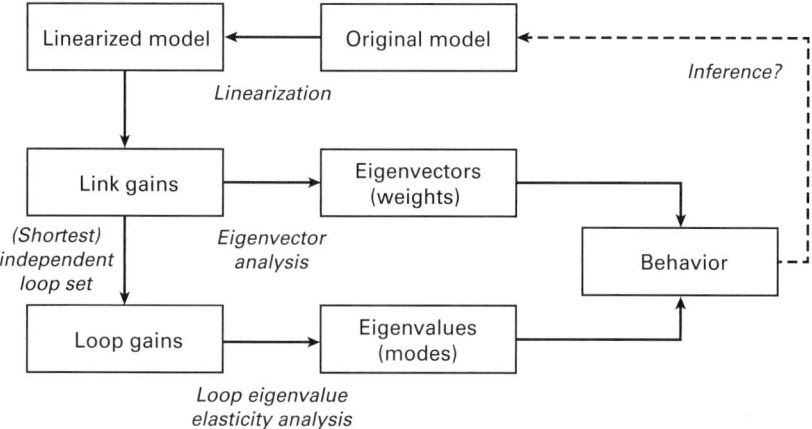

Figure 7.3
Schematic representation of eigenvalue elasticity analysis, or EEA, process. (Adapted from Saleh et al. 2010.)

as was demonstrated by Kampmann and Oliva (2006), it appears to be one of the areas where eigenvalue analysis shows the most promise.

In the context of unwanted instabilities (oscillations), effective policies are normally defined as those that either increase the damping of oscillatory behavior modes by making the real part more negative or decrease the frequency of oscillation. The LEEA can aid in finding the changes that have those desired effects and explain why the effect occurs in terms of the changes in feedback loop gains they imply. Another perspective, afforded by the DDWA, is to make changes that reduce the weights w of the oscillatory behavior modes in a particular system variable; that is, reduce the amplitude of the variable's oscillations. Another aspect addressed by the DDWA is the degree to which external disturbances (from the exogenous variables) can be absorbed and dampened by the system. I have chosen to relegate this aspect to subsequent work.

Finally, as discussed above, interpretation of the results from these analyses is not necessarily straightforward, in particular because the methods could be used for different purposes, such as identifying structural explanations of behavior or policy design. As such, the outcomes of these analyses have not been standardized. In the examples below, I will use different representations of the outputs (i.e., the eigenvalues and eigenvectors, and the loop, link, and parameter influences on them) that have proven useful to explain observed behavior in terms of system structure (feedback loops) and develop policy recommendations.

Detailed Example

Both the LEEA and DDWA methods have been implemented in Mathematica notebooks and are available in this book's electronic supplement, with the example models in Vensim and text parsing routines that generate the appropriate Mathematica files from a Vensim model file; the tools are also available online (Oliva and Kampmann 2010). Names of files available in the electronic supplement will be listed using a fixed-width font (e.g., `model.mdl`).

The model I use to illustrate the method is Nathan Forrester's macroeconomic model, developed in his PhD dissertation (1982). The model serves this purpose well, both because it is close to linear and because the main emphasis of the model is to understand macroeconomic instabilities such as business cycles or longer cycles and develop policies to stabilize these cycles. The purpose of the original model was to investigate various suggestions for fiscal and monetary policies to stabilize the economy. The model, which is shown in figure 7.4, represents the relationships that are all part of standard macroeconomic theory, such as the consumption multiplier, the permanent income hypothesis, the Phillips curve, and the investment accelerator. For the reader unfamiliar with macroeconomics, detailed description of this theory can be found in any standard textbook, such as Dornbusch, Fischer, and Startz (2010).

The model (`NF_model.mdl`) has a total of 10 levels. Four of these (LU, SED, LED, AY, and PY) are first-order delays of unemployment, short- and long-term expected demand, aggregate production, and disposable income, respectively. Moreover, employment (EMP) and capital stock (K) are also effectively first-order adjustments to desired employment (DEMP) and desired capital (DK), respectively. Thus, only the price level (P), the inventory level (IV), and the money supply (M) are not first-order delays. (The latter is not active in the base run of the model.)

The model used in our study is identical to the one listed in Forrester's dissertation, with the following three minor exceptions.

• All smooth functions have been replaced by the explicit formulation of the stock adjustment process.

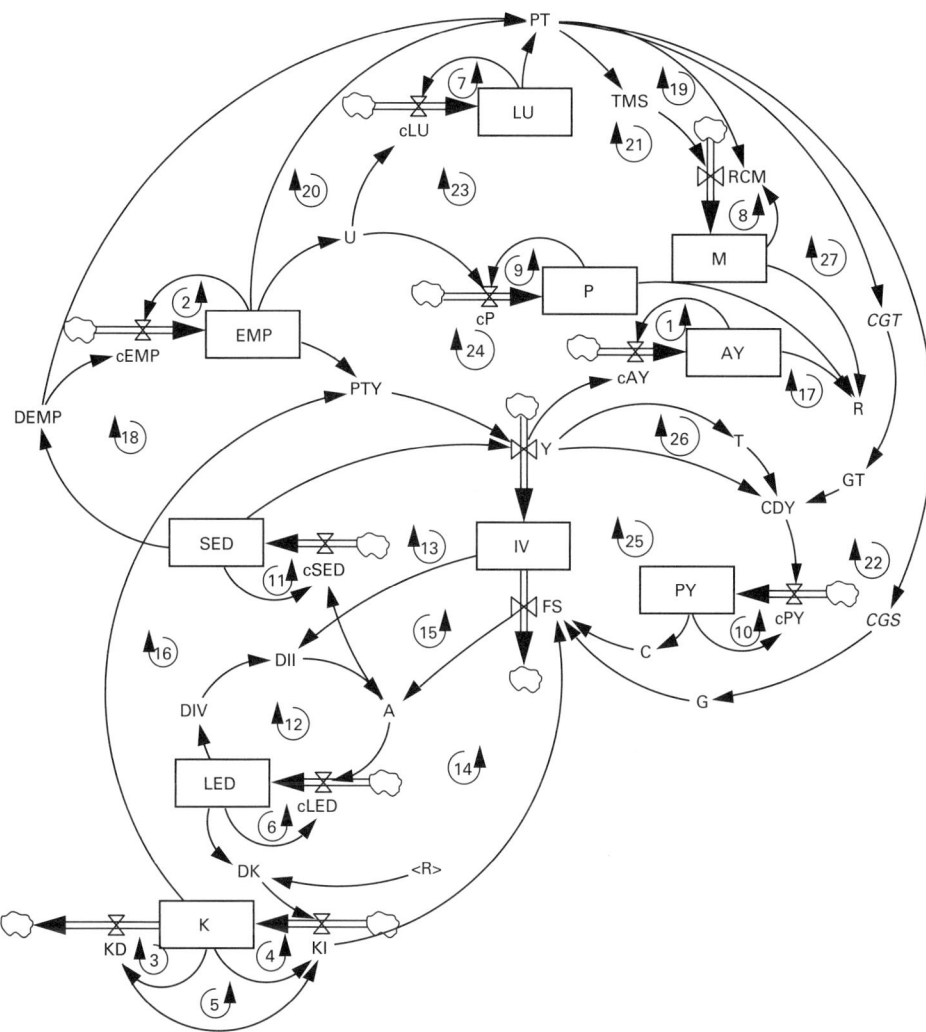

Figure 7.4
Diagram of the Forrester macroeconomic model.

- Stock values have been scaled by a factor of 10e8 to avoid very large negative exponents, which the software may otherwise truncate when doing the analysis.
- Instead of random noise, the IV stock is initialized about 3% below its equilibrium level. This generates smooth trajectories that are easier to interpret and understand.[6]

For our base simulation (`Base.vdf`) all policy levers are inactive, and the behavior of the model is the same as reported in chapter 4 of Forrester's dissertation. Figure 7.5 shows the characteristic response of the system, in this case aggregate output (Y) and the inventory (IV), capital (K), and employment (EMP) stocks. The inventory and employment stocks show a damped oscillation with a period of approximately four years, and the capital stock shows a dampened oscillation with a period of about 30 years (not visible in the figure). The peaks and troughs of inventory stock lag the peaks and troughs of output by approximately two and a half years. As such, the model appears to do a good job of replicating the salient features of the business and the capital cycles.

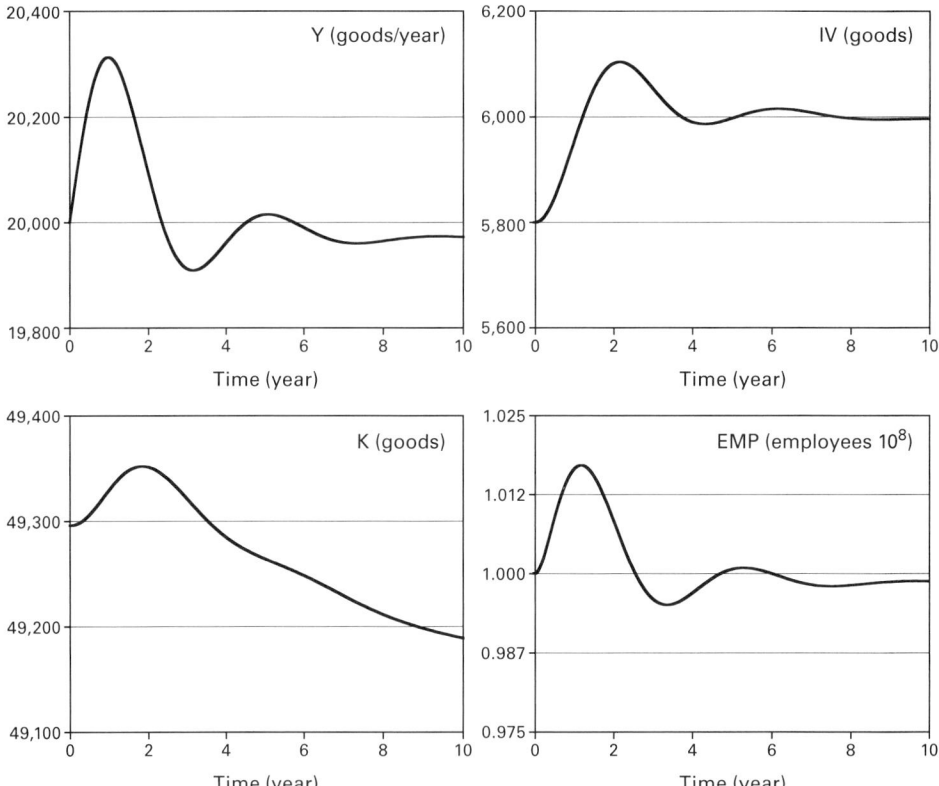

Figure 7.5
Base run of the Forrester model.

LEEA

Base Model   The Mathematica implementation of the LEEA utility requires two inputs to perform the analysis of a model: (1) a Mathematica version of the model, and (2) a data file with the values of the system state variables at all the points in time that have been chosen for performing the analysis. The electronic supplement (`Appendix.pdf`) provides detailed instructions for preparing these files from our sample model (`NF_model.mdl`), as well as executing the analysis. The supplement also lists the current limitations of the implementation.

The first two sections of the LEEA notebook (`LEEA.nb`) contain the instructions to use the utility and the commands to import the data. The following five sections are the computational core of the utility. Section 3 derives the graph representation of the model structure (see Oliva 2004); section 4 identifies a *shortest independent loop set* (SILS) (Oliva 2004) that will be used as the base description of the loops in the model; sections 5 and 6 symbolically derive the link and loop gains as well as the *Jacobian* matrix for the model; and section 7 calculates the loop elasticities for the time periods in the data table. While it is possible to explore the intermediate steps in each of these sections, the sections are not intended for user inspection. Instead, the last two sections present the analysis output in easy-to-interpret formats.

Section 8 reports the evolution of eigenvalues through the different time instances where they were evaluated. The real and imaginary parts of the eigenvalues can be inspected in tabular and graphical form, and it is also possible to obtain a plot with the eigenvalues in the complex plane for each time frame.

Tables 7.1 and 7.2 report the real and imaginary parts of the 10 eigenvalues (one per independent state variable) of our sample model across the 10 annual evaluations. There are several things to note in these tables. First, one of the eigenvalues is 0 throughout the simulation. This is consistent with the fact that one of the stocks in the model (M) does not change in the base run. Second, all eigenvalues have a negative real part, which means that the system is dampened and all perturbations will eventually die off. This is consistent with the behavior observed in the two main stocks (figure 7.5). Third, there are two pairs of complex eigenvalues {3,4} and {7,8}, each representing a different frequency of oscillation. The first pair represents an oscillatory behavior mode with a period of 4.27 years ($2\pi / \text{Im}[\lambda] = 2\pi / 1.47$) that corresponds to the business cycle, and the other corresponds to the capital cycle with a period of almost 30 years ($2\pi / 0.21$).

Table 7.1
System eigenvalues (real part), evaluated annually

Time/Eigenvalue	1	2	3	4	5	6	7	8	9	10
0	-16.000	-3.208	-0.571	-0.571	-0.400	-0.398	-0.160	-0.160	-0.022	0.000
1	-16.000	-3.217	-0.488	-0.488	-0.406	-0.400	-0.227	-0.227	-0.029	0.000
2	-16.000	-3.212	-0.537	-0.537	-0.401	-0.400	-0.188	-0.188	-0.025	0.000
3	-16.000	-3.208	-0.579	-0.579	-0.400	-0.398	-0.154	-0.154	-0.021	0.000
4	-16.000	-3.208	-0.578	-0.578	-0.400	-0.398	-0.154	-0.154	-0.021	0.000
5	-16.000	-3.209	-0.569	-0.569	-0.400	-0.398	-0.162	-0.162	-0.022	0.000
6	-16.000	-3.208	-0.570	-0.570	-0.400	-0.398	-0.161	-0.161	-0.022	0.000
7	-16.000	-3.208	-0.575	-0.575	-0.400	-0.398	-0.157	-0.157	-0.022	0.000
8	-16.000	-3.208	-0.575	-0.575	-0.400	-0.398	-0.157	-0.157	-0.022	0.000
9	-16.000	-3.208	-0.574	-0.574	-0.400	-0.398	-0.158	-0.158	-0.022	0.000

Table 7.2
System eigenvalues (imaginary part), evaluated annually

Time/Eigenvalue	1	2	3	4	5	6	7	8	9	10
0	0.000	0.000	1.469	-1.469	0.000	0.000	0.210	-0.210	0.000	0.000
1	0.000	0.000	1.455	-1.455	0.000	0.000	0.232	-0.232	0.000	0.000
2	0.000	0.000	1.462	-1.462	0.000	0.000	0.222	-0.222	0.000	0.000
3	0.000	0.000	1.472	-1.472	0.000	0.000	0.205	-0.205	0.000	0.000
4	0.000	0.000	1.471	-1.471	0.000	0.000	0.205	-0.205	0.000	0.000
5	0.000	0.000	1.469	-1.469	0.000	0.000	0.210	-0.210	0.000	0.000
6	0.000	0.000	1.469	-1.469	0.000	0.000	0.209	-0.209	0.000	0.000
7	0.000	0.000	1.470	-1.470	0.000	0.000	0.207	-0.207	0.000	0.000
8	0.000	0.000	1.470	-1.470	0.000	0.000	0.207	-0.207	0.000	0.000
9	0.000	0.000	1.470	-1.470	0.000	0.000	0.208	-0.208	0.000	0.000

Finally, it should be noted that all eigenvalues are very stable throughout the simulation, meaning that there are no significant transitions in the model. This stability simplifies the analysis of the linkages between structure and behavior, as there are no significant changes of loop dominance in the trajectories of the base case. Kampmann and Oliva (2006) present a case of a model with significant changes in loop dominance and illustrate how the tools are useful in that context.

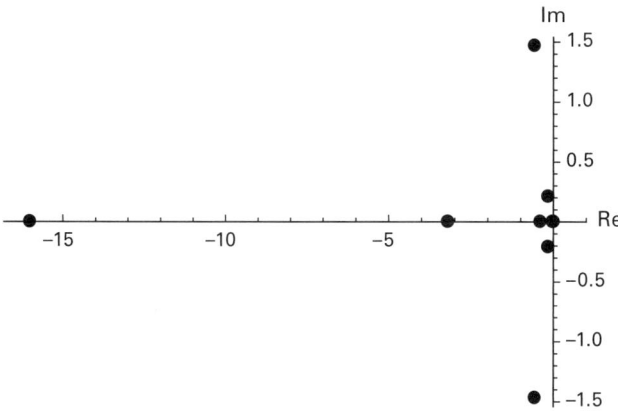

Figure 7.6
System eigenvalues at time 5.

Since eigenvalues are fairly stable throughout the simulation, I focus on reporting the results of the analysis of loop dominance at time 5, the midpoint of the simulation. The utility, however, is capable of instantaneously generating similar reporting for each of the instances when the computations were realized. Figure 7.6 shows the system eigenvalues in the complex plain.

Section 9 of the LEEA utility reports the impact of the feedback loop structure on the reference modes represented by each of the eigenvalues. For our sample model, the utility identified 27 loops in the SILS (Oliva 2004); see table 7.3.

The utility offers the option to report the loop gains for all the loops in the SILS. The gain values, however, are contingent on the magnitudes of the variables traversed by the loop, and it becomes difficult to make meaningful comparisons between loops. The influence metric reported in the following subsection addresses this shortcoming. Nonetheless, it should be noted that in the base case, loop 8, loops 19 through 23, and loop 27 have a gain of zero as the policy trigger (PT) is inactive and there are no changes in the M stock.

As discussed above, the focus of the analysis will be on the model's oscillatory behavior modes; that is, the two pairs of complex eigenvalues. Figure 7.7 shows the LEEA utility's output for the loop influence on eigenvalue 3—the behavior mode with a 4.27 years period

Table 7.3
A shortest independent loop set

Loop 1	AY>cAY
Loop 2	EMP>cEMP
Loop 3	K>KD
Loop 4	K>KI
Loop 5	K>KD>KI
Loop 6	LED>cLED
Loop 7	LU>cLU
Loop 8	M>RCM
Loop 9	P>cP
Loop 10	PY>cPY
Loop 11	SED>cSED
Loop 12	LED>DIV>DII>A>cLED
Loop 13	IV>DII>A>cSED>SED>Y
Loop 14	LED>DK>KI>FS>A>cLED
Loop 15	IV>DII>A>cLED>LED>DK>KI>FS
Loop 16	IV>DII>A>cLED>LED>DK>KI>K>PTY>Y
Loop 17	AY>R>DK>KI>K>PTY>Y>cAY
Loop 18	EMP>PTY>Y>IV>DII>A>cSED>SED>DEMP>cEMP
Loop 19	M>R>DK>KI>FS>A>cSED>SED>DEMP>PT>RCM
Loop 20	EMP>PT>RCM>M>R>DK>KI>FS>A>cSED>SED>DEMP>cEMP
Loop 21	M>R>DK>KI>FS>A>cSED>SED>DEMP>PT>TMS>RCM
Loop 22	SED>DEMP>PT>CGS>G>FS>A>cSED
Loop 23	EMP>U>cLU>LU>PT>CGS>G>FS>A>cSED>SED>DEMP>cEMP
Loop 24	EMP>U>cP>P>R>DK>KI>FS>A>cSED>SED>DEMP>cEMP
Loop 25	PY>C>FS>A>cSED>SED>Y>CDY>cPY
Loop 26	PY>C>FS>A>cSED>SED>Y>T>CDY>cPY
Loop 27	PY>C>FS>A>cSED>SED>DEMP>PT>CGT>GT>CDY>cPY

representing the business cycle. The utility has an option to control the number of loops to be included in this report; in this case, the figure includes the top 10 most influential loops as measured by the absolute value of the influence metric (see equation [5])—regardless of its sign, loops with large-influence metrics are more influential. To simplify the analysis, the utility sorts the loops in descending order of influence and reports the influence measure.

It is also of interest to determine whether a particular loop acts as a stabilizing or destabilizing influence on the behavior mode. The sign

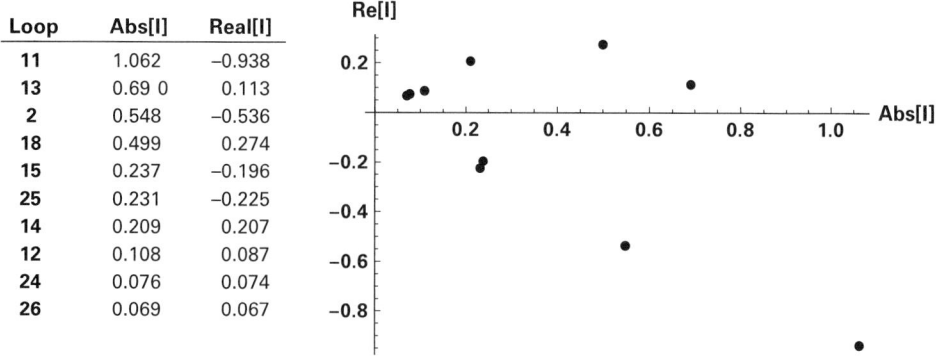

Loop	Abs[I]	Real[I]
11	1.062	−0.938
13	0.69 0	0.113
2	0.548	−0.536
18	0.499	0.274
15	0.237	−0.196
25	0.231	−0.225
14	0.209	0.207
12	0.108	0.087
24	0.076	0.074
26	0.069	0.067

Figure 7.7
Loop influence (top 10 loops) on eigenvalue 3 at time 5.

of the real part of the influence metric determines this role; that is, a negative real part implies a stabilizing (dampening) effect on the behavior mode, while a positive real part implies a destabilizing (exponential growth) behavior mode. The real part of the influence metric is also reported in the output table. To facilitate interpretation of the role of each loop and its relationship to other influential loops, the utility generates a scatter plot of the influence metrics placing the absolute value of the influence in the x axis and the real part in the y axis. The further from the origin a loop is, the more influential it is. Points below (above) the x axis represent stabilizing (destabilizing) loops.

From figure 7.7, it is clear that the business cycle is destabilized by loops 13 and 18 (the rapid adjustment of inventory and the interaction between inventory and employment) and is stabilized by loops 11 and 2 (the slow adjustments of expected demand and inventory). Each of the stabilizing loops has links within the destabilizing loops.

Similar analysis (see figure 7.8) reveals that the capital cycle (the oscillatory behavior mode with a period of 30 years represented by eigenvalue 7) is destabilized by loops 15, 25, and 11 (the multiplier effects of capital and consumption, and the short-term response to estimated demand), and it is stabilized by loops 6, 10 and 13 (the smoothing process to adjust long-term estimated demand, the adjustment of permanent income, and the short-term adjustment of inventory).

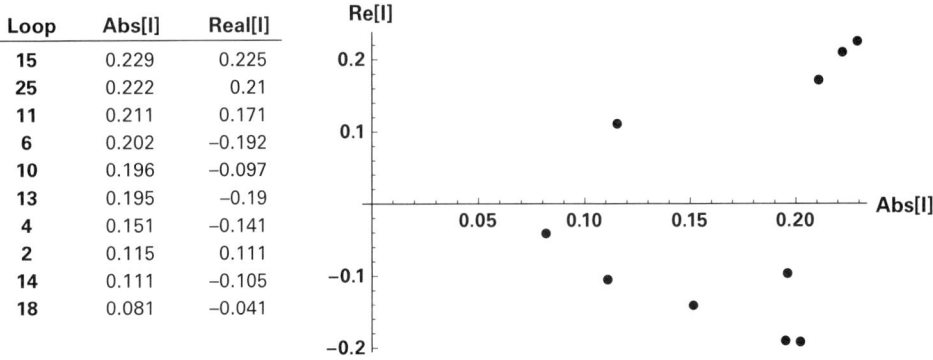

Loop	Abs[l]	Real[l]
15	0.229	0.225
25	0.222	0.21
11	0.211	0.171
6	0.202	−0.192
10	0.196	−0.097
13	0.195	−0.19
4	0.151	−0.141
2	0.115	0.111
14	0.111	−0.105
18	0.081	−0.041

Figure 7.8
Loop influence (top 10 loops) on eigenvalue 7 at time 5.

This explanation of the two oscillatory behavior modes is consistent with the explanation provided by Forrester in his thesis (1982), and also by the explanation obtained from analyzing similar models that include the labor and capital interactions (Mass 1975; Oliva and Kampmann 2006). However, it should be noted that all these insights and a full structural explanation of the behavior of the system were generated out of a single run of the model, as opposed to the exhaustive exploration through sensitivity analysis.

From this analysis one could derive policy recommendations within the structure of the system. That is, the analysis reveals what loops need to be weakened or strengthened in order to bring more stability to the system. It is easy to identify the parameters that control the gain for each of the loops. For example, decreasing the parameter values of time to smooth short-term demand (tssd) and time to adjust employment (tae) would increase the gain of loops 11 and 2, respectively. By increasing the gain of these two loops, which have a stabilizing influence in the business cycle, one would further dampen the business cycle.

The above analysis, however, has the disadvantage that it only identifies leverage points within the active structure of the system. That is, LEEA cannot "see" beyond the active elements of the system, and none of influential loops are part of what Forrester described as part of the policy levers—not surprising, since in this initial simulation the gain on those loops was set to zero.

The ability of the method to assess the full model behavior at once, however, sets up the opportunity to assess several potential policy recommendations at once. While Forrester (1982) tested each of his five policies individually, given this tool one can assess them all at once and get the full benefit of understanding their interactions. I turn to this possibility in the following section.

Full Model   A full version of the Forrester model, including the activation of the policy triggers, is also available in the electronic supplement (NF_model_full.mdl). The model is structurally the same as the model used in the base case, but switches and parameter have been updated to activate feedback loops 8, 19 through 23, and 27. As with the base case, the model is initialized in disequilibrium—IV stock 3% below its equilibrium level. Figure 7.9 shows the behavior of aggregate output and the inventory, capital, and employment stocks under the full model (full.vdf) and compares each to the base case simulation (base.vdf).

With all the intervention policies active at the same time, the frequency of the system's response increases, and the dampening ratio decreases, relative to the base case. That is, the combination of all five policies acting simultaneously makes the system respond faster and more aggressively to deviations from equilibrium. While the faster response prevents the capital stock (K) from deviating from equilibrium as much as in the base case, as a result of this aggressiveness, the system overreacts to those deviations, and now the stocks in the business cycle (IV and EMP) take longer to reach equilibrium. Aggregate production (Y) follows more closely the response of the business cycle in almost the same phase as the employment stock.

As discussed above, to conduct a meaningful policy analysis, it is necessary to specify a set of criteria for what constitutes a successful policy change. I will avoid this challenge, and instead focus on eliminating unwanted instabilities (oscillations), by either increasing the damping of oscillatory behavior modes by making the real part more negative or decreasing the frequency of oscillation. From this perspective, the combination of policies introduced in the full model seems to be effective in increasing the dampening and reducing the frequency of the capital cycle, but has the opposite effect on the business cycle. The LEEA analysis can help us understand why these tradeoffs are

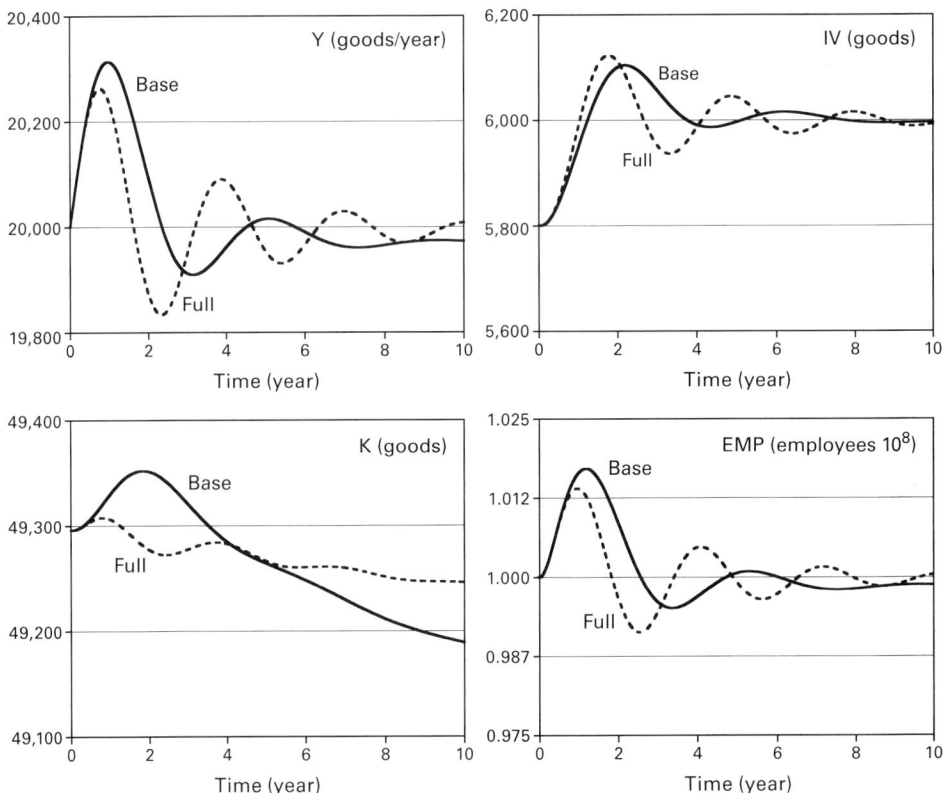

Figure 7.9
Full run of the Forrester model.

taking place and focus on the subset of policies that might yield a better balance of these tradeoffs.

After preparing the Vensim files in the appropriate format (NF_ model_full.mdl → nf_model_full.nb and full.vdf → full. tab), I ran them through the LEEA Mathematica utility, performing the linearization and full analysis of elasticities at one-year intervals.

Again, the system eigenvalues are stable through the simulation, thus I focus on a single period. Table 7.4 reports the eigenvalues of the system at time 5. As this time the M stock is active in the system, all eigenvalues are nonzero. All real parts of the eigenvalues remain negative, and thus the system remains dampened. More interestingly, the system now shows three complex pairs of eigenvalues representing three separate oscillatory behavior modes for the system. Eigenvalue

Table 7.4
System eigenvalues (full model) at time 5

Eigenvalue	1	2	3	4	5	6	7	8	9	10
Real	-16.104	-16.104	-3.055	-0.440	-0.440	-0.454	-0.385	-0.250	-0.250	-0.007
Imaginary	1.054	-1.054	0.000	1.993	-1.993	0.000	0.000	0.167	-0.167	0.000

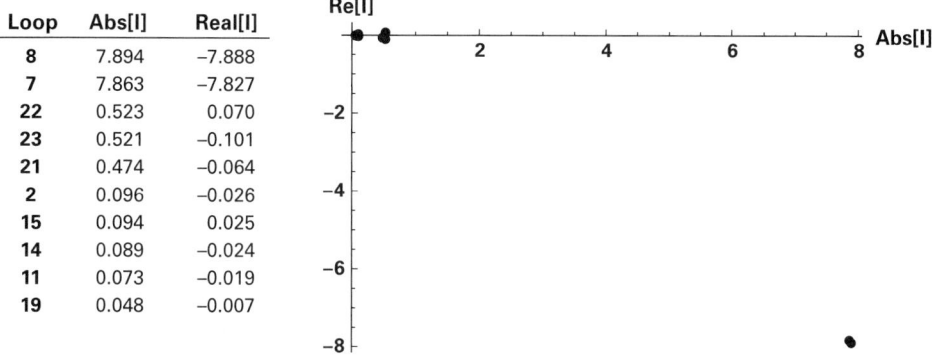

Loop	Abs[l]	Real[l]
8	7.894	−7.888
7	7.863	−7.827
22	0.523	0.070
23	0.521	−0.101
21	0.474	−0.064
2	0.096	−0.026
15	0.094	0.025
14	0.089	−0.024
11	0.073	−0.019
19	0.048	−0.007

Figure 7.10
Loop influence (top 10 loops) on eigenvalue 1 at time 5 (full model).

pairs {1,2}, {4,5}, and {8,9} denote oscillatory modes with periods $(2\pi/$
$\mathrm{Im}[\lambda])$ of 5.96, 3.15, and 37.6 years, respectively.

Analysis of the loops influencing these reference modes reveals the impact of the implemented policies. Figure 7.10 shows the most influential loops on eigenvalue 1 (period 5.96 years). Only loops 8 and 7 (long-term labor adjustment and monetary adjustment) have a significant influence on this behavior mode—both loops are stabilizing. Loop 8, as mentioned above, was not active in the base simulation, and while loop 7 was active, long-term unemployment (LU) had no further effect on the model, as the policy trigger (PT) was not activated. Thus, this behavior mode is strictly the result of the policies introduced. This is in itself an interesting finding, since in addition to affecting the two existing oscillatory behavior modes (see discussion below), the policy implementation introduces an oscillatory pattern with a period of six years. The two influential loops in this reference mode are first-order delays on stocks that determine the intensity of the policy trigger (LU) and the speed of the policy adjustment (M), and their gain is determined by identical time constants.

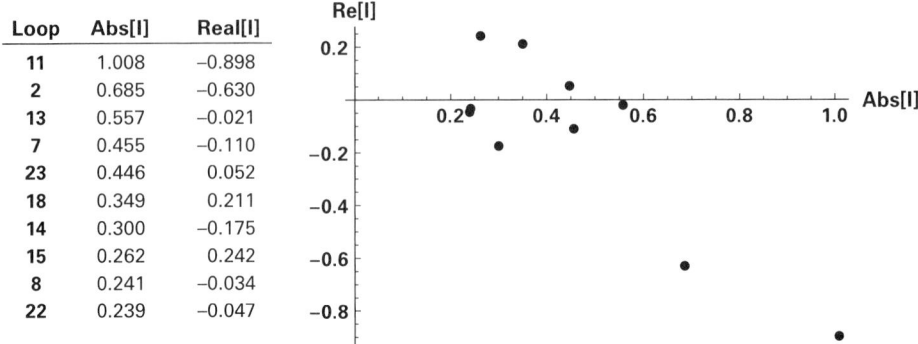

Loop	Abs[l]	Real[l]
11	1.008	−0.898
2	0.685	−0.630
13	0.557	−0.021
7	0.455	−0.110
23	0.446	0.052
18	0.349	0.211
14	0.300	−0.175
15	0.262	0.242
8	0.241	−0.034
22	0.239	−0.047

Figure 7.11
Loop influence (top 10 loops) on eigenvalue 4 at time 5 (full model).

This oscillatory behavior mode represented by eigenvalues {4,5} (period 3.15 years) is most influenced by two stabilizing loops (11 and 2; see figure 7.11). These are the same two loops identified as dampening the business cycle on the base simulation, and they have almost the same effect on the eigenvalue. Comparing this influence scatter plot with the one for the business cycle in the base case (figure 7.7), two major differences become apparent. First, loop 13, the short-term adjustment of inventory, changes from destabilizing to marginally stabilizing. Both the magnitude and the destabilizing effect (the size of the real part) of this loop have been reduced with the introduction of the policies. It is interesting to note that loop 13 has no active links to the 7 loops introduced with the policy recommendations, but its change in relative influence is the result of other loop interactions. Second, loop 23, the countercyclical government spending policy, is now a significant destabilizing influence in the business cycle. While the policy increases aggregate demand (A) in order to affect demand expectations and long-term employment (EMP) toward faster equilibrium, the policy has a short-term effect of depleting the existing inventory (IV)—through final sales (FS)—that further exacerbates the disequilibrium between aggregate production and demand, causing inventories to drop.

This behavior mode from eigenvalues {8,9} is the same as the behavior mode captured by eigenvalues {7,8} in the base simulation. Figure 7.12 shows the 10 most influential loops on this reference mode. Comparing this scatter plot with the one from the base case (figure 7.8), it

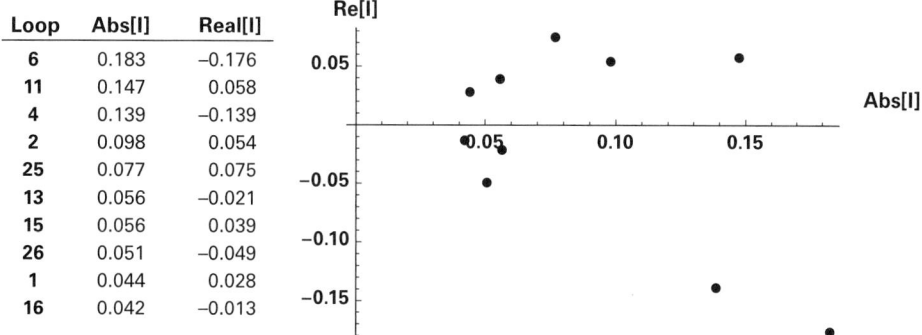

Loop	Abs[l]	Real[l]
6	0.183	−0.176
11	0.147	0.058
4	0.139	−0.139
2	0.098	0.054
25	0.077	0.075
13	0.056	−0.021
15	0.056	0.039
26	0.051	−0.049
1	0.044	0.028
16	0.042	−0.013

Figure 7.12
Loop influence (top 10 loops) on eigenvalue 8 at time 5 (full model).

is clear that the destabilizing influence of loops 15 (capital adjustment), 25 (demand from disposable income), and 11 (short-term demand expectations) has been significantly reduced—the real part of the influence metric of these three loops has been reduced by 24%, 35%, and 70% of their original values. The diminished role of loops 25 and 11 implies that stabilizing loops 10 and 13 also lose their influence, as loop 10 shares links with loop 25 and loop 13 interacts with loop 11. Loops 6 (capital investment) and 4 (long-term demand expectations) retain their influence in the stabilizing effect on this behavior mode, since the two first-order loops dampen the response of all capital acquisition. The net effect of the elimination of the destabilizing loop is the reduction of frequency of the capital cycle (the period increases from 30 to 37 years).

Again, this analysis of the full model can be used to design the policy that would improve the results, looking, for instance, to reduce the strength of countercyclical government spending to mitigate its destabilizing effect on the business cycle while retaining the stabilizing effect of all policies in the capital cycle. The fact that the gains of all the loops that are being analyzed can be independently set (since this is what defines the ILS) immediately focuses our attention on the parameter that uniquely affects the gain of the countercyclical government spending loop, i.e., the strength of countercyclical government spending (SCGS).

A quick test reveals that reducing the strength of SCGS does indeed have the desired effect in the system, reducing the frequency of the oscillations in the inventory stock without affecting the oscillations in

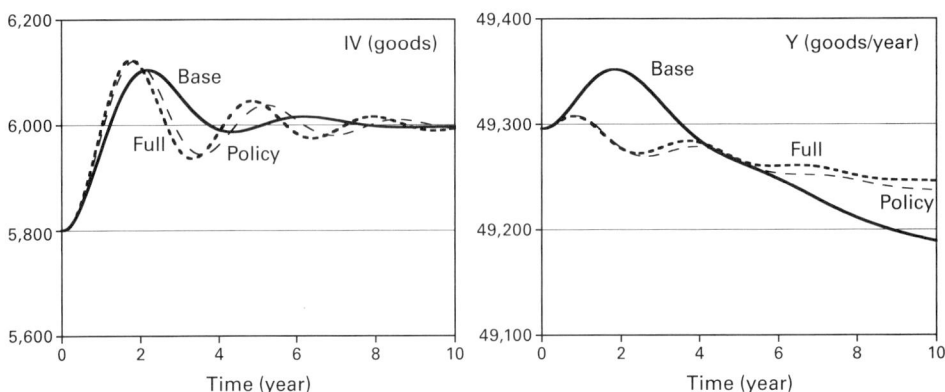

Figure 7.13
Effect of policy to reduce the strength of strength of countercyclical government spending (SCGS) from 1.00 to 0.25.

the capital stock. Figure 7.13 shows the effect of changing SCGS from 1.00 to 0.25 on the behavior of these two stocks. While the effect of the parameter change is in the right direction, it is clear that its leverage on the actual behavior of the stocks of interest is very limited, since a reduction of 75% of the strength of this policy only marginally reduces the frequency of the oscillations in the business cycle. An exploration of the projection of each of the eigenvalues in the stocks of interest and an assessment of the impact of the parameter values in this projection have been found to be much more effective for policy design. The next section addresses the details of the dynamic decomposition weight analysis (DDWA).

DDWA Full Model

Unlike the LEEA utility, the Mathematica implementation of the DDWA utility (DDWA.nb) only requires the Mathematica version of the model,[7] since the elasticities of all parameters are estimated from the model's initial conditions. The electronic supplement provides detailed instructions for preparing this input file.

The first two sections of the DDWA notebook contain the instructions to use the utility and some functions required by the utility. The following three sections are identical to corresponding sections in the LEEA utility. Section 3 derives the graph representation of the model structure. Sections 4 and 5 symbolically derive the edge and loop gains as well as the *Jacobian* matrix for the model.[8] Section 6 performs the computations for the dynamic decomposition weight (DDW; see equation [3] above) and reports graphs of the decomposition of the behavior of each of the stocks decomposed to each of the behavior modes represented by the eigenvalues. The graphs are available in absolute stock values, or normalized by dividing by the constant term of equation (3), thus making the contribution of each eigenvalue comparable.

Section 7 performs the core computations to determine the elasticities of the weights (w in equation [3]) to parameters values and links, and section 8 provides different reporting options for these computations. There are three options to report the parameter and link elasticity tables: (1) reporting by stock, (2) reporting by behavior mode, and (3) reporting with eigenvalue elasticity. The first option allows the user to focus on a particular stock, and a table with the elasticity of all DDWs of that stock (the ws in equation [3]) to all parameters (links) is displayed. The second option allows the user to focus on a particular behavior model (eigenvalue) and a table of the elasticity of all DDWs of that eigenvalue to all parameters (links) is displayed. The final option reports, for a selected stock and behavior mode, the weight elasticity as well as the elasticities of the real and imaginary parts of an eigenvalue to each parameter (link).

From the LEEA of the full model, we are interested in reducing the frequency of oscillation of the business cycle (-0.44+1.99i).[9] The stocks involved in the feedback loops responsible for this behavior mode are inventory (IV), short-term expectation of demand (SED), and employment (EMP; loop 18). Table 7.5 reports, in descending order, the elasticities of the DDW on the inventory stock and the real and imaginary parts of the business cycle eigenvalues to all the model parameters.

Quick inspection of the table reveals the reason for the limited impact of the changes in SCGS on the frequency of interest discussed in the previous section. The parameter has only a marginal effect of the DDW on the inventory stock (elasticity = -0.0625, ranked 12th in the table), and the elasticity of the eigenvalue to this parameter is only marginal. Parameters higher in the table will have a more dramatic impact on the specific stock-behavior mode combination. As an example, figure 7.14 shows the effect of increasing the length of the time to adjust inventory (TAI), the top parameter in the list, from 0.4 to 0.8 on the inventory and capital stocks.

Clearly, the tool cannot judge the feasibility of implementation of the different policy changes, but the comprehensive assessment of the impact of all parameters (links) on all stocks and reference modes

Table 7.5
Elasticity of business cycle eigenvalue and DDW on inventory to model parameters

Parameter	Weight	Re[λ]	Im[λ]
tai	1.225	1.139	-0.438
tssd	1.090	-1.492	-0.398
ee	0.254	0.698	-0.363
tsld	-0.223	0.371	0.096
eyvm	-0.183	-0.171	0.237
ep	0.183	0.171	-0.237
alk	-0.136	-0.377	0.152
tae	-0.123	-0.737	-0.229
iem	0.111	0.165	-0.130
lr	-0.105	-0.130	0.120
scms	-0.072	-0.005	0.108
scgs	-0.063	-0.012	0.105
egs	-0.063	-0.012	0.105
tak	0.059	0.225	-0.059
nic	0.053	-0.215	-0.022
sgyt	-0.043	-0.192	0.026
tsu	-0.032	-0.132	-0.001
fcu	-0.031	0.629	0.125
alpha	0.030	0.012	-0.083
tam	-0.027	-0.070	0.012
stm	0.027	0.070	-0.012
tsay	0.025	0.049	-0.013
yem	-0.019	-0.059	0.011
ey	-0.017	-0.536	0.057
scgt	-0.014	-0.099	-0.001
egt	-0.014	-0.099	-0.001
spc	-0.012	-0.101	0.001
scmg	-0.007	-0.001	0.011
tsy	-0.007	0.014	0.014
nru	-0.003	-0.089	-0.011
sdc	0.000	0.000	0.000

allows for a very rapid identification of the parameters with the highest leverage on a particular behavior mode of a specific stock.

Discussion and Conclusion

The two methods described in this chapter, LEEA and DDWA, are predicated on the idea that it is possible to linearize, around a particular

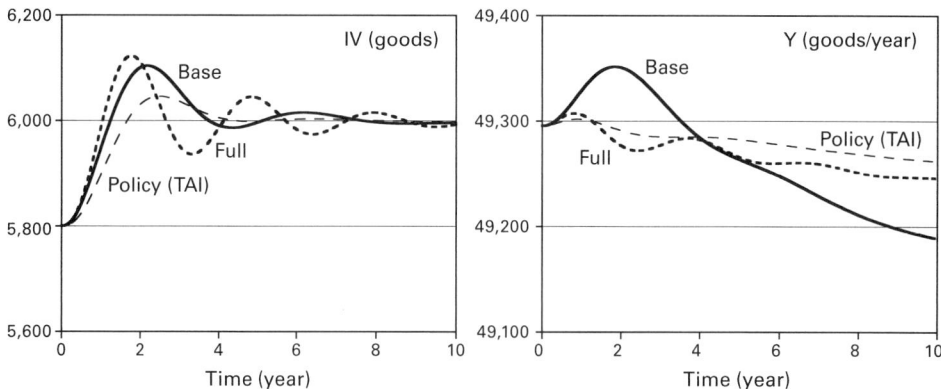

Figure 7.14
Effect of policy to increase the time to adjust inventory (tai) from 0.4 to 0.8.

operating point, a nonlinear model. Once that linearization takes place, it is possible to characterize the system behavior in terms of the eigenvalues of the system matrix. LEEA identifies the feedback loops that are responsible for each of the behavior modes represented by each eigenvalue. As such, it is a powerful tool to formally establish the relationships between model structure and observed behavior. This linkage between model structure and behavior is critical to dynamic modelers, not only in that we now have certainty on what are the structural elements responsible for the behavior, but also as a general map of the feedback loops that are crucial for policy analysis.

DDWA uses the system matrix eigenvectors to identify a closed form projection (weight) of each behavior mode on the state variables. By assessing the elasticity of these weights to model parameters (links), it is possible to formally identify the parameters (policies) with the highest leverage on a particular behavior mode of a state variable.

While the linearization of the model represents an approximation, our experience from having analyzed dozens of models is that the approximation is quite valid whenever the model is reasonably close to the linearization of the operating point and the proposed applications (identification of dominant structure and high leverage policies) do not require the numerical precision that is affected by this approximation.

The benefits of the method are significant. First, it is a formal analysis and linkage of the model structure and behavior. The closed form solutions used by these methods make the analyses traceable, programmable, and replicable. This means that even novice modelers can benefit from the power of the insights generated by the methods. Second, the methods are comprehensive in that they assess the overall behavior of the system and its structure. The simultaneous evaluation of all model structure and behavior allows assessment of how different pieces of structure behave in the context of the overall system—that is, with other structural pieces in place, something that is clearly lost with partial model simulations. Third, the methods are efficient; the analysis of a single simulation reveals model structural insights that used to take hundreds of simulations and sensitivity analysis to develop. Finally, the methods are effective in that they have consistently replicated previous analysis, behavior narratives, and policy design analysis.

The methods, however, have some clear limitations and disadvantages. The first limitation, and probably the most significant in terms of hampering the broad adoption of the methods, is that the interpretation of the analysis output requires some basic understanding of control theory and linear algebra. Second, the linearization step requires that all table functions must be analytical and continuously differentiable (C^1). Future model parsers could eventually address this limitation, and while in most cases this only requires an additional step to express the table function, this requirement reduces the flexibility and spontaneity of testing different formulations in the model. Third, although there are no known computational limitations, the calculation of eigenvalues and eigenvectors, especially through multiple linearization points, might be computationally intensive. Finally, as currently implemented, the parser does not support macros, arrays, and most dynamic functions.

The above limitations still make the analysis an "expensive" undertaking. In a recent analysis of a previously built, large (13 stocks, 44 auxiliaries, 33 parameters, 34 loops in the ILS) model with random inputs, it took ~90 seconds to run the utilities to perform the LEEA and DDWA and about 45 minutes to interpret the results and prepare a report (Oliva 2014). However, preparing the model for the utilities (i.e., formulating SMOOTH functions explicitly, eliminating MIN, MAX and IF_THEN_ELSE statements, and replacing table functions with analytical forms) took more than eight hours. While eight hours is significantly shorter time than the weeks it took the author to develop an intuition for the model behavior, this kind of overhead makes these analyses a post-modeling exercise, rather than an integral part of the model-building and -learning process. While the "model preparation" stage would have been much shorter if the software limitations had been considered when building the model, adherence to these requirements would have limited the developer's appetite for testing the model and exploring alternative formulations. Clearly, reducing the overhead imposed by the current limitations of the experimental tools here presented is a major leverage point for the broader adoption of these methods.[10] Adoption of formal analysis of models' behavior will not only make modelers and analysts more efficient, but would also improve the overall quality of dynamic modeling work.

Challenge

The following is a series of challenges for the reader to develop a better intuition of the analyses output as well as a way to explore the different reporting options of the utilities.

- Loops 17 and 24 are the only loops that contain the interest rate (R) that are active in the base case. What is the role of these loops in that base run? What behavior modes do they affect?
- How does the role of R change in the full model? What behavior models is it affecting?
- What parameter changes (policies) would you introduce to augment (diminish) the impact of the interest rate (R) on the business cycle? On the capital cycle?

Answers to these questions and a description of a strategy on how to go about answering them are available in the `Challenge.pdf` document in the electronic supplement.

Notes

The development of these tools and approaches to EEA has been the result of a long-term collaboration with my friend Christian E. Kampmann. He could not participate in the preparation of this chapter, but I gratefully acknowledge his contributions to these ideas. All errors and omissions are my own. I also thank Burak Güneralp, Alejandro Serrano, and the handbook editors for comments that greatly improved the manuscript.

1. Since behavior modes are based upon the eigenvalues, the terms "behavior mode" and "eigenvalue" are used interchangeably in the following.

2. See Duggan and Oliva (2013) for extended bibliography of this research stream.

3. See Kampmann and Oliva (2006) for a discussion of when such an approximation is appropriate and useful.

4. While the elasticity measure has the advantage that it is a dimensionless measure and hence independent of the choice of units in the model, the influence measure has the dimension $time^{-1}$, and so depends upon the chosen time unit. However, it is still independent of the choice of the other units in the model.

5. The characteristic polynomial is defined as $P(\lambda)=\det(\lambda\mathbf{I}\text{-}\mathbf{A})$, where $\mathbf{I}$ is the identity matrix. The eigenvalues of $\mathbf{A}$ are the roots of $P(\lambda)=0$.

6. In Forrester's original model, the aggregated output (Y) and potential output (PTY) are modified with an additive random noise. The net effect of this noise is to perturb the system and prevent it from reaching equilibrium (the system is heavily dampened). While the EEA methods can work through these perturbations (see Oliva 2014), removing them allows us to understand the transient behavior from the point of linearization as if the system were not perturbed. This is a valid approach, since the eigenvalues of the system are very stable throughout the simulation horizon (see following section).

7. See LEEA section above for a description of the utility to translate a Vensim® model into a Mathematica® version suitable for these analysis.

8. This is clearly a replication of computational effort. I plan to integrate these two analyses into a single utility in the near future.

9. Note that the value of the eigenvalue reported in table 7.5 is slightly different from the value reported in table 7.4. This is because table 7.4 reports the eigenvalues at time 5, whereas the analysis in table 7.5 is based on the values of the eigenvalues at time 0. The easiest way to perform the DDW analysis for that particular point in time would be to initialize the model at the values the full simulation shows at that point in time. This is something that should be addressed once the two analyses are incorporated into a single platform.

10. All Mathematica® notebooks are open for inspection of algorithms and partial output, and the translator Perl code is available for downloading on the utility's website.

References

Chen, C.-T. 1970. *Introduction to linear system theory*. New York: Holt, Rinehart and Winston.

Dornbusch, R., S. Fischer, and R. Startz. 2010. *Macroeconomics*, 11th ed. The McGraw-Hill Series Economics. New York: McGraw-Hill.

Duggan, J., and R. Oliva. 2013. Methods for identifying structural dominance. *System Dynamics Review*, virtual issue. Available at: http://onlinelibrary.wiley.com/journal/10.1002/(ISSN)1099-1727/homepage/VirtualIssuesPage.html.

Forrester, J. W. 1961. *Industrial dynamics*. Cambridge, MA: Productivity Press.

Forrester, J. W. 1969. *Urban dynamics*. Cambridge, MA: Productivity Press.

Forrester, J. W. 1971. *World dynamics*. Cambridge, MA: Wright-Allen Press and MIT Press.

Forrester, N. B. 1982. A dynamic synthesis of basic macroeconomic theory: Implications for stabilization policy analysis. PhD thesis, Massachusetts Institute of Technology.

Gonçalves, P. 2009. Behavior modes, pathways and overall trajectories: Eigenvector and eigenvalue analysis of dynamic systems. *System Dynamics Review* 25 (1): 35–62.

Güneralp, B. 2006. Towards coherent loop dominance analysis: Progress in eigenvalue elasticity analysis. *System Dynamics Review* 22 (3): 263–289.

Kampmann, C. E. 2012. Feedback loop gains and system behavior (1996). *System Dynamics Review* 28 (4): 370–395.

Kampmann, C. E., and R. Oliva. 2006. Loop eigenvalue elasticity analysis: Three case studies. *System Dynamics Review* 22 (2): 141–162.

Kampmann, C. E., and R. Oliva. 2008. Structural dominance analysis and theory building in system dynamics. *Systems Research and Behavioral Science* 25 (4): 505–519.

Kampmann, C. E., and R. Oliva. 2009. Analytical methods for structural dominance analysis in system dynamics. In *Encyclopedia of complexity and systems science*, ed. R. Meyers, 8948–8967. New York: Springer.

Luenberger, D. G. 1979. *Introduction to dynamic systems: Theory, models and applications.* New York: Wiley.

Mass, N. J. 1975. *Economic cycles: An analysis of underlying causes.* Cambridge, MA: Productivity Press.

Mojtahedzadeh, M. T. 2008. Do parallel lines meet? How can pathway participation metrics and eigenvalue analysis produce similar results? *System Dynamics Review* 24 (4): 451–478.

Mojtahedzadeh, M. T., D. F. Andersen, and G. P. Richardson. 2004. Using *Digest* to implement the pathway participation method for detecting influential system structure. *System Dynamics Review* 20 (1): 1–20.

Oliva, R. 2004. Model structure analysis through graph theory: Partition heuristics and feedback structure decomposition. *System Dynamics Review* 20 (4): 313–336.

Oliva, R. 2014. Structural dominance in large and stochastic system dynamics models. Paper read at International System Dynamics Conference, July 2014, at Delft, The Netherlands.

Oliva, R., and C. E. Kampmann. 2010. Toolset for eigenvalue elasticity analysis. Available from http://iops.tamu.edu/faculty/roliva/research/sd/leea/toolset.html. Accessed April 15, 2014.

Saleh, M. M., R. Oliva, C. E. Kampmann, and P. I. Davidsen. 2010. A comprehensive analytical approach for policy analysis of system dynamics models. *European Journal of Operational Research* 203 (3): 673–683.

Sterman, John D. 2000. *Business dynamics: Systems thinking and modeling for a complex world.* New York: Irwin McGraw-Hill.

III Decision Support and Optimization

8 An Introduction to Deterministic and Stochastic Optimization

Erling Moxnes

In *Industrial Dynamics*, Jay W. Forrester expressed skepticism at the preoccupation with optimum solutions in management science:

> For most of the great management problems, mathematical methods fall far short of being able to find the "best" solution. The misleading objective of trying only for an optimum solution often results in simplifying the problem until it is devoid of practical interest. (1961, 3)

While this is still a useful warning, there are areas where both the philosophy and practice of optimization could be useful for system dynamics (SD) analysis. Graham and Ariza (2003) give an overview of earlier articles dealing with system dynamics and optimization and point out that improved methods for optimization and easy access to computing power have opened up new possibilities for analysis. (For an earlier overview, see Dangerfield and Roberts [1996]).

This chapter explores what optimization can offer to the analysis of dynamic systems. Likewise, it gives some examples of how optimization can be helped by simulation. The analysis relies on references to SD and optimization literature. A new optimization package for stochastic optimization in policy space, SOPS,[1] is used to work out examples.

Hopefully the chapter gives hints for future policy analysis. It should also help reduce communication barriers and thus help prevent unwarranted disagreements between practitioners of SD and optimization. Since both types of analysis have their strengths and shortcomings, optimization and simulation should be seen as complements. Hence, in many cases a combination of simulation and optimization could be the most effective way to improve real-life policies, as indicated by a laboratory experiment with this question in mind (Brekke and Moxnes 2003).

This chapter deals with the following uses of optimization:
- Exploring and refining policies in nonlinear, dynamic systems
  - that are deterministic
  - that are influenced by exogenous variables
  - that are influenced by stochastic disturbances
  - where measurements are uncertain
  - where model parameters are uncertain
- Establishing benchmarks in laboratory experiments
- Judging the rationality of descriptive theory
- Testing policy sensitivity to model uncertainty
- Capturing near-optimizing behavior in simulation models

After a section giving general background information, the chapter proceeds according to the above list.

Optimization in Dynamic Systems

The focus here is on dynamic systems of moderate size where SD offers principles for model building (Forrester 1968) and guidance on practical model building of social systems (Sterman 2000).

Optimization is about maximizing a criterion under restrictions defined by a system. Hence, optimization searches for increased net benefits by using available resources more wisely. This is a problem-oriented perspective similar to that of SD philosophy (Forrester 1961). The following three equations give a first definition of what an optimization problem for a continuous dynamic system may look like. The criterion or objective function can be written

$$J = E\left[\int_{t=0}^{T} e^{-rt} U\{\pi(x,u)\} dt \right],$$ (1)

where J expresses the expected discounted utility of a flow of net benefits over a time horizon T. Here E denotes expectation in case of randomness, r denotes a constant discount rate, U is a utility function, π is a function that expresses the flow of net benefits (or profits) over time, x is a vector of state variables (here usually referred to as stocks), and u represents a vector of decisions.

Restrictions are represented by a set of coupled differential equations, corresponding to a typical SD model

$$dx / dt = f(\psi, x, u, w),$$ (2)

where w denotes random variation over time (stochastic disturbances). Uncertainty in model parameters ψ will also be considered.

Finally, maximization of the criterion under the above restrictions leads to an optimal feedback policy

$$u = u(x) \tag{3}$$

for the steady-state case with *infinite horizon* (T→∞) where perfect information is available about stocks x and about model parameters ψ.

Policies are not always as simple as in equation (3). In case the time horizon is finite, the feedback policy will be a function of time as well. Finite horizon is not considered here. Predictable exogenous influences may also imply that time enters the policy. Errors in state measurements $y=x+v$, where v is random noise, imply that optimal policies also depend on past measurements. With uncertainty about model parameters ψ, policies may also have to consider how they influence the potential for learning about ψ.

The three basic equations define the stochastic dynamic optimization (or programming) problem. They may also be seen to be in agreement with the philosophy of SD analysis: a model (equation [2]) is used to explore policies (equation [3]) that improve system performance (equation [1]).

Optimization could be thought of as an ideal and hence impossible task. In order to truly optimize utility, there is no end to model and criterion complexity. The needs for detailed information are overwhelming. This is similar to prediction, which in principle is also an impossible task because of all the factors that could influence the variable to be predicted. However, if one thinks of optimization techniques and theory as aids to policy improvement, optimization has much in common with standard SD analysis. Policy improvement is also likely to be the practical concern of decision makers.

To deal with overwhelming detail, both optimization and SD rely on simplifications of model structure. Influential SD models typically have from 2 to 10 stocks, and nonlinearity is typically represented, while uncertainty tends to be neglected. Dynamic optimization models tend to have less than 4 state variables, and linearization is frequently resorted to, while uncertainty is sometimes central.

Optimization requires an explicit criterion that captures a tradeoff between benefits and costs. Formulating such a criterion is a challenging task. One main reason for this is that in order to optimize in dynamic models, the criterion must aggregate over space and time;

optimization can only take place with respect to one aggregate variable. A second reason is that criteria typically involve comparisons (or weighing) of (incommensurable) apples and oranges. A third reason is that optimal policies at times can be very sensitive to criterion formulations and parameter values. With these types of concerns in mind, the economist Tjalling Koopmans wrote: "The problem of optimal growth is too complicated, or at least too unfamiliar, for one to feel comfortable in making an *entirely* a priori choice of an optimality criterion before one knows the implications of alternative choices. One may wish to choose between principles on the basis of the results of their application" (1965, 226).

For dynamic models, the obvious way to inspect "results" is to simulate policies to see their consequences in terms of developments over time. Such simulations give much richer impressions of results than aggregated criterion values (Moxnes 2014). Traditionally, SD analysts have tended to formulate goals verbally and vary policies manually, before judging simulated behavior. Thus, they have circumvented Koopmans' problem while giving up on finding optimal policies.

In chapter 2 of *Industrial Dynamics*, Forrester defines problems as "characteristic and undesirable modes of behavior" (Forrester 1961, 21). He later defends this approach by saying: "In evaluating criteria one seldom encounters a zero-sum game. There is a free lunch. Most systems operate so badly that one can initially expect to find policies that improve some or most measures of performance without degrading others" (Forrester 1994, 251). Thus, in such cases the challenge is to find policies that make the modes of behavior more desirable, such as dampening cycles in supply chains.

When judging the behavior of an SD model, one is of course free to also consider tradeoffs between costs and benefits. Actually, this is one of the benefits Forrester sees in structurally rich SD models: "Criteria for accepting a proposed policy will usually depend on multiple measures of behavior. For example, in a business problem, one must balance the effect of a policy change on market share, profitability, inventory investment, stability of employment, available finances, and long-run versus short-run results" (Forrester 1994, 251).

In the above quotes, it appears that Forrester contradicts himself. However the two positions can be understood if one splits the optimization problem in two parts: finding the *goal* to pursue and *controlling* the system to reach that goal. Inventory management illustrates the point: determining desired inventory level involves a tradeoff between the costs of holding inventories and the benefits of being able to serve customers with short delivery delays. Controlling the inventory level involves policies for hiring, training, and utilization of production capacity. In the latter case, implementation costs may not differ much between good and bad policies. If this is the case, the split is justifiable. However, in general, the split is not perfect, while it may be practical when defining and prioritizing subproblems.

Optimization poses more methodological challenges to analysts than simulation. Therefore, optimization typically requires model simplifications in addition to criterion simplifications. Sometimes this is useful, since simple models can yield clear qualitative insights and help establish guiding principles. The next sections benefit from such insights. On the other hand, when simplifying, overlooked complexity is left to decision makers to deal with. In such cases, optimal policies found in simple optimization models could be tested in more complex simulation models and be compared to policies based on simulation experimentation, experience, and intuition. In case of uncertainty, Monte Carlo simulations could be used to test policies.

Failure to capture all aspects of reality implies that no policy is truly optimal; policies are only optimal in a restricted sense. In the optimization literature, the term optimal policy denotes the very best policy, the policy that maximizes the criterion for given and simplified (restricted) criteria (equation [1]) and models (equation [2]). This literature tends not to use the term optimal for heuristics (practical policies) that restrict the solution space. However, from a practical point of view, there is no principally important difference between restricting models and restricting policies.[2] Hence, in this chapter, the term optimal is also used to denote results following from criterion maximization when policies are restricted (heuristics). In this case, optimization procedures identify the set of policy parameters that maximizes the criterion value. With this definition of optimal, policies are not truly optimal because criteria (equation [1]), models (equation [2]), as well as policies (equation [3]) are simplified (restricted).

In an attempt to build a stronger bridge between optimization and simulation, this chapter makes use of "stochastic optimization in policy space" using a program called SOPS (Moxnes 2012b, 2005, 2003). Simulation models are formulated and tested in Powersim Studio before they are transported into SOPS. Resembling a manual trial-and-error search for policy improvement, SOPS searches automatically for parameters in flexible policy functions (equation [3]) that maximize a criterion (equation [1]) given a dynamic simulation model (equation [2]). Hence, in SOPS policies are written $u=f(x,\theta)$, where θ represents a vector of parameters.

The method can tackle more complex problems than nonlinear dynamic programming and optimization methods requiring linear models and quadratic criteria (also known as LQ problems). This is because SOPS allows for policy simplification as an alternative or complement to model and criterion simplifications.

In addition to complex dynamics (Forrester 1971; Sterman 1989a), numerous studies have shown that people have difficulties dealing with uncertainty (Tversky and Kahneman 1974). Hence, one should suspect that problems combining dynamics and uncertainty are in particular need for formal analysis. This combination will be in focus in this chapter, as indicated by equations (1)–(3). To deal with stochasticity, the SOPS program makes use of Monte Carlo simulations and search for policies that maximize the average (expected) criterion value over M Monte Carlo simulations of different stochastic and unknown futures. In this way, the problem is transformed from a stochastic dynamic programming problem to a deterministic nonlinear search problem where parameters in policy functions are adjusted to maximize the average criterion value. Since a large number of simulations is needed, the Powersim Studio model is automatically converted to C++ code before SOPS performs the policy parameter search. This speeds up calculations by a factor of about thousand.

The search for parameters in policy space is not a new idea. Dangerfield and Roberts (1996) review the use of this technique for deterministic problems in the field of SD. The method has been described and compared to other methods, though it is difficult to find examples of applications (Keloharju 1983; Polyak 1987; Gaivoronski 1988; Walters 1986; Bertsekas and Tsitsiklis 1996).

Challenge 1: When using optimization in policy space, SOPS, what is it that the optimization routine varies in order to maximize the criterion?

Challenge 2: Discuss the difference between model simplification and policy simplification in light of resulting policy recommendations.

Deterministic Nonlinear Dynamic Systems

This section deals with two situations: first, the case without exogenous influences, and second, the case with predictable exogenous influences.

No Predictable Exogenous Influences

First consider a continuous, dynamic system that is not influenced by exogenous variables, where the time horizon is infinite, $T=\infty$, and where measurements are perfect and model, criterion, and parameters are known perfectly. This system is a continuous Markov chain, where all information about the system at any point in time is contained in the current values of the state variables (stocks). For such a system described by equations (1) and (2), it can be proven that the optimal policy is a stationary nonlinear function of all state variables—as stated in equation (3) (Bertsekas 1987, 184).

This means that in principle, all policies in these types of systems are functions of all stocks. This will typically imply a restructuring of the system in terms of information channels for decisions. With this restructuring in place, there will exist an optimal policy. Hence, in this ideal case, there is in principle no room for the doubt that Coyle expresses: "The only drawback is, however, that there is always a nagging doubt that, had one tried only one more experiment, something even better would have been found" (1999, 429). However, doubt returns if one is not able to find or approximate closely the perfect nonlinear policy function.

Consistent with optimization theory, restructuring is also a maxim in SD philosophy. However, restructuring could extend beyond the nonlinear policy function. An example is steering of large ships. Moving the large rudder of such a ship takes great force and much energy. To reduce this need for force and energy, one can mount a small rudder on the big rudder. This way the big rudder is controlled similar to a small boat. Such a restructuring does not come out of optimization, nor does it suggest itself when using simulation. It follows from creative "thinking outside the box." Once a creative idea is in place, however, the model is augmented and can be analyzed by both simulation and optimization.

As an example of a complex, deterministic, nonlinear dynamic problem, consider an inverted pendulum. Such a pendulum is illustrated in figure 8.1. It is mounted on a cart, which is controlled by a force F in the direction that the wheels are pointing. The pendulum can swing in the same direction, and its deviation from an upright direction is denoted by the angle α. The inverted pendulum is not just a toy problem; it represents the core challenge for the control of a bicycle, a Segway, and even a rocket. The problem is representative of feedback control problems and is similar to dynamic management problems.

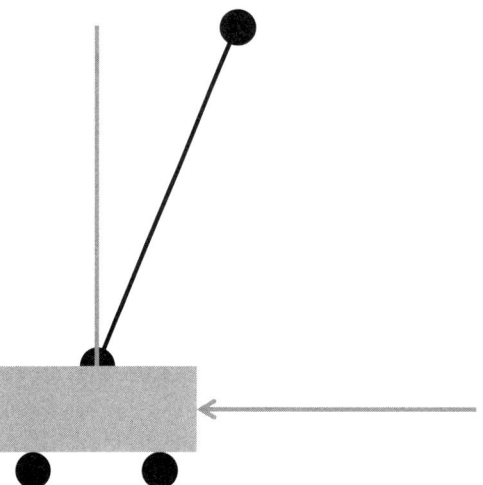

Figure 8.1
Inverted pendulum.

Figure 8.2 shows a stock and flow diagram for the inverted pendulum. The criterion, equation (1), is the accumulated sum of weighted (CW_c) costs, consisting of squared angle deviations from zero, squared deviations from the desired position of the cart, and the square of force F. The model, equation (2), has four stocks: the angle (α), the angle velocity, the cart position, and the cart velocity. Force F is determined by a linear policy, equation (3), of all four stocks, with weights (W_s) for each stock level. Hence, while the model has the nonlinear structure known from physics,[3] the policy is simplified and not the nonlinear function that theory prescribes. Hence, the policy is a heuristic that restricts the policy space. Angle and position can be easily measured, and angle velocity and cart velocity can be found from angle and position, respectively, by comparing neighboring observations in time. Perfect measurements are assumed.

Finding an optimal policy for F consists of two distinct problems. First, a set of feasible initial policy parameters (the weights W_s) must be found that are sufficiently near the global optimum to allow the optimization routine to find the global optimum. Second, one must search for policy parameters that minimize the accumulated costs.

For the inverted pendulum, different methods can be used to find initial policy parameters. First, it simplifies matters if the model is such that the pendulum stops moving if it reaches a horizontal position.

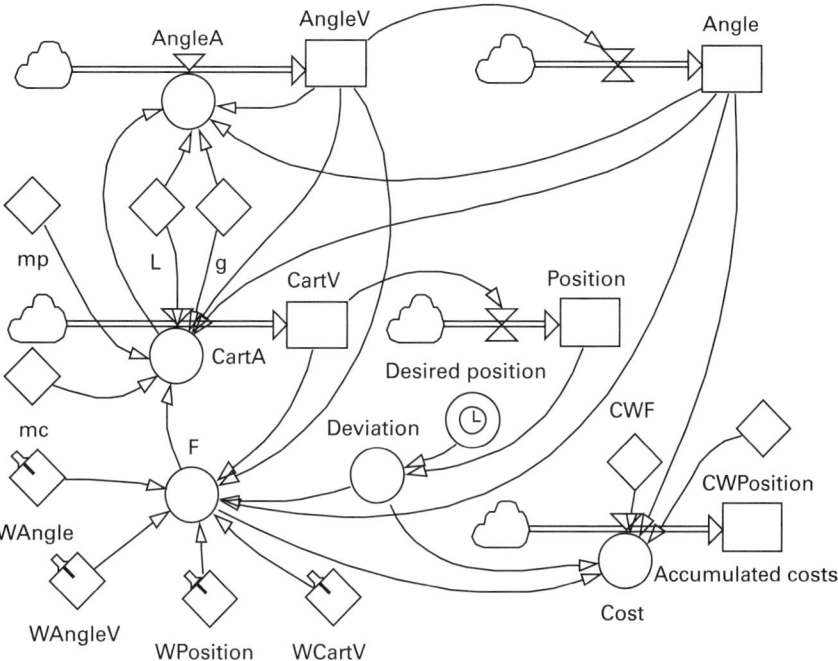

Figure 8.2
Stock and flow diagram for inverted pendulum.

Second, control theory or control engineering can guide the initial choice of parameters, either by use of S-transforms or optimal control theory. Both methods require linearization, and state-space optimization typically relies on a quadratic criterion. For most system dynamicists, this would require new and quite time-consuming skills to learn.

Third, for system dynamicists, feedback loop analysis can help find initial policy parameters. Eigenvalue analysis is one type of loop analysis that requires linearization and advanced skills. However, simple reasoning about loop polarities and gains may suffice. For instance, the reinforcing loop through angle and angle acceleration will bring the pendulum to the ground unless this loop is dominated by a control mechanism. Therefore, the control loop through angle and force F must have higher gain and opposite polarity to the reinforcing loop. With a large and positive value for the weight (WAngle) on the angle in the policy function, while the three other weights are set to zero, simulations show slowly increasing oscillations. This turns out to be a good starting point for optimization.

Fourth, one can also perform a random search for feasible policy parameters. In SOPS this is done by performing repeated criterion calculations for randomly chosen initial policy parameter values. With four policy parameters and wide ranges for each parameter, a very large number of searches may be needed to find a feasible parameter set. However, if one makes use of prior information from loop analysis, the range is narrowed, and fewer searches are needed to identify a set of feasible parameters.

The second problem is to find optimal policy parameter values that minimize costs. SOPS uses a gradient search (or, alternatively, a more robust and slower eclectic routine) to find optimal policy parameters. Since the search routine may stop prematurely or at distinct local minima, multiple searches are performed from different starting points for the policy parameters. After discarding searches that end up with clearly inferior criterion values, the remaining searches are used to calculate standard deviations for the obtained criterion values and policy parameters.

Table 8.1 shows optimal policy parameters for different cost weights (CWF) on force F in the criterion. This weight reflects the costs of using force (equipment and operating costs). The table shows that all four parameters are determined with high accuracy; accumulated costs typically vary with a standard deviation less than 0.0001. Lower weights on force lead to more use of force and lower accumulated costs.

So an interesting question is: would it be possible to reach similarly good results using a manual search for policy parameters in the simulation model? For a given criterion, it seems impossible to reach the same accuracy by manual search. If the criterion is uncertain, a manual

Table 8.1
Criterion values and policy parameters for different weights (CWF) on force (standard deviations based on multiple searches from different sets of initial policy parameters), initial Angle=1.0 radian, T=30s.

CWF	10^{-2}	10^{-4}	10^{-6}
Accumulated Costs	3.81	0.55	0.22
WAngle	36 (0.0)	90 (1.0)	334 (3.9)
WAngleV	11 (0.0)	12 (0.1)	27 (3.0)
WPosition	-1.1 (0.0)	-3.1 (0.0)	2.4 (0.7)
WCartV	2.5 (0.0)	5.9 (0.0)	16 (1.8)

search is more likely to produce policy parameters that will be considered satisfactory. Table 8.1 shows that for widely differing weights on force CWF, optimization results in a wide range of policy parameters to choose from. However, with four stocks to consider, finding appropriate relative weights remains a challenge for manual search.

It is not straightforward to quantify criterion weights on use of force and on deviations between desired and actual angles and positions. It is a comparison of apples and oranges. Further uncertainty is caused by the structure of the criterion. While the above criterion sums up costs due to deviations from desired angles and positions, it does not state any preference for the frequency of oscillations around the goals. It takes simulations to reveal that tight control also implies high frequencies. Simulations also reveal that a positive value of Wposition in the case with CWF=10^{-6} leads to a slow drift in position away from its desired level, implying that the time horizon should have been longer.

Since the underlying model is nonlinear, a linear policy is not optimal. Table 8.2 shows that the parameters of the linear policy vary with the initial angle, and thus with how much the angle deviates from zero. For an initial angle of 0.2 radians, the control is more aggressive than for an angle of 0.6 or 1.0 radian. Interestingly, if the policy for an initial angle of 0.2 radians is used in the case of an initial angle of 1.0 radian, the system becomes unstable and the pendulum falls to the ground.

Because the linear policy function introduces a restriction on the optimal policy, a next step would be to explore nonlinear policy functions. With four stocks, the truly optimal policy is a function in four dimensions, and probably a quite complex one. SOPS could be used to approximate such a function; however, this will not be attempted here.

Table 8.2
Criterion values and policy parameters for different initial values for angle (standard deviations based on multiple initial policy parameters); CWF =10^{-4}, T=30s.

Initial Angle	0.2	0.6	1.0
Accumulated Costs	0.01	0.11	0.55
WAngle	128 (7.1)	90 (0.2)	90 (1.0)
WAngleV	25 (2.7)	17 (0.1)	12 (0.1)
WPosition	-9.2 (0.9)	-6.2 (0.0)	-3.1 (0.0)
WCartV	14 (1.9)	9.6 (0.1)	5.9 (0.0)

Nonlinear policy functions will be explored later for a simpler model where nonlinearity also seems more important.

Feedback policies represent rules to control systems toward goals; as such, they also serve to counteract effects of unexpected environmental variation. Above, policies were found for an initial angle different from the goal. This deviation for the angle could be seen as caused by some random disturbance just before time zero. Thus, the feedback policy should be expected to work similarly well to correct for unexpected exogenous disturbances in the future. It follows from table 8.2 that in the case of a simplified linear policy, the choice of policy parameters must be adjusted to expected variation in future random disturbances. This is because of model nonlinearities. This gives some intuition behind stochastic optimization in nonlinear models where policies are sensitive to the amount of unexpected disturbance. An example follows later.

Predictable Exogenous Influences

Predictable exogenous developments make it possible to prepare before changes take place. There are different ways to prepare. First, one could simplify and maximize the criterion under the earlier assumption that the policy is a function of all stocks, $u = u(x)$. In this case, the parameters of the policy function would in most cases adjust to reflect the predictable exogenous influence. Hence the method may seem to work, but this procedure is only likely to give miniscule improvement. With exogenous variables, the process is no longer a Markov chain where all information is contained in the stock values.

A second method, which is theoretically correct, is to allow the feedback policy to be a nonlinear function of time t (time as a stock) in addition to all other stocks x, $u = u(x,t)$. Dependent on exogenous developments, this function could become highly nonlinear.

As a third alternative, one could model the process generating the exogenous variable and let the policy be a function of the stock variables z for the this process. Then the policy could be written $u = u(x,z)$. This method brings the system back to the standard form, except that the order of the system has increased. Often it will be a challenge to identify low order models that represent exogenous influences with good accuracy.

Fourth, in the deterministic case, with perfectly predictable disturbances, a much simpler approach is to find a path for force over time, $u = u(t)$. The weakness of this fourth approach is that the optimal path is only useful for one particular exogenous development, and it is not robust in case of additional unpredictable random disturbances. Since the equilibrium for the inverted pendulum is unstable, lack of feedback control implies that the pendulum will always fall to the ground with such a policy. For systems that are not dominated by reinforcing loops, and where randomness is minimal and exogenous influences are predictable, this method may produce useful results. Using SOPS, $u(t)$ for the problem dealt with below, was found with accumulated costs only 12% higher than those found for the next method, feedforward. As predicted, however, the pendulum eventually fell to the ground. Feedback control is essential.

A fifth approach is to combine feedback with a simplified and explicit feedforward policy (alternatives two through four are implicit feedforward policies). Simplified feedforward means that known future movements in the exogenous variable are allowed to influence selected variables ahead of time. This way, imbalances are minimized because the system is allowed time to prepare for what is coming. In the following, feedback and feedback combined with simplified feedforward are compared. Results are shown before the details of the feedforward policy are discussed.

Assume that the desired position of the cart steps up from zero to 10 meters at time 4. Table 8.3 shows the optimized parameters when this event is seen as unexpected (only feedback is possible) and when the change in desired position is known in advance (feedforward is added). Use of feedforward in the latter case reduces accumulated costs by 70% from 1.15 to 0.34.

Table 8.3
Criterion values and policy parameters optimized for a 10-meter shift in the desired position for an unexpected and an expected change (standard deviations); CWF $=10^{-4}$, T=30s.

	Unexpected (Feedback)	Expected (Feedforward)
Accumulated Costs	1.15	0.34
WAngle	241 (1.1)	55 (1.8)
WAngleV	42 (0.3)	21 (0.4)
WPosition	-17 (0.1)	-2.8 (0.1)
WCartV	28 (0.2)	4.8 (0.2)
Time constant		0.9 (0.01)
Time maximum		2.8 (0.03)

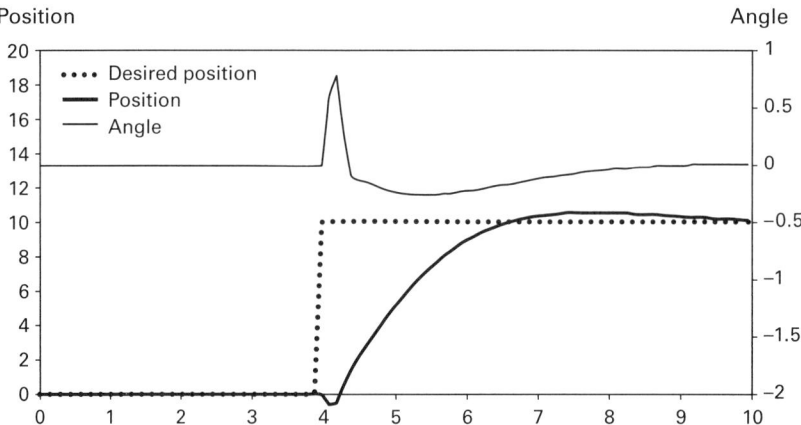

Figure 8.3
Development in angle and position as a response to a step increase in desired position.

Figure 8.3 shows that the feedback policy is a worse-before-better policy. First the cart moves in the wrong direction, the pendulum starts falling in the desired direction, and the cart moves quickly in order to prevent the pendulum from falling further. The cart succeeds to the extent that the angle becomes negative, and toward the end the cart decelerates in order to bring the pendulum back to an upright position. As a result of all this, the position changes from zero to 10 meters. Since the event is unexpected, the temporary gaps between desired and actual position and between desired and actual angle become quite large. Amplifications in the feedback loops are large (see table 8.3).

In figure 8.4, the feedback rule for force is influenced by the deviation between actual position and a modified desired position, which is a feedforward version of the desired position (the criterion is still influenced by the actual desired position). The modified desire puts more and more weight on the known future desired position as the step is approached.[4] Figure 8.4 shows that the modified desire increases quite fast with a time constant of 0.9 seconds and maximum is reached after 2.8 seconds; that is, 1.2 seconds before the desired position steps up. As a result, the actual position increases smoothly and nearly symmetrically around the step in desired position. Maximum temporary deviations from desires are smaller than for the case without feedforward. Amplifications in the feedback loops are lower (see table 8.3).

In this example, the simplified feedforward policy seems intuitive, and it leads to a considerable percentage reduction in costs. Simple trial

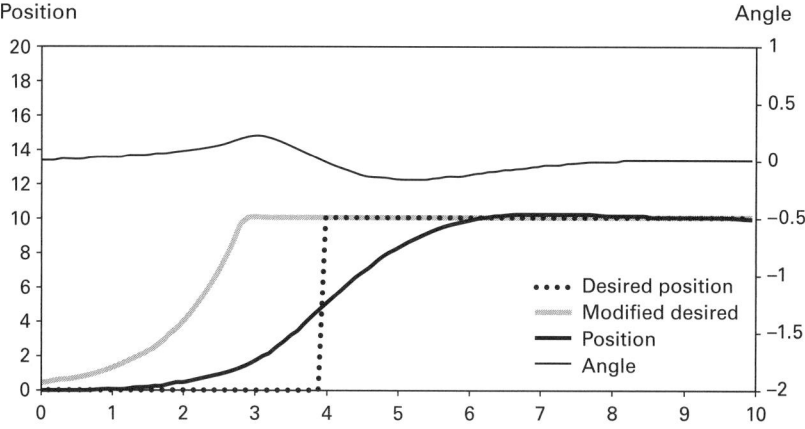

Figure 8.4
Development in angle and position as a response to an expected step increase in desired position.

and error may have been sufficient to find parameters in the feedforward policy. However, table 8.2 shows that the feedforward policy also has quite large implications for the four parameters of the feedback policy. This result would be more difficult to establish by trial and error. In general, simplified feedforward strategies must be based on understanding of underlying model structure. A shifting goal is probably one of the easier examples one could think of.

Feedforward is also important in systems where disturbances can be detected before they hit controlled systems. For instance, dynamically positioned oil-drilling vessels are influenced by observable and hence predictable waves. Even in cases where disturbances are not directly observable, they could enter policy functions. This is the case when the disturbance states can be estimated from variables they influence, and when the disturbance is produced by a known dynamic process (autocorrelated).

For management, the lesson is that one should start reacting to predictable future events before they impact the system. How much earlier depends on how long it takes to adapt to new conditions. That is neither a new nor a surprising insight. However, laboratory experiments suggest that the importance of delays and inertia is often ignored (Sterman 1989b; Brehmer 1989; Moxnes and Jensen 2009). What optimization may contribute to is a better understanding of how early one needs to react to a predicted event, how far into the future forecasts need to reach, and how large a cost savings an early start could contribute to.

Challenge 3: In deterministic, infinite-horizon, dynamic systems, the optimal policy is in principle a function of all stocks. Try to give an intuitive explanation for this, and think about cases where some stocks do not matter.

Stochastic Nonlinear Dynamic Systems

In infinite horizon problems with linear systems and quadratic criteria (LQ), unpredictable random disturbances are of no consequence for optimal policies. Randomness is only important for policies when models are nonlinear or when criteria are asymmetric, for instance due to risk aversion. Similar to the deterministic case, the optimal policy is still a nonlinear function of all stocks (Bertsekas 1987).

In this section, a nonlinear fishery problem is used to illustrate optimization. What is the effect of random disturbances on optimal harvesting strategies and fishing capacity? Should capacity be big enough to take advantage of exceptionally large fish stocks randomly distributed in time? Historically, there are numerous examples of overfishing and of overbuilding of fishing capacity (Schrank 2003). Governmental subsidies of capacity when capacity is already overexpanded is perhaps the best example of poor intuitive understanding of the policy problem. Laboratory experiments show not only overshooting investments in capacity but also efforts to sustain overcapacity (Moxnes 1998, 2012a).

Figure 8.5 shows a model of a fishery where one and the same agency makes decisions about both harvesting and capacity. This could be a privately owned fish stock or a fishery governed by use of yearly quotas and capacity licenses. The criterion, equation (1), is the expected value of the bank account at the end of a 50-year period (approximately infinite horizon). This is the same as maximizing the net present value of the cash flow. The cash flow is made up of revenues[5] minus costs[6] and minus interest payments.[7] Effort is given by harvest divided by catch per unit effort (CPUE).[8] Since randomness can lead to many different and for the decision maker unknown futures, policies are tested against a large number of future scenarios (stationary processes). The expected criterion value is approximated by the average value of the bank account over all simulated futures, here produced by 100 Monte Carlo runs.

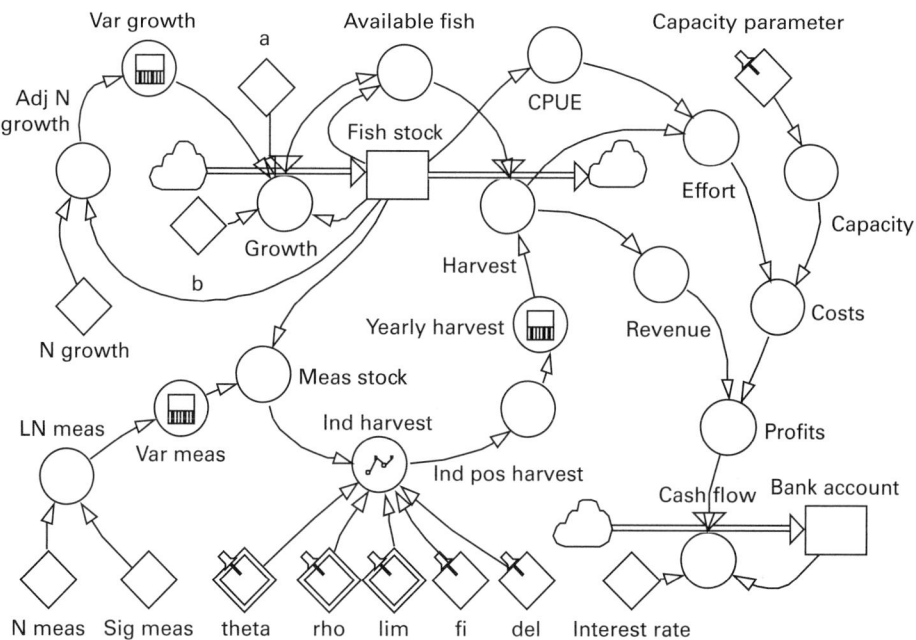

Figure 8.5
Stock and flow diagram for fishery management.

The model, equation (2), is made up of the fish stock with its inflow of net growth and outflow of harvest. To ensure that the fish stock never goes negative, the harvest is the minimum of the decided yearly harvest and the available fish (fish stock plus growth in the next time step of 0.1 years). Growth is given by a surplus growth curve[9] plus random variation in growth.[10] The randomness updates only once a year.[11] Maximum growth (maximum sustainable yield, or MSY) is 0.8 million tons per year. The fish stock is measured once a year, correctly or with error.[12]

The task is to simultaneously find two policies (equation [3]): one constant-capacity parameter and a harvesting strategy, first for perfect measurements of the fish stock. The harvesting strategy is defined by six gridpoint values for indicated harvest (a vector theta) for six different sizes of the measured fish stock (0, 1, ... 5 million tons). In Powersim Studio the strategy is formulated using a special function SOPSPOLI-CYGRID, which is recognized by the optimization program SOPS. The function interpolates between the grid points. Actual decisions about yearly harvests are made only once a year and are restricted to non-negative numbers.

Since the criterion is not symmetric and the model is nonlinear, optimization theory predicts that the optimal policy will be nonlinear. Figure 8.6 shows the optimal harvesting policy found by SOPS. For fish stocks below 0.8 million tons, there is no fishing. Above that level, harvest increases along a concave curve. Optimal capacity is 0.85 million tons per year, slightly higher than MSY. The figure shows that capacity is not utilized fully for fish stocks below 3.1 million tons. Above that level, capacity is utilized more than 100%. For large fish stocks, unit operating costs increase somewhat because the effect of increased utilization dominates the beneficial effect of higher CPUE. However, for all fish stocks, unit operating costs are well below the fish price minus the fixed capacity costs (NOK 4/kg). Hence, the fishery produces a large resource rent (which would stimulate competitors to enter the fishery in case of open access).

Even without risk aversion in the criterion, the optimal policy suggests a stable equilibrium point at a fish stock of about 2.6 million tons, well above the stock size yielding MSY. This protects against episodes where a combination of a high catch and low growth due to randomness causes the fish stock to go very low. In case of such an episode, it would take many years to rebuild the stock, with low cash flows in the

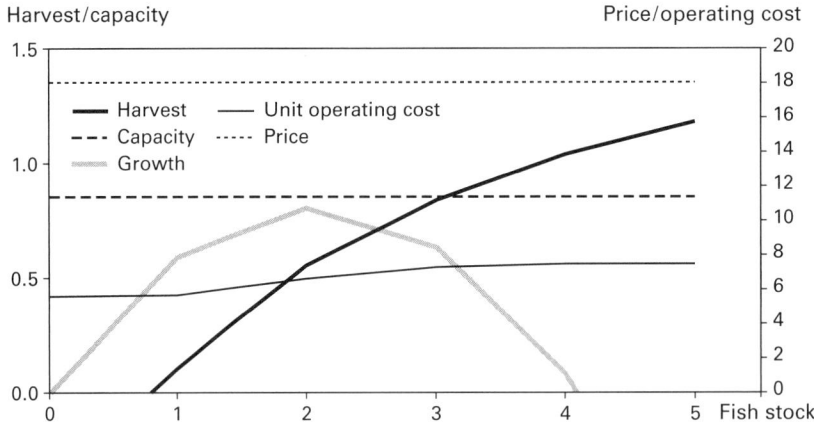

Figure 8.6
Optimal capacity and harvesting strategy for the case with random variation in growth together with expected fish growth. Thin lines show fish price and unit operating costs.

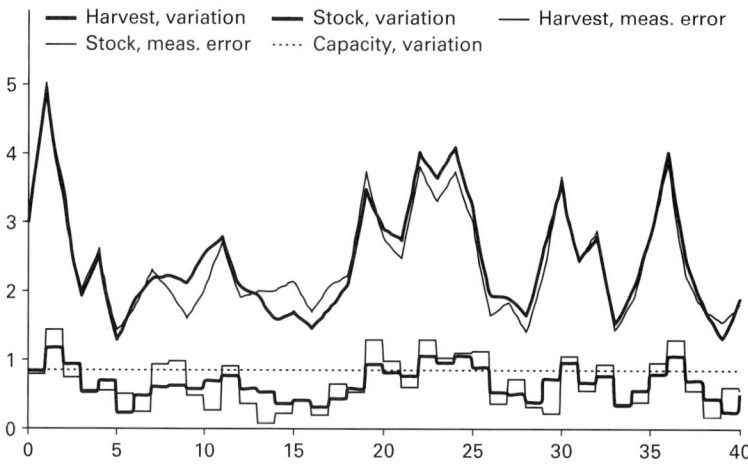

Figure 8.7
Simulated harvest and fish stock developments for the case with random variation in growth (thick lines) and in the case with random variation in both growth and stock measurements (thin lines). The dotted line shows capacity for variation in growth.

intervening period. Figure 8.7 shows that with the given assumptions about randomness, the fish stock never goes below 1.28 million tons over a randomly chosen 40-year period.

The optimal capacity restricts harvest at large fish stocks. For a stock size of about 4 million tons, which is observed only five times in figure 8.7, harvest is merely 1.04 million tons. It could have been nearly 2 million tons per year and still only reduced the fish stock to the stock level giving MSY. Such a situation can easily cause decision makers to conclude that capacity is much too low. However, the optimal solution represents a tradeoff between expected costs and revenues.

While the loss of revenue due to limited capacity is obvious, high costs require deeper analysis. Marginal capacity that is only utilized every eighth year (5 out of 40) is eight times as costly as capacity used every year. As capacity utilization increases, marginal operating costs increase rapidly. Furthermore, much of the fish that is not caught in one year with high fish stocks survives and allows for higher harvests in ensuing years. The optimization routine considers all these factors and produces results that stimulate analysis and understanding.

The above optimization is clearly more demanding than a manual adjustment of policy parameters in a deterministic simulation model. Optimization allows for the use of grid functions with many parameters. As long as optimal policies are reasonably smooth, interpolations between grid points give good accuracy and, importantly, do not restrict policies to predetermined and guessed-at functional forms.

Also note that adjusting policy parameters for each of a few selected scenarios leads to nonoptimal policies. When adjusting to one particular scenario, one makes the false assumption that the future is known with certainty. The average policy found from several such scenarios is still a nonoptimal policy (Rockafellar and Wets 1991).

Also note that in case of randomness, a formally stated criterion seems more needed than in the deterministic case. In the deterministic case, inspection of time developments can often give better ideas about the goodness of a policy than the numerical value of a simplified and hard-to-understand criterion. With randomness, as illustrated by figure 8.7, it is far more demanding to judge the goodness of different policies. However, inspection of simulated developments is still important to spot problems such as instabilities that are not considered by the criterion.

With the optimal policy in place, the expected future value of the bank account for the case with random variation in growth is 745 billion NOK. The optimal policy for the case without random variation is 789 billion NOK. Hence, in this case the cost of variation is 6% of the expected value.

Challenge 4: Compare the SOPS method and a method where optimal policies are found for each and every Monte Carlo future (deterministic) and where the final policy recommendation is the average of these policies.

Measurement Error

Random errors in measurements of stock values complicate optimization considerably. In the LQ case, the separation theorem applies. This means that one may keep the feedback policy found for the system without measurement error. Since stocks are not perfectly measured, information must come from optimal state estimates, for instance produced by a Kalman filter. For problems that are not LQ, the separation theorem no longer works in its simple form. In this case the optimal policy is a function of the entire conditional probability distribution for the state variables (Bertsekas 1987). According to Bertsekas, "there may be no computationally efficient way to solve the problem" (127). Moxnes (2003) investigates six different approximations, of which a quite simple and still relatively efficient one will be used here. As before, harvest u and constant capacity K are determined simultaneously. Hence, harvest will reflect the optimal value of K, however, harvest will not be a direct function of K.

Instead of the traditional two-step process of finding state estimates and feedback policies separately, Moxnes suggested that the two steps could be combined into finding one nonlinear function linking observations y and harvest u, provided the system is observable with the existing set of measurements. For a time-discrete system, using all the information that is available for state estimation, the policy for harvest can be written:

$$u_t = f(y_t, y_{t-1}, ..., y_0, u_{t-1}, u_{t\ 2}, ..., u_0). \tag{4}$$

As time progresses, the policy space expands. Under the assumption that this policy is followed consistently at all times, harvest u_{t-1} can be expressed by measurements y_{t-1}, y_{t-2}, ... and harvests u_{t-2}, u_{t-3}, ... Similarly, u_{t-2} can be expressed by even earlier measurements and harvest decisions, and so on.

$$u_t = f(y_t, y_{t-1}, ..., y_0). \tag{5}$$

Going back to the continuous case, the policy can be written,

$$u = f(y, y^{t-1}, y^{t-2}, ..., y^{t0}), \tag{6}$$

where terms y^{t-1} denote current records of earlier values of measurements (delayed measurements). This is still a complex problem, since the ideal policy function will be a nonlinear function of all previous measurements. However, in practice there are limits to how interesting old measurements are and how frequently measurements should be stored. If the system changes quickly relative to the time interval between measurements, old measurements are quickly outdated. For the purpose of reducing measurement error, it helps to combine many observations; however, the marginal value of additional observations also drops fairly quickly. Hence, in many cases the most recent measurement and/or a first-order delay of the measurement will do a relatively good job.

Interestingly, the latter approach is often used in SD models as a descriptive theory of how people make use of information. In addition to capturing delays in information handling, perception delays are also motivated by the need to smooth observations. In the following, a harvesting strategy that is simply a function of current measurements will be used to illustrate.

Figure 8.5 shows how measurement error combined with random variation in fish growth influences the fishery model. Using the harvesting strategy and capacity found for the case with variation in growth only, measurement error causes the expected criterion value to drop by NOK 118 billion from NOK 745 to 627 billion; that is, by 16%. Using SOPS to adjust both harvesting strategy and capacity to reflect measurement error, the expected criterion value increases from NOK 627 to 686 billion, an increase of NOK 59 billion or 9%.

The maximum improvement in the expected criterion that could be obtained by perfect measurement accuracy is NOK 118 billion. Since marginal assessment costs increase rapidly with accuracy, the optimal net gain from increased accuracy is likely to be considerably lower than NOK 118 billion. Adapting the strategy to the current level of accuracy saves NOK 59 billion. Considering the small effort involved in improving the strategy, the payoff is huge (adding older measurements could lead to a moderate further improvement). Whether the same improvement can be obtained in reality is an interesting question that requires empirical research. The answer depends on the quality of current policies.

Harvest

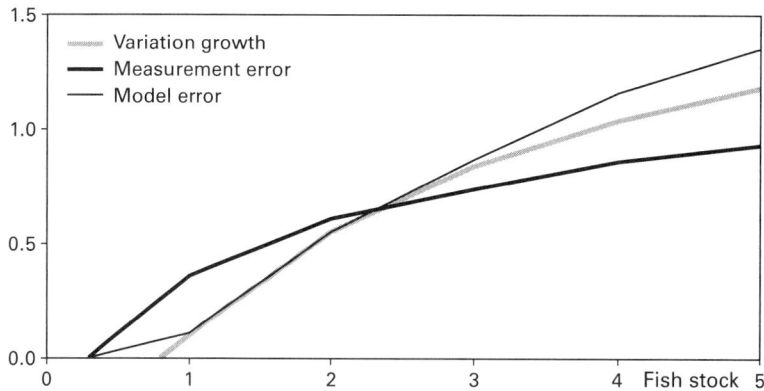

Figure 8.8
Harvesting strategies for variation in growth and variation in growth combined with measurement error and with model error.

Figure 8.8 shows the harvesting strategy for measurement error. Compared to the strategy for variation in growth only, the strategy for measurement error is less sensitive to changes in the measurements of the fish stock. The strategy can be said to filter observations in space as opposed to filtering in time (e.g., Kalman filtering). Since the measurement error strategy has the higher harvest for low stock estimates, one should expect that measurement error leads to greater chances of low fish stocks and depletion. However, figure 8.7 suggests that the effect on the fish stock is similar in the two cases. Stocks develop almost identically, while harvesting varies more in the case of measurement error.

The optimal fishing capacity is 0.81 million tons per year, close to MSY and 5% lower than in the case with variation in growth only.

Challenge 5: Compare optimal policies for systems with measurement error and Forrester's use of first-order delays to model perceived information.

Model Error

Models are not perfect descriptions of decision problems; there is model error (or uncertainty or ambiguity). There is parameter uncertainty and uncertainty about model structure. However, structure could be described by parameters defining nonlinearities such that parameter uncertainty could be seen to encompass structural uncertainty as well. In its general form this problem is very complicated, and, according to Bertsekas, "suboptimal controllers are thus called for" (1987, 163). Again, optimization theory can be useful in the search for approximations.

There are three versions of this problem that are dealt with in this section: persistent error in the policy period (when the policy has its main effects), exogenous learning in the policy period, and endogenous learning in the policy period.

Persistent Error in Policy Period

If environments and hence parameters change over time, the preceding and most relevant historical period will always be too short to obtain very precise parameter estimates. If only modest learning is going on during the period for which policies are explored, the optimization problem boils down to finding the best policy in light of probability distributions for uncertain parameters ψ in equation (2). With new data, say once a year, policies can be updated. However, each new year, one makes the somewhat inconsistent assumption that there will be no learning and no updating.

To illustrate how this problem can be dealt with, consider uncertainty in the two parameters a and b in the formulation for the surplus growth curve in the fishery model (see earlier endnote 9). Using SOPS with the same assumptions as for the case with random variation in growth, the two parameters are drawn from normal distributions before each Monte Carlo run. Expected parameter values are as before, and standard deviations are 20% of the expected parameter values. With a at plus one and b at minus one standard deviation (high case), MSY becomes 1.44 million tons per year. With the reversed assumptions (low case), MSY becomes 0.43 million tons per year. Hence, the uncertainty in the growth curve is assumed to be considerable.

The optimal harvesting policy for model error (and no measurement error) is shown in figure 8.8. At large fish stocks, harvesting is higher than in the case with variation in growth only. This enables large harvests in case uncertain parameters yield high MSY and high carrying capacity. In case uncertain parameters yield low MSY and low carrying capacity, fish stocks are not likely to reach levels above 3 million tons, and the higher harvests are of little relevance. At low fish stocks harvesting is slightly higher. This enables profitable fishing in case MSY and carrying capacity are low, while it is of little concern if MSY and carrying capacity are high. Fishing capacity is somewhat higher than in the case with variation in growth only, 0.91 million tons per year.

It is not easy to estimate the loss due to uncertainty in parameters. The following method gives a rough idea. Using the obtained optimal policy for the case with uncertainty in a and b, simulations show that for the above "high case" the expected criterion value becomes 1,876 billion NOK and for the "low case," it becomes 150 billion NOK. The difference is great, because the criterion values represent two different fisheries. The important question is: what is lost by using the optimal policy for the case with uncertainty in the two parameters a and b? To answer this question, SOPS can be used to find optimal policies under the assumption of perfect information about the parameters. In the "high case" an optimal policy based on perfect information leads to an expected criterion value of 2,063 billion NOK; the loss due to model uncertainty is 9%. In the "low case," the criterion value for perfect information is 255 billion NOK and the loss is 41%. These numbers indicate that higher parameter accuracy can have considerable value.

However, as for measurement error, due to high marginal costs of model accuracy, there are limits to economical model improvement.

Exogenous Learning in Policy Period

Exogenous learning denotes learning that is independent of the effects decisions have on the system. Thus, learning is supposed to follow from research and path-independent observations. If decisions made today have longlasting consequences, uncertainty about model parameters could be revealed over that period. If so, policies should take advantage of this fact by introducing flexibility to be able to adapt to the new information. This is referred to as recourse in optimization literature.

In the above fishery example with model error it seems risky to invest in large amounts of long-lasting fishing capacity in case one finds out after a few years that MSY is much lower than expected. Capacity would be excessive and costly. However, as in many real situations, idle capacity could be sold or used for other purposes. Hence, with exogenous learning, better policy recommendations would follow if capacity were modeled as a stock that can change by flows for investments, sales, or scrapping. This requires policies for these flows, and these policies should ideally depend on updated (recursive) estimates of primarily MSY. If so, estimated MSY would appear as an extra stock in the model. This stock should influence harvest decisions as well, together with fish stock measurements. If used fishing capacity can be sold for a good price, flexibility is occurring naturally, and one can invest more aggressively than otherwise, before uncertainty is revealed.

Flexibility is not always occurring naturally. Consider climate change. The amount of greenhouse gases in the atmosphere is inflexible; it takes time to change the rate of emissions, and natural removal is only a small fraction of the stock each year. If, sometime into the future, one learns that the effect of greenhouse gases on climate (climate sensitivity) is larger than expected, future climate costs would be high. Based on optimization with exogenous learning, lack of flexibility implies that near-term emissions should be kept lower than otherwise. Optimization also implies that future flexibility has value and that flexibility should be provided by research and development of alternatives to fossil energy. Having technological options available eliminates time needed for research and development and thus increases future flexibility. Hence, optimization with exogenous learning can be used to understand and to quantify the value of flexibility for different assumptions about current and future uncertainty in climate sensitivity.

Endogenous Learning in Policy Period

Endogenous learning occurs when speed of learning is influenced by the decisions made in a system; learning is sensitive to the path taken. This is known as the *dual problem*, where one has to make a tradeoff between optimizing short-term profits based on current knowledge and deviations from optimal short-term policies in order to speed up learning to make better future decisions. This is a very complicated problem, for which formal optimization has less to offer than practical decision-making based on awareness of the nature of the problem.

Again the fishery problem serves to illustrate. Figure 8.9 shows two different surplus growth curves. Assume that the thin line represents current parameter estimates and that fishing is going on according to an "optimal" fishing strategy, with fish stocks somewhat higher than the stock size that yields MSY, in accordance with the optimal strategy in figure 8.6. This operation will over time produce data points in or close to the gray area in figure 8.9. These data will tend to support current knowledge. However, these are also the data one would obtain if the true growth curve were represented by the thick curve. In order to find out which curve is the more correct one, harvest must be changed away from what the current "optimal" strategy prescribes. If harvest is reduced in this case, learning will come at a cost. If increased, learning will be immediately profitable.

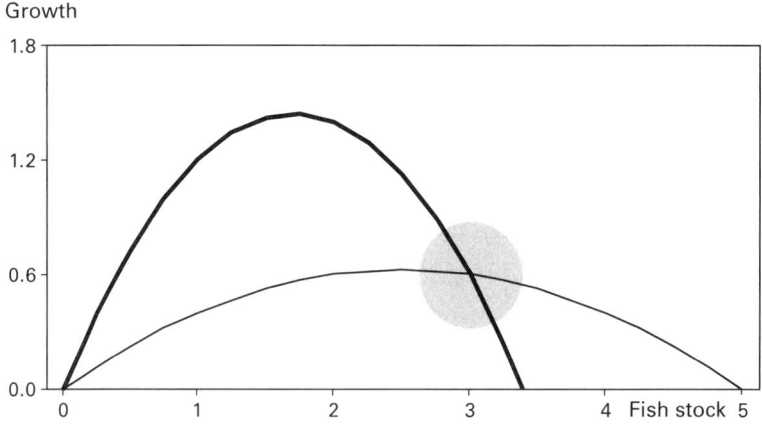

Figure 8.9
Two growth curves with different carrying capacities (CC) and different maximum sustainable yields (MSY).

For practical reasons, one must separate the two questions. Prior knowledge, estimation techniques, and risk aversion should play a role when determining in what direction and how much to deviate from current "optimal" policies in order to learn. Then new knowledge is used to reformulate management policies before new deviations from "optimal" policies are considered. Most important seems to be awareness of the dual problem.

Different Uses of Optimization

Earlier sections have dealt with the main area for use of optimization: improving policies in complex nonlinear dynamic systems under uncertainty. Here four other areas where optimization may complement SD analysis are discussed.

Establishing Benchmarks in Laboratory Experiments

For laboratory experiments where subjects get full information, optimal policies may be found and can serve as benchmarks to judge the goodness of subject decisions. Full information may include probability distributions instead of exact parameters, in which case benchmarks require methods that can deal with uncertainty. In cases with less than full information, optimal policies assuming full information only define upper limits for expected subject performance. Benchmarks can be developed both for strategies and for criterion values.

Judging the Rationality of Descriptive Theory

Assessing the rationality of descriptive decision rules is complicated because decision-maker objective functions (equation [1]) and the conditions they operate under (equation [2]) are not well known. Still, speculation about people's motives and intended rationality is useful when developing or revising descriptive theory of behavior. By testing descriptive theories under increasingly complex assumptions about objectives and conditions, it should be possible to identify under what assumptions decision rules could be described as rational or not. Morecroft (1985) describes how this task can be approached with simulation. Optimization could be a useful complement to simulation for this type of study, particularly in the field of economics, where rational decision-making is a central topic.

Testing Policy Sensitivity to Model Uncertainty

Simulation models are often tested for behavior sensitivity to uncertain model parameters. Similarly, one can test the behavior sensitivity to policy parameters. In contrast, policy sensitivity analysis measures how the optimal policy changes when model parameters change.

Policy sensitivity analysis can be very valuable because it is directed at the main outcome of model studies: policy recommendations. What parts of a model are particularly important for the policy recommendations and what parts are of little importance? Where should uncertainty be reduced in order to make policy recommendation more robust and criterion values higher? For this purpose, optimization is very useful, because it produces policies that are consistent with model parameter values. Doing policy sensitivity manually leaves the analyst with uncertainty about accuracy and consistency. In addition, there is a risk of bias due to subjectivity and wishful thinking.

To illustrate, Moxnes (2005) uses SOPS to test policy sensitivity for harvesting policies in fishery models. He finds that policies can be quite insensitive to whether one uses a simple aggregate one-stock model or a much more complex model with age classes. This conclusion is surprising in light of the much greater level of detail in age-class models. Assumptions about increasing marginal operating costs turned out to be more important for policies. This is a mechanism that is left out of most models used to recommend fishing quotas. Interestingly, in this case, policy sensitivity analysis indicates the value of interdisciplinary research.

These examples also illustrate that policy sensitivity analysis can help guide model simplification. Model structures that have some importance for model behavior but minimal importance for policy recommendations could be removed. Simplification is important to facilitate diffusion of insights; few decision makers are likely to use optimal policy functions without a minimum of understanding. In contrast, ordinary behavior sensitivity tests could encourage inclusion of unnecessary model complexity in order to replicate historical behavior modes with high accuracy.

Modeling Near-Optimizing Behavior in Simulation Models

A totally different reason for making use of optimization in SD studies is the occasional need to capture fast, near-optimizing behavior in simulation models. In simple cases, one can find analytical solutions to simple optimization problems and insert the solution into the simulation model. For instance, in a simplified market model, one could assume that consumers maximize utility by adjusting demand instantaneously to price. This can be done by setting supply equal to demand, and then reading out the equilibrium price from the demand curve. When modeling decision-makers' choices between various options, logit models could be used to determine indicated optimal shares for each of the options. Actual share could follow after delays. In more complex cases with binding constraints, linear programming or nonlinear programming could be used to find optimal resource allocations or indicated optimal allocations.

Conclusion

For both analysts and decision makers, it seems fruitful to see optimization and SD as complementary. Optimization can be useful to SD practitioners in the following ways. Optimization may inspire system dynamicists to become more conscious about tradeoffs between benefits and costs and about how to deal with stochasticity. Results in the optimization literature give general insights about the formulation of policies for different problems. Optimization also inspires policy sensitivity analysis as a more problem-oriented test than parameter-sensitivity tests.

Simulation and SD philosophy can be useful to practitioners of optimization for the following reasons. Simulation models allow for more dynamic complexity than typical optimization models, and SD can be used to develop and explore models before they are used for optimization. Because optimization typically requires quite drastic simplifications of both criteria and restrictions (models), it seems wise to test obtained policies in more complete simulation models. Here optimal policies can be compared to existing policies and more intuitively based policy suggestions. Perhaps counterintuitively, rather than using optimization for fine-tuning of policies, optimization could be a provider of policy suggestions to be tested and adjusted in simulation models.

Optimization in policy space is a method that helps bridge the gap between simulation and optimization. Using this method, it is possible to find simplified policies to complex problems. For decision makers, this should be more attractive than exact solutions to simplified problems, since the latter approach may come to leave the most demanding part of the analysis to the decision maker. Optimization in policy space is intuitively appealing in that it can be seen as an automated version of the type of policy testing one can do manually in simulation models. With a tool like SOPS, it has also become easier to search for optimal policies once a simulation model has been developed. As for other optimization methods, policies obtained by use of SOPS should be checked by simulation because of the difficulties in formulating criteria.

Acknowledgments

I am grateful to Arne Kråkenes and the team of programmers at Powersim Software for making SOPS a user-friendly commercial product. Thanks to the Research Council of Norway for supporting the early development of the optimization software as part of a project on fishery management, and thanks to the System Dynamics Group at the University of Bergen for supporting further development of the software and for providing an inspiring environment for research.

Notes

1. The user manual and the SOPS software is available from powersim.com/main/products-services/sops/.

2. That simplifying models and simplifying policies are related is illustrated by the effect of using optimization methods that require models are linearized and criteria quadratic. Such a model simplification restricts optimal feedback policies to linear ones.

3. Angle and position are integrations of corresponding velocities. Velocities are integrals of acceleration (Newton's first law) and accelerations are given by force divided by mass (Newton's second law). Angle acceleration is given as:.

$d\alpha_v / dt = (g \sin(\alpha) - a_v \cos(\alpha)) / L,$

where g is acceleration of gravity, a_v denotes cart acceleration, and L is the length of the pendulum. Cart acceleration is given as:

$dv / dt = (-m_p g \sin(\alpha)\cos(\alpha) + m_p L \alpha_v^2 \sin(\alpha) + F) / (m_c + m_p \sin^2(\alpha)).$

Equations for F and cost are weighted sums, where the weight for angel (α) in cost is set equal to 1. Deviation is the difference between desired position and actual position. The implicit goal for α is zero degrees.

The equations do not show the logical operations that ensure the pendulum cannot fall below a horizontal position. Once the pendulum reaches horizontal position, cost becomes zero and accumulated costs stop updating. This is realistic in this case and makes it easier to find optimal policies for force.

4. The exact formulation is inspired by a folding integral for a first-order linear process. The feedforward desired position (D) is given by the formula $D = S \min(1, e^{-(t_m - t)/\tau})$, where S is the step size, t_m is the time when $D = S$, and τ is the time constant. The two parameters are found by optimization together with the parameters of the feedback policy.

5. Harvest times a fixed export price for fish of NOK 18/kg.

6. Fixed unit costs of capacity of NOK 4/kg/year are multiplied by capacity. Unit operating costs vary with utilization $8(0.7 + 0.3*(\text{effort/capacity})^3)$ NOK/unit effort and are multiplied by effort.

7. In order to compare to a risk-free investment option, the interest rate is set to 4% per year.

8. CPUE is an increasing function of the fish stock; CPUE=(fish stock/2.5 million metric tons)$^{0.6}$.

9. Growth curve is given as $a*$'fish stock' + $b*$'fish stock'^{2}, where a=0.98 and b=-0.19.

10. Randomness is given as 'Fish stock'*'N growth'*(1-e$^{-\text{'Fish stock'}/3}$), where 'N growth' is a normally distributed random variable.

11. Assuming that randomness is updated only once a year means that autocorrelation is introduced. The stocks needed to model autocorrelation are not used to influence the policy. However, this is not a problem here, since the fish stock is only measured once a year and quotas are also set only once a year.

12. Measurement error is given as 'Fish stock' times a log-normally distributed randomness with mean 1.0.

References

Bertsekas, D. P. 1987. *Dynamic programming: Deterministic and stochastic models*. New Jersey: Prentice-Hall.

Bertsekas, D. P., and J. N. Tsitsiklis. 1996. *Neuro-dynamic programming*. Belmont, MA: Athena Scientific.

Brehmer, B. 1989. Feedback delays and control in complex dynamic systems. In *Computer based management of complex systems*, ed. P. Milling and E. Zahn. Berlin: Springer Verlag.

Brekke, K. A., and E. Moxnes. 2003. Do numerical simulation and optimization results improve management? Experimental evidence. *Journal of Economic Behavior & Organization* 50 (1): 117–131.

Coyle, R. G. 1999. Simulation by repeated optimisation. *Journal of the Operational Research Society* 50 (4): 429–438.

Dangerfield, B., and C. Roberts. 1996. An overview of strategy and tactics in system dynamics optimization. *Journal of the Operational Research Society* 47 (3): 405–423.

Forrester, J. W. 1961. *Industrial dynamics*. Cambridge: MIT Press.

Forrester, J. W. 1968. *Principles of systems*. Cambridge: Wright-Allen Press.

Forrester, J. W. 1971. Counterintuitive behavior of social systems. *Technology Review* 73: 52–68.

Forrester, J. W. 1994. System dynamics, systems thinking, and soft OR. *System Dynamics Review* 10 (2–3): 245–256.

Gaivoronski, Alexei. 1988. Stochastic quasigradient methods and their implementation. In *Numerical techniques for stochastic optimization*, ed. Y. Ermoliev and R. J.-B. Wets. Berlin: Springer-Verlag.

Graham, A. K., and C. A. Ariza. 2003. Dynamic, hard and strategic questions: Using optimization to answer a marketing resource allocation question. *System Dynamics Review* 19 (1): 27–46.

Keloharju, R. 1983. Relativity dynamics. In *Acta Academiae Oeconomicae Helsingiensis*, Series A. Helsinki School of Economics.

Koopmans, T. C. 1965. On the concept of optimal economic growth. *Academiae Scientiarum Scripta Varia* 28 (1): 225–287.

Morecroft, J. D. W. 1985. Rationality in the analysis of behavioral simulation models. *Management Science* 31 (7): 900–916.

Moxnes, E. 1998. Not only the tragedy of the commons, misperceptions of bioeconomics. *Management Science* 44 (9): 1234–1248.

Moxnes, E. 2003. Uncertain measurements of renewable resources: Approximations, harvest policies, and value of accuracy. *Journal of Environmental Economics and Management* 45 (1): 85–108.

Moxnes, E. 2005. Policy sensitivity analysis: Simple versus complex fishery models. *System Dynamics Review* 21 (2): 123–145.

Moxnes, E. 2012a. Individual transferable quotas versus auctioned seasonal quotas, an experimental investigation. *Marine Policy* 36: 339–349.

Moxnes, E. 2012b. *SOPS, stochastic optimization in policy space, user's manual version 1.0.* Bergen: University of Bergen and Powersim Software ASA.

Moxnes, E. 2014. Discounting, climate and sustainability. *Ecological Economics* 102: 158–166.

Moxnes, E., and L. C. Jensen. 2009. Drunker than intended; misperceptions and information treatments. *Drug and Alcohol Dependence* 105: 63–70.

Polyak, B. T. 1987. *Introduction to optimization*. New York: Optimization Software, Publications Division.

Rockafellar, R. T., and R. J. B. Wets. 1991. Scenarios and policy aggregation in optimization under uncertainty. *Mathematics of Operations Research* 16 (1): 119–147.

Schrank, W. E. 2003. Introducing fisheries subsidies. *FAO Fisheries Technical Paper*. No. 437, Rome: FAO. 52p.

Sterman, J. D. 1989a. Misperceptions of feedback in dynamic decision making. *Organizational Behavior and Human Decision Processes* 43 (3): 301–335.

Sterman, J. D. 1989b. Modeling managerial behavior: Misperceptions of feedback in a dynamic decision making experiment. *Management Science* 35 (3): 321–339.

Sterman, J. D. 2000. *Business dynamics: Systems thinking and modeling for a complex world.* Boston: Irwin/McGraw-Hill.

Tversky, A., and D. Kahneman. 1974. Judgment under uncertainty: Heuristics and biases. *Science* 185: 1124–1131.

Walters, C. J. 1986. *Adaptive management of renewable resources.* New York: Macmillan.

9 Addressing Dynamic Decision Problems Using Decision Analysis and Simulation

Nathaniel D. Osgood, Karen Yee, Wenyi An, and Winfried Grassmann

A wide range of real-world planning problems involve planning in the context both of substantial dynamic uncertainty—uncertainty sufficient to affect which decision is most desirable—and substantial dynamic complexity—complexity that prevents straightforward calculation of scenario outcomes. Decision analysis (Pratt, Raiffa, and Schlaifer 1995) is an established tool well suited to dynamic decision problems, but offers little to assist outcome assessment for dynamically complex scenarios. By contrast, while dynamic modeling is well suited to complex decision problems, policy selection using traditional dynamic models can be misguided when used in the context of uncertainty over time. Within such contexts, traditional fixed ("static") policy decisions—choices to be made regardless of evolution of uncertain factors—tend to fare poorly compared to *adaptive decisions*—decisions that learn from how uncertainties play out over time.

This chapter describes an integrated conceptual framework encompassing simulation and decision analysis that allows for addressing such dynamically complex decision problems, and describes an implementation of that framework for a popular system dynamics modeling package. The section below discusses motivations for the approach. The following section introduces decision trees using an example related to control of West Nile virus, while the subsequent section ("Realizing the Hybrid Framework") discusses the unified framework and the interfaces that support it. The section "Advantages of the Hybrid Approach" discusses the advantages conferred by this approach. The section "An Example Hybrid Framework Implementation" discusses one implementation of the framework, including how models created in the popular modeling package Vensim—including the example model discussed here—can be prepared to work with the decision tree and details on the use of the decision tree software.

The chapter concludes with a discussion of the interpretation of the hybrid decision tree software output and a summary and discussion of important avenues for future work.

Simulation Modeling

Dynamically complex problems are often associated with long delays and feedback loops between causes and effects, as well as nonlinearities that confound reasoning about the combined impact of policy choices whose results have only been studied in isolation—if at all. The dynamic complexity associated with such systems complicates evaluation of policy impact, and frequently means that decision making in such contexts further involves negotiating conflicting multiple goals and interests. Such problems are seen across a wide variety of areas. In an example from zoonotic infection control that will follow us throughout this chapter, the prevalence of a vector-borne disease such as West Nile virus at a given point in time depends upon complex contact dynamics between and within multiple reservoir and vector species as well as contact with humans. Such a system exhibits classic hallmarks of complex systems (Sterman 2000): nonlinearities associated with the transmission process; reinforcing feedbacks associated with viral transmission; balancing feedbacks associated with risk perception and draining the pool of susceptibles in all species; delays associated with organism development, viral maturation, intervention rollout, risk perception, and so on. The impact of interventions such as advisories, source reduction, vaccination, and larvaciding are all shaped in important ways by such factors.

Drawing on an example from a very different area—municipal planning—the well-being of a city is shaped by complex nonlinear interactions of factors that include—among many others—employment, residential, and nonresidential real-estate, city government finances, recreational spaces, and transportation infrastructure. Long delays are associated with the aging of a workforce, turnover in residential and commercial space, and with intervention outcomes associated with securing funding, project planning and construction, impacts of changing property taxes, and so on.

Simulation modeling is a useful methodology to promote greater insight into a system through a dynamic rather than static lens. Among other benefits, modeling can be used to understand the impact of different corrective actions and alternatives for change within a system through asking "what-if" questions. Simulation modeling can enhance the decision-making process by capturing the intricacy of the environment in a holistic fashion, with an improved understanding into the possible long-term and unintended effects of policies intervening within such systems (Sterman 2006).

Dynamic Decision Problems

When defining the scope of a simulation model, there is a need to determine which represented components will be "endogenous" (with their evolution simulated within the simulation model, and subject to changes due to other components of the model) and which remain "exogenous" (subject to either the assumption that they adhere to a fixed value, or where their evolution occurs in a predefined manner; Sterman 2000).

Exogenous factors commonly include many considerations that are not under the direct control over the system being modeled, but which influence it. For a municipal planning model, this might include aspects of weather such as rainfall or temperature, the price of oil, the rate of global economic growth, and so on. With respect to our primary example—control of West Nile virus infections in Canada's prairies—there are acute dependencies on external temperature. Higher temperatures can very significantly accelerate both mosquito reproduction and virus maturation within a mosquito's gut. In temperate climes—such as the Canadian prairies—elevated temperatures also tend to enhance human participation in outdoor recreation and lessen the likelihood of use of protective clothing. Similarly, municipal planning may be strongly affected by the relative growth of local and regional industry and resource revenues.

We can broadly distinguish between two types of dynamic modeling problems, according to the relative significance of exogenous factors for our policy planning:

- **Problems in which exogenous conditions are stable, reliable, or have small enough impact on the behavior of the system that we can plan our choices with confidence against an "expected" or "typical" course of events.** For such situations, a traditional simulation model—created, for example, in a system dynamics, agent-based modeling, or hybrid (Borschev 2013) framework—will aid us in grappling with the ramifications of our choices across the system, and we need only consider a single or a small number of scenarios for the exogenous factors that impinge upon our scenario.

- **Problems for which we are unable to accurately anticipate the evolution of exogenous factors, and where those factors are likely to have important impacts for the relative impact of interventions.** In this case, we have a stiffer challenge: not only do we need to consider the complex ramifications of our choices on system dynamics when considering one scenario involving exogenous factors, but we need to consider many possible future scenarios. Tradeoffs between our choices may be quite different depending on how those exogenous factors outside of our control evolve over time.

We term the second class of such problems *dynamically complex decision problems*; in this case, decisions are to be made over time in the context of unfolding uncertainty, with each sequence of uncertain events and choices leading to a set of consequences. These problems generally recommend a policy that will respond robustly and *incrementally* to a broad range of possible futures, rather than "putting all of our eggs in the same basket" using a single preselected strategy designed with reference to one or a small number of typical or expected futures.

Within dynamically complex decision problems, we are often torn between two principles, each exemplifying aphorisms:

- **Intervene early ("A stich in time saves nine").** Applications of this principle seek to head problems off by investing early—but face the risk that such investments will prove unnecessary. To draw on our example, an early application of interventions such as source reduction or larvaciding may head off human cases of West Nile disease if the weather turns unusually hot, but could squander precious resources if typical or cooler temperature conditions prevail. For municipal planning, similar early investments may be made with respect to infrastructure such as transit systems, roadways, and water treatment facilities.
- **Focus on learning ("Wait and see").** Observing this principle, we seek to learn which investments to make based on ongoing observations of evolving external conditions. While this can allow for tailoring our policy to the specifics of the conditions that present themselves, we run the risk that our commitment to a certain form of action will come too late to be effective. For our example, we may seek to observe temperature patterns over time, and invest only when and if the weather grows sufficiently hot or when elevated mosquito populations are observed. But by the time we roll out our intervention, large numbers of West Nile cases may already have appeared, including very serious and costly

neurological cases. For municipal planning, waiting too late could mean a significant strain on infrastructure, such as congested roadways or overcrowded schools.

Phrased in the language of decision analysis, we seek an optimal *decision rule* (Pratt, Raiffa and Schlaifer 1995) that will balance the tradeoffs associated with the principles above. By definition, a decision rule should specify the decisions to be made over time in response to any given possible observed sequence of (uncertain) events. Within dynamically complex decision problems, the net benefit conferred by a specific decision at a certain point in time may depend strongly on the preceding events. For example, the desirability of a given intervention (e.g., wait and see, source reduction, larvaciding, adulticiding, issuing an advisory, or mass vaccination) to address West Nile virus might depend on "learning" by observation regarding the mosquito population, temperature, and rainfall (all of which might impact future growth of that population), and factors such as the observed prevalence of West Nile virus among mosquitoes and birds until that point (Yee, Osgood, and An 2011). Conversely, the relative desirability of making a given decision at one point in time will also reflect the occurrence of future decision points to which one might defer an intervention, or which might be used to "correct course" with respect to that intervention.

Reflecting this dependence of intervention benefit and cross intervention tradeoffs on uncertainties and knowledge of future decision points, it is rare for the most competitive or effective policy to be a *static* policy—a "stay the course" plan that hews to particular predefined sequence of decisions (actions) regardless of eventualities. We can instead gain advantages by employing *adaptive* policies, whereby contingent plans are established that explicitly adjust the decisions and actions to undertake at a particular point in time based on how uncertainties have unfolded until that point, and the anticipation of future decision points. Because an adaptive policy offers the flexibility to respond in a different fashion by exercising different policies to different observed uncertainties, it will have a much higher likelihood of being robust in the context of different possible eventualities. For example, in some jurisdictions, public health authorities will meet weekly to reflect on the evidence to date regarding West Nile virus risk and to decide whether to take steps toward interventions. In such situations, the quality of planning on the part of authorities could be substantially enhanced by models allowing authorities to understand the tradeoffs between possible decisions in light of the observed environmental patterns and the implications of an intervention now for the impact of interventions and vulnerabilities in the future.

The Computational Challenge of Evaluating Adaptive Decision Rules

The Intractability of Traditional Policy-by-Policy Evaluation of Adaptive Decision Rules
While adaptive policies are attractive from the standpoint of robustness under divergent eventualities, the process of identifying such policies can pose daunting challenges when undertaken in a traditional fashion of policy-by-policy enumeration and exploration. Because the decisions to be undertaken at a given point in time will in general depend on the observed events, the number of candidate policies (i.e., the count of possible *decision rules*) that must be explicitly considered is tremendously large.

Consider, for example, the incremental planning associated with West Nile virus decision making. Given the pronounced impact of temperature on rates of viral maturation, mosquito maturation, and egg-laying, as well as on human outdoor activity and risk behavior, it is natural to consider a policy in which the actions undertaken will depend on the observed temperatures. Consider a policy that adapts one of $d=6$ possible decisions ({no action, source reduction, larvaciding, adulticiding, issuing advisories, vaccination}) on a month-by-month basis for each of $n=3$ months, based on one of $c=3$ unknown ("chance") temperature outcomes for that month ({mean temperatures $< 22.5°C$, $\geq 22.5°C$ but $<30°C$, or $\geq 30°C$}). A straightforward, classic approach to the evaluation of possible decision rules would exhaustively enumerate each rule, evaluate its expected (possibly risk-weighted) outcome in light of possible future eventualities, and select the decision rule offering the most desirable outcome. Unfortunately, while conceptually simple for even modest-sized trees, this naïve approach is computationally intractable. For example, the count of possible policies requiring explicit evaluation in a tree with simple, regular structure evaluates to approximately 10^{31}, as given by the formula $d^{\left(\frac{c^{n+1}-1}{c-1}\right)}$ (proof is provided in Osgood 2005). It is notable that the count of policies to be evaluated exhibits a double exponential. This reflects a double combinatorial explosion resulting from trying to characterize the set of all possible responses to all possible ways that an uncertain situation could play out. As a result of this combinatorial explosion, identifying the optimal policy via traditional, systematic policy-by-policy evaluation is prohibitively expensive for all but the most trivial trees.

The Common Substructure of Decision Rules   Fortunately, one of the reasons for the combinatorial explosion of the number of possible adaptive decision rules with the rise of the count of decision points and uncertainties—the high degree of common substructure exhibited by the possible decision rules—is simultaneously the key to addressing it.

Employing the tools and algorithms of decision analysis permits us to overcome such challenges, while at the same time realizing many additional advantages. Specifically, using decision trees permits us to characterize structured policy space, encapsulate most decision-influencing uncertainty, take advantage of sophisticated statistical tools, perform multistrategy sensitivity analyses, and—most critically here—rapidly identify robust strategies for a tremendously large set of possible decision rules using the process of backward induction. In essence, performing a simple linear analysis across a single decision tree will allow us to simultaneously evaluate the vast number of possible decision rules represented by that decision tree. The ability to perform such analysis in a much faster fashion reflects the tremendous amount of common substructure shared across distinct decision rules—for example, the fact that many adaptive policies will share the same response to situations with consistently low temperatures early in the summer. To better understand this substructure and how we can take advantage of it to efficiently analyze the set of all possible policies, we now provide a brief introduction to decision trees, referring interested readers to the more in-depth coverage available in textbooks devoted to the subject (e.g., Pratt, Raiffa and Schlaifer 1995).

Introduction to Decision Trees

Tree Structure
The decision tree is a central decision analytic tool that systematically organizes and diagrammatically depicts decisions and uncertainties over time, specifying the associated consequences of each possible sequence of decisions and events over time. Figure 9.1 shows an example of a decision tree. Decision and event nodes are depicted in a left-to-right sequence that mirrors their chronological ordering. In our example, decisions are made chronologically and in a uniform way over the course of time, reflecting the occurrence of regular

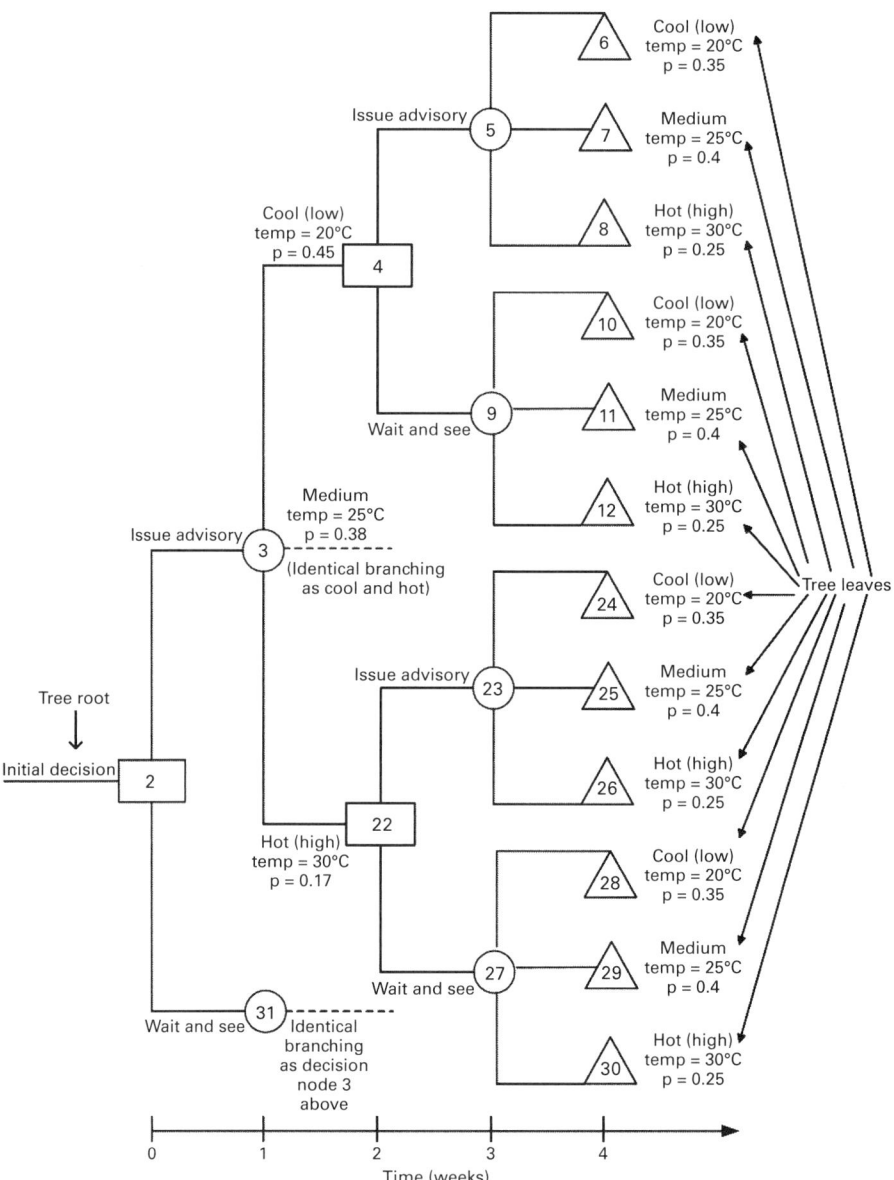

Figure 9.1
Decision tree showing two possible interventions at decision nodes (squares).

policy-setting meetings among decision makers. While not required by the decision tree formalism, this regular structure is a good match for our needs here. We explain the elements of the decision tree and its interpretation next.

Decision Nodes   A decision point in the decision tree is visually represented with a small, square box termed a decision node; possible decisions associated with that decision node are represented with one or more edges connecting that decision node to its children in the tree. The decisions associated with a given decision node are assumed to be mutually exclusive (with only one such decision being made at a time) and collectively exhaustive (thus, one such decision will always be made—even if the decision is simply a passive "wait and see" decision to defer intervention). Within the tree in figure 9.1, possible decisions include "wait and see" or "issue advisory" (a public health advisory that warns individuals to comply with personal protective practices).

Event Nodes   The occurrence of an uncertain event in the decision tree is visually represented by a small circle in the tree called an event (or "chance") node. As with the decision node, the event is treated as being realized at a point in time, and possible outcomes of an event are represented by edges emanating from the right side of the event node. The set of possible outcomes also represents a mutually exclusive and collectively exhaustive set of possibilities. In contrast, however, the decision maker has no choice as to the outcome of an event node. Instead, that outcome is dependent on external factors beyond their control (e.g., temperature, rainfall, level of economic growth, oil price, interest rates, etc.). Each outcome for a given event node is associated with a given probability of occurrence conditional on that node being reached. Hence, the sum of the probabilities for all the outcome branches must equal one. Within figure 9.1, three possible temperature ranges (temperatures closest to 20, 25, and 30°C) are associated with each event, leading to specific consequences or outcomes (triangles). In our stylized example developed below, the probability for each of the two possible outcomes of an event node varies by two-week period. In other cases, the likelihood of each event might remain invariant over time, with the spacing between such events being associated with finer or coarser time periods.

Terminal Nodes and Scenarios   The "leaves" at the first right of the tree are termed terminal nodes, and are visually denoted by triangles. These nodes have no "children" within the tree. Each terminal node of the tree is associated with a numerical value representing the outcome (or "consequence") of a given scenario in terms of a metric specific to the tree. For the purposes of this chapter, we use the term "scenario" to describe a specific sequence of decisions and events over time. Such a scenario represents a unique path from the "root" of the tree (with the initial node with the earliest occurrence time) to a specific consequence.

A Simple Tree

To aid the reader in understanding the relationship between policies (decision rules), the decision tree, and the associated combinatorial challenges, we consider here a stylized and reduced-scale version of the West Nile virus planning problem. Specifically, while the structures used can support an arbitrarily large number of alternative decisions and events, although with greater computational cost, we consider the simple case of dichotomous decisions (the choice of either "wait and see" or "issuing an advisory"), and just two temperature ranges— temperatures closer to 20°C ("low") or to 30°C ("high"). While later sections will examine the operation of a far deeper hybrid tree, for our tree explored below (figure 9.2), we further limit ourselves to considering just one month of time, with decisions occurring every two weeks (starting weeks 0 and 2), and two temperature ranges at each event node (starting weeks 1 and 3), and with consequences (outcomes) at each terminal node (at week 4).

Associated Decision Rules   Recall from above that a decision rule describes the sequence of decisions that will be made in response to any possible sequence of events, and the number of decision rules will in general rise rapidly with the depth of the tree. Table 9.1 lists the set of possible decision rules for the stylized two-week West Nile virus decision-making problem. Because the decision at the beginning of week 0 is not preceded by any events, it is not conditional on any event occurrence. By contrast, the decision at the beginning of week 2 may be conditional on the temperature observed during week 1. It can be observed that there are quite a number of decision rules despite the very simple structure of the decision tree, which includes just two decisions and two events along any one scenario.

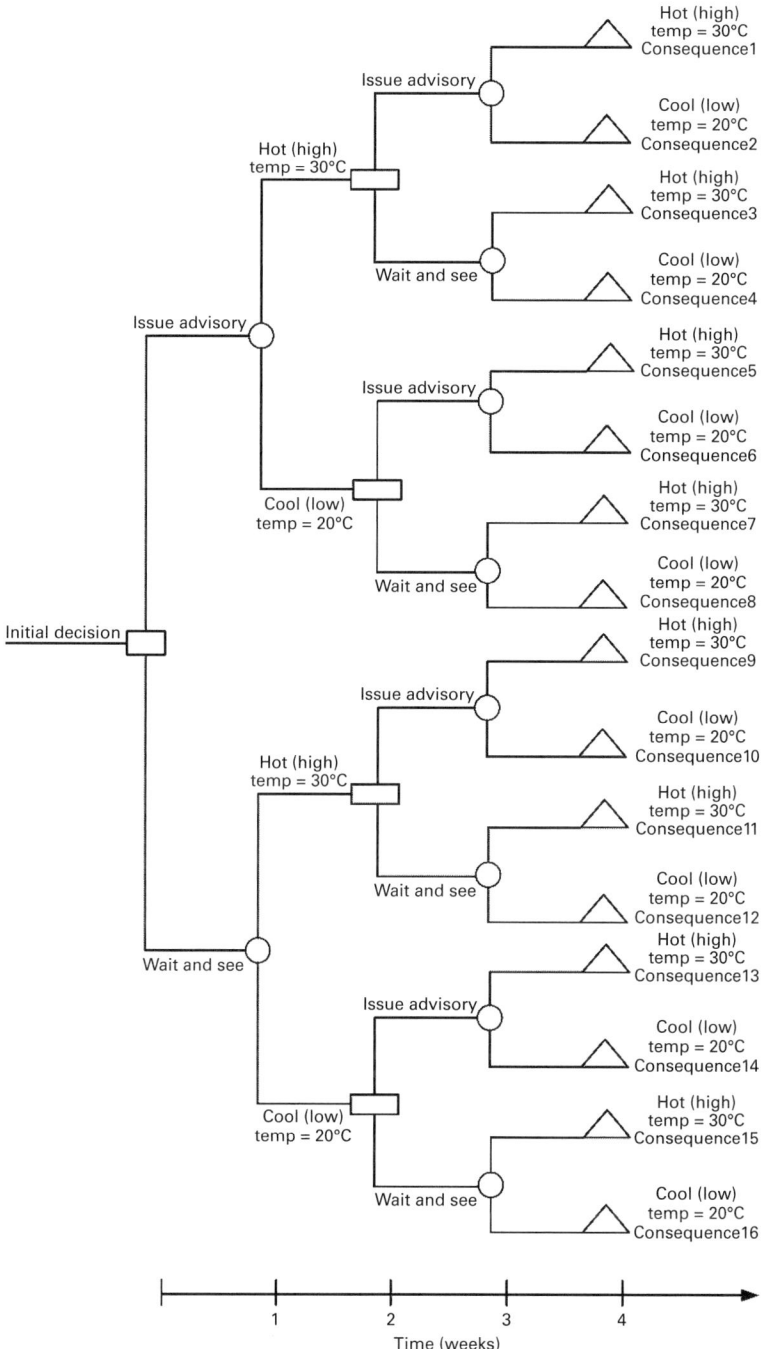

Figure 9.2
Schematic structure of an example five-layer decision tree.

Table 9.1
Enumeration of possible decision rules for the example shown in figure 9.2

	Decision Rule (Expressed Informally)		
Rule ID	Decision at the Beginning of Week 1	Decision at the Beginning of Week 2	Associated Terminal Nodes
1	Wait and see	Wait and see	11,12,15,16
2	Wait and see	If temperature in week 1 was low, issue advisory, otherwise temperature is high and wait and see	11,12,13,14
3	Wait and see	If temperature in week 1 was high, issue advisory, otherwise temperature is low and wait and see	9,10,15,16
4	Wait and see	Issue an advisory	9,10,13,14
5	Issue advisory	Wait and see	3,4,7,8
6	Issue advisory	If temperature in week 1 was low, issue advisory, otherwise temperature is high and wait and see	3,4,5,6
7	Issue advisory	If temperature in week 1 was high, issue advisory, otherwise temperature is low and wait and see	1,2,7,8
8	Issue advisory	Issue advisory	1,2,5,6

For this example, we must choose between eight decision rules, reflecting the fact that each decision rule must choose at two points in time (the beginning of weeks 0 and 2), with each such decision point involving a choice between two possible choices ("wait and see" versus "issue advisory"), and the fact that the second decision can be made in response to two possible eventualities ("Low temperature observed during week 1" versus "High temperature observed during week 3").

Relationship Between Decision Rules and Terminal Nodes   Understanding the relationship between decision rules and terminal nodes is central to understanding the tradeoffs and analysis presented in this paper. In all but the most degenerate trees, a particular decision rule will be associated with many terminal nodes. Each of these terminal nodes represents a scenario that could occur if the decision rule were in place. The path to a given terminal node captures both the sequence of events and the decision rule's particular responses to those events that define the scenario. For example, under decision rule 3, different patterns of events could lead to scenarios associated with any of the terminal nodes labeled 9, 10, 15, and 16 (figure 9.3). The fact that there are four such terminal nodes in this simple example reflects the fact that there are two possible events (one in week 1, one in week 3), each with two possible outcomes. To thoroughly evaluate the effectiveness of a decision rule—and to help ensure that it is robust under a wide range of eventualities—it is important to calculate the consequences of the various scenarios with which it can be associated.

Each decision rule will typically be associated with several terminal nodes, and the relationship is many-to-many: a particular terminal node t within a decision tree will also, in general, be associated with several decision rules. The decision rules that include terminal node t happen to be identical in their response to the particular sequence of events represented on the path to terminal node t, but may differ dramatically (and *must* differ at least in some regards) with respect to the decisions they would make for other sequences of events. For example, terminal node 15 in figure 9.2 could be reached in either decision rule 1 or decision rule 3. This reflects the fact that both decision rule 1 and decision rule 3 forego any intervention in response to a scenario in which low early temperature is observed. Evaluating the single scenario associated with the terminal node helps contribute to evaluating (i.e., tells us something about) the behavior of *all* of the associated decision rules.

Conclusion   This section introduced a small example that illustrates the many-to-many relationship between terminal nodes and decision rules. Because decision rules are distinct if they differ in response to *any* possible eventuality, the number of possible decision rules rises super-exponentially with the size of the tree—exponentially faster than the number of terminal nodes or the size of the tree. However, the fact that a given terminal node is associated with multiple decision rules— and that evaluating the outcome of that terminal node can contribute to evaluation of multiple decision rules—suggests that significant economies can be secured in evaluating decision rules. We turn next to a procedure that exploits such economies to secure tremendous performance advantages over the naïve approach mentioned above.

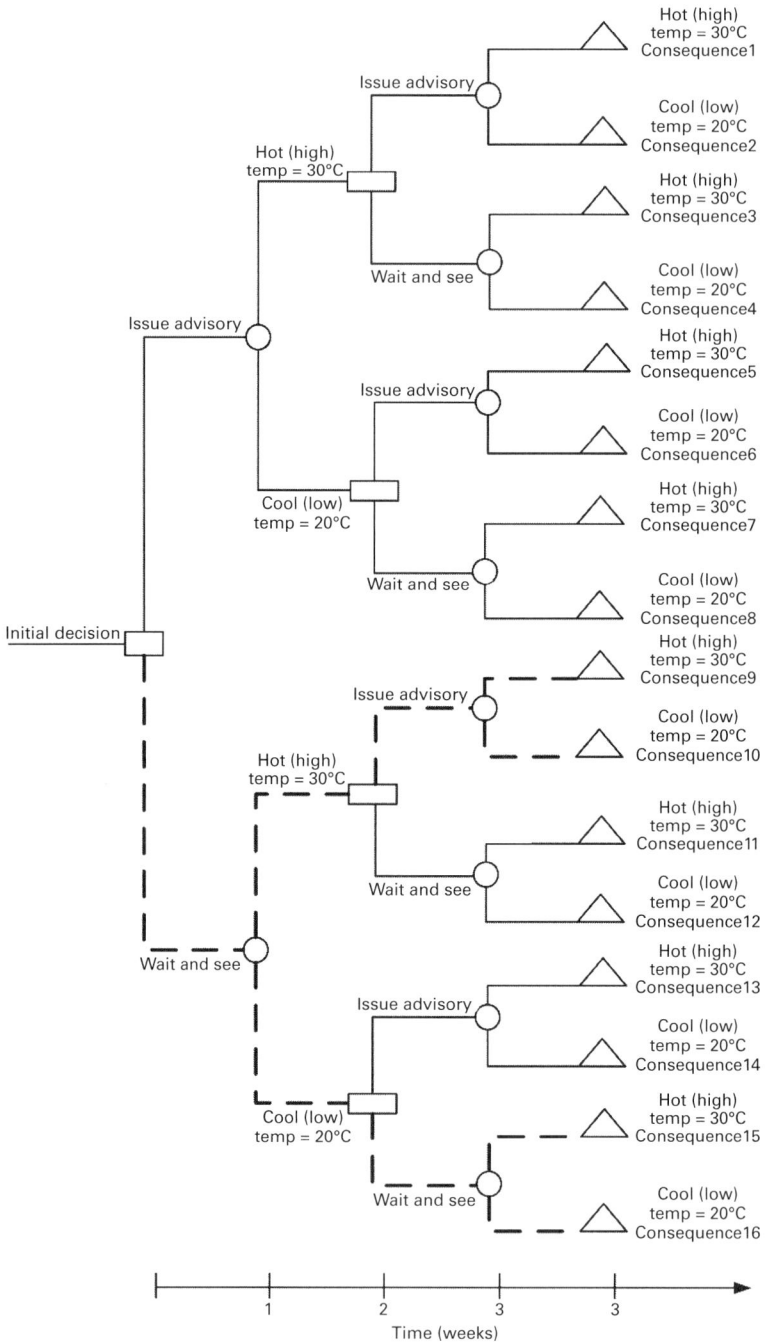

Figure 9.3
Representation of a specific decision rule in the decision tree.

Backward Induction

The process of identifying a decision strategy that is optimal given the assumptions of the models involves making use of a procedure called backward induction. This procedure is a variant of dynamic programming that avoids explicit enumeration of decision rules for evaluation on a case-by-case basis. Instead, it involves taking advantage of the characteristics noted in the section "The Common Substructure of Decision Rules"—whereby evaluation of a single node of the tree contributes to evaluation of multiple decision rules—through use of a single evaluation process across the tree that allows for identification of the preferred decision rule. This process involves recursive computation of values for each node of the tree, starting from the leaves and proceeding "backwards" (in terms of the chronological placement of the nodes involved) to the root of the tree. The rollback procedure employed for a given node is dependent on the type of node:

- **Terminal node.** The value for a given terminal node is simply the value of the consequence of the scenario associated with that terminal node, at the moment in time with which that terminal node is associated within the tree.
- **Event node.** Each event node is assigned a numerical value given by the *expected value* of outcome branches arising from it (i.e., the sum of the values associated with each child, weighted by the probability of that occurrence of that child).
- **Decision node.** Because the choice of decision at a given decision point (node) is assumed to be under the control of those using the decision tree, the selection of the decision can be made such that it chooses whichever decision is maximally beneficial. As a result, the value assigned to a given decision node is the best value associated with each of its children. Depending on the outcome metric (whether a larger value is favorable or unfavorable), this will be either the highest or lowest numerical value from among its possible decisions (child nodes).

The decision tree provides a high-level construct by which a manager can diagrammatically depict a sequence of decisions and uncertain events as they unfold over time, gain insight into how these interact with one another, and determine the optimal decision strategy (i.e., secure advice on what choice to make given each possible eventuality, in light of the model structure specified).

The result of this procedure—whose computational effort scales linearly with the number of nodes in the decision tree—is a situation where each and every decision node is associated with a "recommended" outcome (i.e., one that is optimal of the chosen outcome metric). Given that a decision rule is a specification of "what to do" (i.e., what decision to make) for every possible eventuality, and that choice points for decisions are precisely the decision nodes, this specification of the recommended outcome of each decision node can be recognized as implicitly specifying a decision rule that is optimal for the assumptions involved.

Summary
This section examined the combinatorial challenges associated with evaluating dynamically complex decision problems and suggested a practical route for such problems to dramatically lessen the overhead associated with policy selection. This approach achieves economy of computational effort using two approaches. First, it provides an efficient way of capturing the tremendously large amount of overlap between the exponentially large set of distinct decision rules by representing such decision rules using a decision tree. Secondly, rather than following the traditional route of enumerating and evaluating each such decision rule in turn to identify the one that achieves the most favorable result, dynamic programming is used in the form of the "backwards induction" to efficiently evaluate the set of such decision rules within a single "rollback" of the tree, which occurs from the terminal nodes back to the root of the tree.

Having identified this conceptual framework for evaluation of dynamically complex decision problems, it remains to be seen how simulation results for the appropriate scenarios can be realistically associated with the decision tree terminal nodes so that the rollback can occur. Next, we describe the use of the software that realizes such an integration.

Realizing the Hybrid Framework

This section discusses the implementation of a hybrid framework for analyzing dynamically complex decision problems by allowing simulation results to be associated with the terminal nodes in a decision tree. Specifically, the value of a given terminal node is obtained by start-to-finish simulation of the scenario associated with that node. Given an association of a value with each such terminal node, an optimal decision rule can then be identified using the rollback procedure.

We begin by discussing the conceptual framework, and then proceed to discuss an implementation of that framework for the Vensim system dynamics modeling platform.

Division of Responsibilities

By systematically examining large numbers of scenarios, the hybrid framework can identify those decisions that appear to give the best results within a given context (i.e., given the events that have occurred and the decisions that have been made in the past, or that can be made in the future). Within this context, the simulation model is responsible for simulating the consequences of a given scenario, while the decision tree is responsible for representing the set of possible decisions and events over time. The generic division of responsibility of the hybrid framework is shown in table 9.2.

The hybrid modeling technique involves performing simulation modeling and sensitivity analyses in the context of a decision tree that encapsulates decision choices, uncertain events, and outcomes. Each of these elements has corresponding components in both the decision tree and in the simulation model, but the direction of influence between the simulation model and the decision tree differs across such elements. The occurrence of a particular event or decision at a specific moment of time within the decision tree is associated with (and is communicated to the simulation model by) imposing a specific value for a parameter in the simulation model at the same point in time. Different possible outcomes of that event or different decisions will be signaled by the decision tree *to* the simulation model by setting that parameter

Table 9.2
Generic division of responsibilities for hybrid approach

Simulation Model	Decision Tree
• Calculates dynamic consequences of a sequence over time of	• Represents possible sequences of
o Events	o Uncertainties (using event nodes)
o Choices	o Decisions (using decision nodes)
• Performs deterministic simulation given events and decisions	• Expresses consequences (outcomes in terminal nodes)
	• Encapsulates "policy space"

within the running simulation to hold different specific values at that point of time. As is the case for decisions and events, the value of a terminal node within the simulation model corresponds to the value of a variable within the simulation model at a particular point in time. However, in contrast, the value of the terminal variable in the decision tree obtains its value *from* the simulation model rather than vice versa. Specifically, the value of a particular outcome is derived from running the simulation model on the corresponding scenario and reading the resulting value of that outcome variable (e.g., "accumulated discounted cost," "West Nile symptomatic cases") from the model at the appropriate point in time. This value summarizes the outcome of the scenario associated with the specific sequence of decisions and events over time that appear on the path from the root of the tree to that terminal node.

Within the approach presented here, the simulation model is systematically run to evaluate the consequences associated with each terminal node—a number of nodes that can be very large, but is typically incomparably smaller than the number of distinct decision rules. To do this for a specific terminal node, the simulation model must simulate the specific sequence of decisions and events over time associated with that terminal node (i.e., on the path from the root to that terminal node) and read the resulting outcome value from the simulation model. Following the evaluation of each such terminal node, the preferred decision rule can then be evaluated using backward induction through dynamic programming to summarize the experimental results.

This bottom-up approach involves the decision tree not merely as a conceptual device, but also as a computational construct. This approach features a natural division of labor: The decision tree explicitly represents event likelihoods and the policy space and can be used to identify the most desirable decisions (given model assumptions) at each decision point within the tree, while the simulation model serves specifically to evaluate the consequences of specific scenarios.

The hybrid framework is most general when employed for continuous-time simulation models, but can be used for discrete-time models as long as there is alignment between the discrete time steps used in the simulation model and those used in the decision tree.

Most of the description in this chapter assumes a deterministic model. In our closing remarks, we briefly discuss the prospect of extending this framework to handle stochastic models.

Details of the Interface Between the Decision Tree and the Simulation Model

General Character of the Interface In order to run the simulation model in the context of different scenarios in the decision tree, the hybrid modeling software controlling both the decision tree and simulation model needs to be able to communicate between the decision tree and the simulation model. This communication needs to go in each of two directions and takes place through an interface defined by variables that are located within the simulation model—variables whose identity and semantics are recognized by both sides. What is communicated in each of these directions, and the specific mechanisms by which that communication is realized within the decision tree, are described below and depicted in figure 9.4.

- The occurrence of decisions and events need to be communicated *from the decision tree to the simulation model.* In virtually all cases, the occurrence of a given choice (e.g., "source reduction," "vaccination," "wait and see"), or an occurrence of an event (e.g., a "hot week," a "cool week") at a point in time must have some specific impact in the simulation—that is, it will result in specific dynamics. As a result, the simulation model needs to "know" when a given decision or event has occurred in the course of simulating a specific scenario (e.g., a hot summer with no interventions undertaken) so that it can take that into account by simulating the system-wide consequences of those decisions and events. The communication of this information concerning decisions and events to the simulation model is performed by setting variables in the simulation model. Specifically, occurrence of an event at a time point t is communicated to the simulation model by setting, at time t during the simulation, a specific simulation model variable (a variable whose job it is to encode events) to a unique value encoding that class of event. Occurrence of a decision at a point t is correspondingly realized by

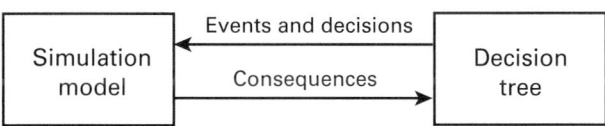

Figure 9.4
Bidirectional communication between simulation model and decision tree.

setting at time *t* during the simulation a distinct specific variable (whose job it is to encode decisions) to a unique value encoding that class of decision (e.g., "hot week").

- The outcome of each scenario is calculated by the simulation model. The results of the scenario are read from a particular designated variable in the simulation model (again a common point "known by" the decision tree) and placed into the decision tree.

- While the model begins at one time, the decision tree itself can start at the same or a later time. If the decision tree starts later than the model, the baseline scenario within the simulation model (i.e., without imposed changes to simulation variables) will play out until the decision tree starts; decisions and events then start to deviate from the baseline starting from the decision tree start time onwards.

Interface Variables   Variables within the simulation model serve as a conduit for communication between the decision tree and the simulation model. Such variables are connected to the remainder of the simulation model in a manner appropriate to their function.

Within the example model (included in the appendix), each variable occurs to the left of a prominent purple block of text that points to and helps visually highlight the interface variable.

We comment briefly on each of these variables.

- **Decision variable.** This variable is set to any floating point number at a certain point in time to indicate the decision being taken at that time. In our example model, the decision variable is entitled "InterventionSelected" and is used to communicate the desired intervention to put into place. While this variable accepts a continuous number, currently there are only a discrete number of choices that can be made within the decision tree. This variable retains its value over time during the simulation, until it is again set by the decision tree (typically via another decision node later in the path to the terminal node). There is, however, a variable ("Last Decision Tree Decision Time") that is used by the decision tree to communicate the (simulation) time of the last decision made; the contents of this variable may be compared to the current time in order to decide whether an intervention of limited duration should remain in place within that interval, or to capture the impact of that intervention over time.

- **Event variable.** An event variable is set to indicate the occurrence of an event at a certain point in time. For the example model, the event variable is entitled *"TemperatureInCentigrade from Decision Tree"* and communicates the range of temperature that occurs (a temperature that falls closer to 20°C, 25°C, or 30°C.) This variable retains its value until it is reset, and admits to representing only discrete choices. There is additionally a variable ("Last Decision Tree Event Time") that can be used by the decision tree to communicate the time of occurrence of an event to the simulation model, and which can be used to impose events whose impact varies as a function of the time that has been in effect.
- **Outcome variable for consequences of the decision tree.** The final variable to be discussed is used to communicate the outcome of scenario simulations to the decision tree—for our example, factors such as accumulated cost, and cumulative severe neurological West Nile virus cases. This is in contrast to the two variables above, which serve to communicate from the decision tree to the simulation model. The hybrid modeling software will associate outcome value x from a particular scenario with the appropriate terminal node. While the hybrid modeling software supports the association of terminal nodes with any simulation model variable, it is important that it be set to some value that dictates relative "goodness" of a given scenario outcome. This is important because the "rollback" algorithm used with the decision tree to identify the best decision rule (specifying the decision to make at any possible decision point that one could reach) will prefer decisions that have a higher value of this variable. For simplicity, in our example model, we consider only the negative of *cost* as an outcome, where both health care and intervention costs are considered. In more realistic models, the outcome variable might be defined as a weighted combination of several distinct outcomes.

Within the simulation model, the variables set by the hybrid software as part of the protocol defined here—the decision and event variables—are commonly used to further set other variables. In our example, the decision variables within the interface between the simulation model and decision tree are used to represent interventions to be put into place. The simulation model contains additional logic to realize the implementation dynamics and effects of these interventions. For example, an *advisory* intervention might take a certain amount of time to come into effect and will affect certain components of human behavior. The model logic located downstream of the hybrid interface decision variable would implement such effects. By contrast, a *larvacide* intervention would have its immediate effects localized to the region of the mosquito lifecycle components related to mosquito larvae and pupae.

Advantages of the Hybrid Approach

There are several advantages of the hybrid approach for dynamically complex decision problems. First, it is a framework geared toward addressing dynamically complex decision problems by facilitating robust planning in the midst of an ongoing process of observation and decision making that captures uncertainties as time progresses, simulates a broad range of future possibilities rather than just a single scenario, and allows for staging of decisions over different time points—including decisions to "wait and see" that exploit the flexibility associated with recognition of opportunities for making and revisiting decisions at decision nodes later in the tree. While we have applied the approach here to contexts where the primary uncertainties represented are meteorological in character, it can be used for a wide range of other planning challenges that combine decision making over time with dynamic complexity—for example, in pandemic situations. This hybrid modeling technique can offer a significant and general approach that can be incorporated into decision making to allow for robust, adaptive response planning. This allows both for more reliable planning, and nimbler response to emergent eventualities.

Second, the hybrid framework provides the capacity to depict in a systematic and structured fashion both the policy space and the set of uncertainties of concern. With the appropriate decision tree software, one could view "risk profiles" (patterns of possible outcome) at multiple levels of the tree for particular consequence nodes, at specific decision and event points, and for given decision rules. Third, the hybrid approach permits observing the impact of event likelihood changes on risk profiles or on the decisions that are deemed most desirable. Finally, the approach most notably offers far higher performance than a straightforward "top down" approach to evaluating decision rules because it obviates the need to enumerate an intractably large set of different decision rules and reduces the needless re-evaluation of particular scenarios. As a result, the backward induction procedure reduces the expense of resources in identifying preferred adaptive strategies.

An Example Hybrid Framework Implementation

This section discusses representing our example model in one implementation of the hybrid framework, and illustrates its use on a sample model. The software framework presented here operates with system dynamics models implemented in the popular system dynamics package Vensim (Eberlein and Peterson 1994); a manual for the hybrid software is included in the supplemental material for the chapter.

Preparing the Simulation Model for Work with the Decision Tree

The section "Realizing the Hybrid Framework" discussed some of the preparations required for the simulation model variables (for decisions, events, and outcomes) to interface with the decision tree. In addition to providing this interface, the simulation model as a whole must be prepared for work with the decision tree software. This process involves four steps, mentioned here only in brief (users interested in details should consult the manual in the supplemental material):

- setting the "SAVEPER" of the model
- reforming and cleaning the model
- establishing decision and event variables as gaming variables
- publishing the model.

Use of the Decision Tree Software

Starting the software, the user sees an entry dialog box. Pressing the "Browse" button to select a .vpm file on the hard drive, the user can then select the appropriate .vpm file representing the dynamic model, and press "Open." The user will subsequently return to the initial dialog box.

Creating a New Tree   The second choice on the welcome dialog box is to create a new tree, which is sought for our example. Pressing the "Create New Tree" button opens an interface (figure 9.5) used to specify the characteristics of trees with regular, uniform structure that consists of alternating layers of decision nodes and event nodes that are followed by the terminal nodes at the far reaches of the tree (the latest points in time). The user can identify the number and character of tree layers desired (top-left panel); the decisions being made (lower-left panel); the outcomes being considered (top-right panel); and the uncertain events playing out over the course of the summer (bottom-right panel). The times given are specified using the time unit used for the simulation model (e.g., days). Trees created by this interface are constrained to have decision nodes sharing the same structure with identical choices; event nodes must have similar structure with each other; and all terminal nodes must be located at an identical depth. These are restrictions of the software *creating* the tree, but not of either the software processing the tree, or of the hybrid approach itself. More general trees can be readily handled by the approach discussed, whereby the user manually describes them using XML and then subsequently runs the tree described by XML with the software.

As described in greater detail in the manual, there are four major regions of this dialog box that specify different aspects of the tree.

In our example depicted in figure 9.5, the interface is used to create a variant of the example tree introduced above. Specifically, we consider here a three-month time horizon where decisions regarding West Nile virus interventions must be made every two weeks (14 days) on

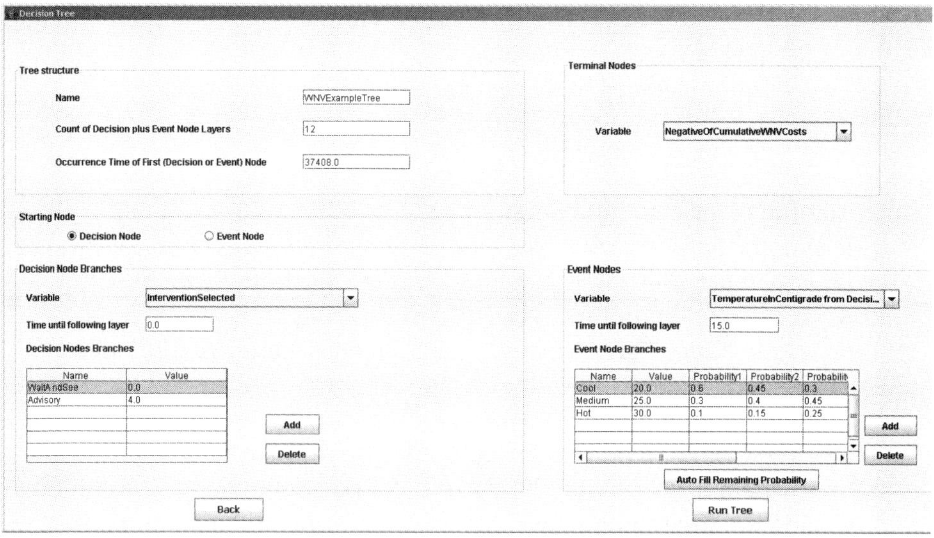

Figure 9.5
Dialog box for creating a new tree.

the basis of temperatures observed until that end time point. For every two-week interval, we consider three possible outcomes for temperature (20°C, 25°C, 30°C; compared to earlier, where only two were depicted), and the dichotomous policy choices ("wait and see" vs. "issue advisory").

As indicated by the tree depth (12), the tree has six alternating levels of decisions and events over time, with one for each two-week interval. Because the probabilities associated with the event outcomes are assumed to be identical at a given point in time (i.e., a given point in the summer[1]), this leads to six levels of probabilities for each event branch. As can be seen in the set of radio buttons labeled "Starting Node," the root of the tree is a decision node (figure 9.5). We wish to treat the decisions as being made every two weeks, with the uncertainty regarding temperature playing out over the ensuing two-week period. To allow the decision to be followed immediately by the evolution of an uncertain temperature regime, we have assigned the time between the decision and event nodes to be negligible (0 days). To allow the decisions to be made every two weeks, the time between an event node and the subsequent decision node (i.e., the time over which that temperature regime plays out and imposes its effects) is the full two-week period. This is a pattern seen in many—but by no means all—decision trees; in other cases, there may be distinct lag between when a decision is made and when the ensuing uncertain events play out.

Evaluating the Decision Trees   Once we have provided the specifications of the tree in the interface, pressing the "Run Tree" button will initiate simulation of the tree.

During tree evaluation, the model is evaluating all the scenarios in the tree, and will then perform backward induction ("Rollback"). The size of the tree will determine the amount of time it takes to execute, and whether the user sees ephemeral output from Vensim (the work in progress dialog box) appearing while the scenarios are being evaluated. Vensim will produce temporary results stored as Vensim data files. If there are existing Vensim data files with that same name, the user may be prompted as to whether they wish to overwrite these files. While such files can be useful as forensic devices (to better understand the variation in model output seen across scenarios), these files can be safely deleted after the simulation is finished, or after later simulations within the same tree have started. If any errors (e.g., numeric errors) appear within the Vensim simulations, the user may also receive a pop-up window from Vensim reporting this fact; in this event, it is strongly recommend that the user investigate the cause underlying such errors.

After a period of time where Vensim is running the simulation, we are presented with a dialog box reporting the value deduced by backward induction for the root node of the entire tree, and allowing the user to specify the name of output files for model results. The most important of these output files is the "*.xml*" file, which can optionally be reused at a later time.

Interpreting Decision Tree Software Output

There are two ways of exploring the rollback information in decision trees created either manually or built using the software described above. First, the hybrid tree visualization interface (figure 9.6) can be used to graphically explore and interact with decision trees created either using the software described above or manually. This approach allows exploring both the structure of the tree (visible on the left panel of figure 9.6) and behavior over time graphs for particular terminal nodes (visible on the upper-left panel of figure 9.6). Second, at the cost of wading through much additional information on the tree itself, and using a less tailored interface, we can explore—and even change—that tree information using general-purpose tools for working with the format used to encode the tree, XML. Both of these approaches to browsing a tree are discussed in the manual included in the supplemental material for this chapter.

Browsing the tree can give important pieces of information. Perhaps the most important is the decision recommended for each decision node by the rollback procedure. Of all the nodes in the decision tree for which this can be assessed, the most important area is often the earliest decision nodes, as this is likely to shape near-term decision making. For trees that start with a decision node, of particular note is the starting point ("root") of the main decision tree. In the example tree described in the previous section, the root decision node has a "best decision" of "WaitAndSee." This is associated with a rollback value (indicating an expected value of outcomes if this choice is selected) of approximately –\$3.6M cumulative costs over the period of simulation (including both intervention and health care costs). Additional browsing in the tree would reveal that this compares favorably to other outcomes.

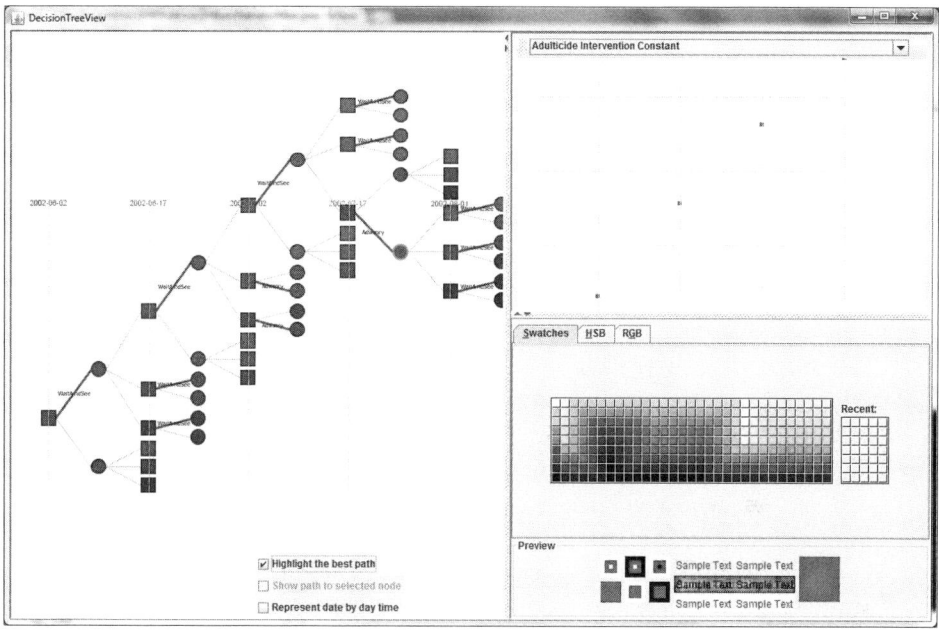

Figure 9.6
Hybrid tree visualization interface.

Further browsing in the tree reveals considerably more texture. For example, there are distinctly different recommendations in place for decisions at different points in time (even given similar recent weather), and for the same point in time, given how uncertainties resolved recently. There are also large differences in outcomes between scenarios (e.g., disparities in the mean costs imposed between hot and cold spells, even early in the summer).

Challenge: Using the tree visualization software, go through the example tree (as output after the rollback) and find cases where different decisions are recommended at the same point in time, depending on the preceding observed events (temperature changes).

Hybridizing Decision Analysis and Stochastic Simulation Models

Endogenous Stochastics and the Hybrid Model
The hybrid technique for address dynamically complex decision problems described in this chapter is well suited for describing stochastics whose dynamics lie external to the scope of the simulation model, but

not stochastic factors whose dynamics are tightly bound up with emergent dynamics being simulated.

For example, evolution of rainfall and temperature clearly lie outside the bounds of the West Nile virus simulation model (being unaffected by both the natural endogenous dynamics and interventions simulated), and are therefore comfortably treated as exogenous events. By contrast, one could readily imagine seeking to incorporate into a hybrid analysis stochastics involving mosquito population dynamics. Recall that the existing West Nile virus model incorporates articulated dynamics associated with the mosquito population, but does so in a deterministic fashion. In truth, the mosquito population is affected by factors lying outside the current scope of the simulation model, including some that exhibit pronounced temporal variability—for example, rainfall. Even if our model captures well the mean dynamics to be expected in a nonlinear model such as that depicted here, fluctuations around the mean could be important, as they could lead to extremes in exposure and health risks. Put another way, the mean outcome of the model simulated over the stochastic ensemble might be very different than the outcome of the model simulated on the mean of the stochastic ensemble. For factors—such as temperature—whose governing dynamics plausibly lie outside of the simulation model, it may be convenient to stochastically characterize those dynamics within the simulation model (for example, using a random walk around some overall trend) in order to capture the path-dependent temporal dynamics involved. While space constraints prevent its discussion here, the authors have characterized extensions to the framework discussed here that can comfortably handle such endogenous stochastics.

Conclusion

This chapter has presented a hybrid framework for addressing dynamically complex decision problems—problems marked both by dynamic complexity (commonly the domain of simulation modeling) and by the need to make choices over time in the context of pronounced uncertainty (commonly the domain of decision analysis). Taken in isolation, neither simulation modeling nor decision analysis can grapple with such problems in a fully adequate fashion. In the context of dynamic complexity, traditional decision analysis techniques experience difficulty in requiring that the user provide reliable estimates for the consequences associated with terminal nodes. Conversely, while there is a strong need to formulate *adaptive* policies to help make incremental decisions in the context of unfolding uncertainty, the space of adaptive policies is far too large to evaluate exhaustively using traditional simulation-based policy analysis, which depends on exhaustive enumeration of each successive decision rule.

The hybrid approach presented in this chapter captures the strengths of each analysis approach, while limiting the other's weaknesses. The hybrid scheme uses decision trees to capture the common substructure of adaptive and static policy space in a way that is graphically insightful, computationally frugal, and permits exploration of policy-oriented sensitivity analyses on the decision trees (a topic not covered here). For each terminal node in the decision tree, the simulation model is used to evaluate the dynamically complex outcomes of the associated scenarios of decisions and events over time. Backward induction (a form of dynamic programming) can then be used to evaluate possible choices to be made for each eventuality represented in the tree.

This chapter further presented one particular computational implementation of the hybrid approach that defines a protocol to allow the decision tree and simulation model to communicate the information required to perform hybrid simulation to one another, and provides a graphical user interface for analyzing dynamically complex decision problems with simulations implemented in the Vensim simulation package.

While the system presented here offers considerable support for dynamically complex decision problems, there are several important avenues for future work. First, the existing framework should be extended to allow for more accessible, graphical exploration of hybrid decision trees by the user. Second, the tree creation framework should be generalized to allow for straightforward creation of trees with more general tree structure, while still maintaining an easily understood interface. Third, while the current framework handles the case of uncertainties exogenous to the simulation model, there is potential for great enhancements to the value of the current system by extending it to handle variants of such problems that consider uncertainty within the simulation model. Planned enhancements to the software system to address endogenous uncertainties will allow that system to address an important larger subclass of problems.

Note

1. This assumption is a notable oversimplification imposed by this particular interface, as the likelihood of occurrence of a given temperature in a particular two-week period of the summer is likely to vary widely conditional on the temperatures experienced in the immediately preceding two-week period. Unfortunately, the existing tree creation interface does not allow for capturing this conditional probability. It should be stressed that this is a limitation of the tree creation interface and *not* of the tree evaluation algorithm itself. If a tree is be manually built or modified to specify such varying probabilities at a given point in time (for our example, conditional on recent temperatures), the software presented here can readily analyze it.

References

Borschev, A. 2013. *The big book of simulation modeling: Multimethod modeling with AnyLogic 6.* AnyLogic North America.

Eberlein, R., and D. Peterson. 1994. Understanding models with Vensim. In *Modeling for learning organizations*, ed. J. D. W. Morecroft and J. D. Sterman. Portland, OR: Productivity Press.

Osgood, N. 2005. Combining system dynamics and decision analysis for rapid strategy selection. In *Proceedings of the 23rd International Conference of the System Dynamics Society.* Boston.

Pratt, J. W., H. Raiffa, and R. Schlaifer. 1995. *Introduction to statistical decision theory.* Cambridge, MA: MIT Press.

Sterman, J. 2000. *Business dynamics: Systems thinking and modeling for a complex world.* Boston: Irwin/McGraw-Hill.

Sterman, John D. 2006. Learning from evidence in a complex world. *American Journal of Public Health* 96 (3): 505–514.

Yee, K., N. Osgood, and W. An. 2011. A hybrid system dynamics and decision analysis tool for effective strategy selection to control for a West Nile virus epidemic. Extended abstract and presentation at System Dynamics Winter Conference, Austin, TX.

10 Using Decision Trees to Value Managerial Real Options

Burcu Tan and Edward G. Anderson Jr.

Traditional discounted cash flow methods fail to account for the value of managerial options inherent in many types of projects that can be represented with dynamic models. For example, a simple net present value analysis does not capture the values of the options to delay, expand, or abandon a project. The state-of-the-art method to value investment decisions that involve managerial options is the real options valuation approach. Real options valuation applies financial options theory to "real options" derived from managerial flexibility; that is, options associated with real assets rather than with financial assets (Myers 1987; Trigeorgis and Mason 1987; Trigeorgis 1988; Dixit and Pindyck 1994).

Traditional real options solution approaches typically rely on models that are highly stylized closed-form formulations (e.g., McDonald and Siegel 1986; Paddock, Siegel, and Smith 1988; Capozza and Li 1994; Kogut and Kulatilaka 1994). Despite their various advantages, however, these solution approaches cannot generally handle complex systems involving stochastic processes that stem from nonlinear feedback structures (Forrester 1961). This concern can be overcome by integrating system dynamics (SD) models into the valuation approach. Thanks to the unique capabilities of SD in modeling complex feedback systems, it is possible to capture the underlying structure that produces the complex dynamics and thereby to improve the accuracy of a valuation when these complexities play a critical role.

Another advantage of integrating SD models into the valuation approach is the flexibility offered by the SD methodology in defining complex feedback systems and separate stochastic effects. This flexibility becomes particularly valuable when the managerial decisions involve multiple and potentially interacting sources of uncertainty. Furthermore, describing the distribution of uncertainty around SD variables is intuitive thanks to the methodology's emphasis upon the use of concrete variables that correspond to "real" phenomena (Sterman 2000). As a result, SD provides clearer insights into the drivers of the effect of a strategic action (Johnson, Taylor, and Ford 2006).

There is a burgeoning literature that recognizes these benefits, particularly in the context of project management. For example, Ford and Sobek (2005) model a product development project and use real options concepts to manage product design risk. Ford and Bhargav (2006) examine the effect of project management quality on the value of flexible strategies. Johnson, Taylor, and Ford (2006) evaluate a large petrochemical project using an SD model. Osgood and Kaufman (2003) and Osgood (2005) develop a decision-tree–based method to perform strategy selection. Tan et al. (2010) extend this methodology by allowing the representation of multiple sources of uncertainty efficiently, with less vulnerability to "the curse of dimensionality." Finally, Tan et al. (2009, 2012) develop a diffusion approximation approach that is more in line with the finance literature, since it can account for the change in risk profiles as managerial options are exercised.

The difficulty in evaluating complex option structures within SD models lies in the necessity to optimize a sequential decision process, which typically involves backward induction (Bertsekas 2005), a technique incompatible with most SD simulation evaluation algorithms. In contrast, decision tree analysis provides an intuitive approach to model sequential decision processes (Clemen 1997) and is compatible with backward induction. The decision tree models are also appealing because they are easy to explain to nonpractitioners.

The complementary strengths of SD and decision analysis in representing stochastic models and decision processes motivate the use of SD-based decision tree methods to evaluate managerial options. In this chapter, we illustrate and contrast the uses of two such methods, the SD-based decision tree approach in Tan et al. (2010) and the diffusion approximation approach in Tan et al. (2012), to evaluate a renewable energy investment project. The methods are based upon first formulating a SD model of the project and then transforming the cash flow data obtained from the model into a decision tree. By leveraging the decision analysis literature, the algorithms enable a backward induction solution approach to evaluate the project. Note that even though the two methods are illustrated in the context of corporate investment projects, they can be used to improve decision making in a variety of domains, from product development projects to public policy interventions (e.g., Hovmand and Ford 2009).

The remainder of the chapter is as follows. The next section provides a motivating example, a solar power plant capital investment project, and briefly describes an SD model for that project. The third section presents the SD-based decision tree algorithm by illustrating the transformation of SD simulation data into a decision tree that incorporates the project's managerial options. Since the diffusion approximation approach heavily relies on finance theory, we omit its illustration within the chapter and refer the interested reader to the appendix at the end of this chapter. The fourth section compares the results obtained under the two methods and introduces a modified example. The discussion section provides a comparison of the two valuation approaches as well as a discussion of some limitations. Finally, the sixth section concludes the chapter.

A Motivating Project and Its SD Model

We will use a hypothetical photovoltaic solar power project as an example to illustrate the methods. Despite increased support for renewable energy technologies such as solar and wind power, developing these technologies has proven problematic, in part because of the difficulty in estimating the return from investing in such projects. One major factor that affects the viability of a solar power plant is the price of electricity, which highly depends on the price of natural gas. Yet natural gas prices are uncertain and are influenced by various geopolitical and macroeconomic factors. The cost of developing a new technology is also uncertain. A significant portion of a solar power plant's cost is the fixed cost of building the plant. As for many other new technologies (Argote 1999), this cost is typically reduced significantly with each doubling of the cumulative capacity installed. However, the steepness of this "learning curve" and its final plateau are generally unknown *ex ante*. For example, estimates for photovoltaic solar power learning curves vary from 15% to 22% cost reduction per doubling of the cumulative capacity installed (National Renewable Energy Laboratory 2006; Hearps and McConnell 2011; Martin 2010). Finally, state and federal regulations also play a key role in determining the profitability of a solar power plant investment project. For example, renewable energy producers in the United States currently receive a 2.3 cent benefit per kilowatt-hour of generation, known as the production tax credit (PTC). The uncertain expiration date of the PTC has been a major factor in solar capacity investment decisions.

The illustrative example in this chapter is based on a hypothetical firm's effort to evaluate an investment opportunity to build a 30-MW photovoltaic solar power plant with the option to expand capacity within the first four years. The firm can also delay the beginning of the project by up to two years. After the solar power plant comes online, which takes a year, the firm has the option to expand by adding 25 MW of capacity with the option to build another 25 MW in the successive year within the first four years from the beginning of the project. This example is kept simple for the sake of clarity in exposition, but it contains sufficient nonlinearity and path dependence to make SD models attractive. The algorithms described can handle this, as well as much more complex models (albeit within certain limitations addressed in the discussion).

Four major uncertainties are captured by the model: natural gas price, the learning curve, the expiration date of the production tax credit, and solar power availability (see table 10.1). Natural gas price uncertainty has several components. The demand growth ratio is the rate that base demand for natural gas is assumed to grow. The base demand is assumed to grow continuously. It is a function of population growth less any potential reduction in energy usage intensity. Initial undiscovered resources of natural gas is another major uncertain variable that determines the gas supply and hence the future gas prices. In addition to these factors, the natural gas supply is exposed to random disruptions whose frequency, size, and duration are uncertain.

The learning curve uncertainty has two components: The steepness of the learning curve determines how fast the cost of capacity drops with doubling of the cumulative capacity installed. The weight of global learning determines how much the firm benefits from the technological improvements elsewhere. Finally, solar power availability is also uncertain, modeled using a pink noise process. Other uncertainties (e.g., the minimum possible capacity cost) could easily be incorporated without increasing the complexity of the solution procedure.

SD-Based Decision Tree Approach

In this section, we describe the SD-based decision tree approach (table 10.2) developed in Tan et al. (2010) and illustrate the method using the motivating example described above. Basic components of a decision

Table 10.1
Uncertainties captured in the model

Uncertain Parameters	
Gas price model parameters	**Units**
- Demand growth ratio	1/Month
- Initial undiscovered resources	Trillion cubic feet
- Supply disruption size	1/Month
- Supply disruption frequency	Dimensionless
- Average length of supply disruptions	Month
- Demand elasticity	Dimensionless
Learning curve parameters	
- Steepness of the learning curve	Dimensionless
- Weight of global learning	Dimensionless
Solar power availability	Dimensionless
Official expiration date of PTC	Month

tree are decision nodes, chance nodes, and terminal nodes. Square nodes (e.g., figure 10.4) are the *decision nodes*, which represent the decisions to be made at a particular time, such as to expand or suspend. Branches leaving a decision node represent the decision alternatives. Circular nodes are the *chance nodes*, which represent the uncertainties underlying the project. Branches leaving a chance node represent possible outcomes of an uncertain event and any continuous uncertainty has to be approximated by a discrete probability distribution. Triangular nodes are the *terminal nodes* that depict the final outcome of a particular scenario after all decisions have been made, all uncertainty has been resolved, and all payoffs are received.

The first step of the algorithm is identifying the managerial ("real") options in the project. At each period, the manager needs to decide whether or not to exercise the available options; hence there is a sequence of decisions to be made throughout the horizon of the project. Each possible sequence of these decisions is called a *decision sequence*. In the example problem, the firm has several managerial options that are valid at different time periods: invest (I), delay (D), expand (E), and suspend further investment (S). The time unit is set to one year, so $t = 1$ represents the end of the first project year. Hence, these options result in the seven decision sequences that can be seen in table 10.3.

Table 10.2
Steps of the SD-based decision tree algorithm (Tan et al. 2010; Tan 2010)

Steps of the Algorithm	
Step 1	Identify the decision sequences
Step 2	Build the deterministic SD model that captures the project dynamics
Step 3	Model the uncertainty by specifying the random variables and their distributions in the SD model (e.g., using the sensitivity simulation tool in Vensim)
Step 4	Run Monte Carlo simulations of the SD model for each decision sequence
Step 5	Obtain the discrete distribution approximations for the first-period cash flow distribution for each decision sequence
Step 6	Obtain the conditional discrete approximations for the remaining periods for each decision sequence
Step 7	Solve the decision tree by backward induction using the risk-adjusted discount rate

Table 10.3
Decision sequences of the example problem

Decision Sequence ID	Decision in 2012 (t=1)	Decision in 2013 (t=2)	Decision in 2014 (t=3)	Decision in 2015 (t=4)
1	I	N/A	S	S
2	I	N/A	S	E
3	I	N/A	E	E
4	I	N/A	E	S
5	D	I	N/A	E
6	D	I	N/A	S
7	D	D	I	N/A

The second step is building the deterministic SD model that captures the project dynamics. In the example problem, the model must capture solar capacity investment, natural gas price, and learning curve dynamics. Then, as step 3, the underlying uncertainty of the project has to be modeled by specifying the random variables and their distributions in the SD model (e.g., by using the sensitivity simulation tool in Vensim). For the example problem, there are multiple sources of uncertainty, as presented earlier in table 10.1.

The fourth step of the algorithm is running Monte Carlo simulations of the SD model for each decision sequence in table 10.3 (it is straightforward to impose these decision sequences in the SD model with a few additional if-then-else–type equations). A Monte Carlo run for a specific decision sequence gives a cash flow distribution for each period. In the example problem, the time horizon is assumed to be 20 periods; hence, we obtain 20 cash flow distributions for each decision sequence in table 10.3, making a total of 140 distributions. For example, the distribution displayed in figure 10.1 is the distribution of the sixth-period cash flows obtained under the decision sequence "invest-sus-pend-suspend" after 1,000 iterations of the Monte Carlo simulation, while figure 10.2 displays some of the cash flow sample paths for the same decision sequence. Note that when the method is used to evaluate managerial options in a decision-making problem that does not hinge on "cash flow," the analyst can use another metric of interest and study the distribution of that metric instead of the cash flow distribution. For example, in their analysis of mandatory arrest policies in domestic violence, Hovmand and Ford (2009) use "the number of domestic violence cases per year" as a metric.

In traditional decision analysis models, uncertainty is modeled by assigning probabilities to each branch, leaving a chance node. These *exogenously* assigned probabilities may represent subjective beliefs about the likelihood of a specific "event" represented by the chance node (e.g., the probability of the PTC being suspended) or they may be risk-neutral probability measures of price uncertainties derived from market data. However, in the SD-based decision tree approach used here, exogenous scenarios such as PTC expiration are incorporated at

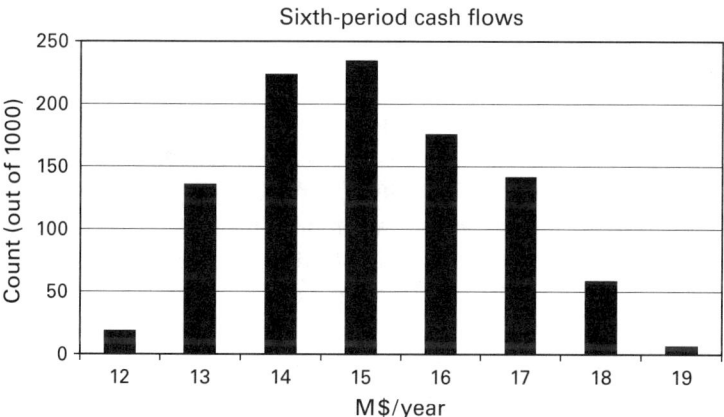

Figure 10.1
The distribution of the sixth-period cash flows under the "I-S-S" decision sequence.

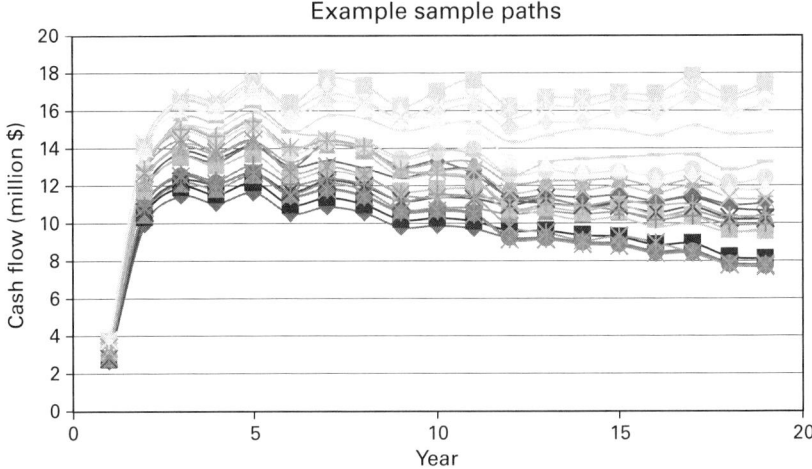

Figure 10.2
Example sample paths obtained under the "I-S-S" decision sequence.

the stage of running the Monte Carlo simulations of the SD model. Hence, instead of building a decision tree that explicitly models every source of uncertainty as separate chance nodes, the method uses a more efficient approach: a chance node represents the distribution of the random cash flow accumulated during a given period t, which is obtained from the Monte Carlo simulation of the SD model for the corresponding decision sequence.

This representation of chance nodes has two advantages. First, it allows multiple uncertainties in a time period to be represented by a single uncertain variable (the cash flow in that time period) without changing the size of the tree and without adding any computational burden. Second, the approach simplifies handling valuation problems that involve path-dependent stochastic processes, such as the investment cost at period t, which is a function of the cumulative capacity investment up to t. Such processes are represented effectively within an SD model. Since the cash flow distributions are obtained through SD simulations, this approach is much more powerful in handling complex, path-dependent stochastic feedback structures compared to the simple stochastic processes used in traditional valuation methods.

The cash flow distributions obtained from the Monte Carlo runs are continuous. In order to place them into the decision tree as chance nodes, one needs to approximate them with a discrete distribution without losing too much precision. We use the bracket median approximation technique (Clemen 1997) to obtain a k-point discrete distribution approximation: first, the distribution is divided into k equally likely intervals. Next, the median of each interval is determined. Then, the continuous distribution is approximated with the k median values that are equally likely to occur (with probability $1/k$). The choice of k is a critical one. If too few intervals are chosen, the accuracy of the approximation will be low, which may lead to the recommendation of a strategy that would not be optimal for the true cash flows. If too many are chosen, the computations involved will become prohibitive. This issue of choosing the appropriate level of detail in a model is common to all applications, and the practitioner needs to resolve this trade-off on a case-by-case basis. As an example, when a project has a relatively short horizon or when managerial flexibility is limited to a certain phase of the project rather than being spread out in the entire horizon (such as an option to delay), dimensionality is less of a problem and it is feasible to use a bigger k. Yet, if the analysis does not demand high fidelity to the tails of the distribution, a bigger k may not be necessary.

The fifth step of the algorithm is obtaining k-point discrete distribution approximations for the *first*-period cash flow distribution of *each decision sequence* by using the bracket median method. In the example problem, three-point discrete approximations are used. That is, depending on the firm's decisions and on how the natural gas price, capacity cost, and tax credit uncertainties evolve, the cash flow at the end of each period may be termed high, medium, or low (each of these could possibly be less than zero). To obtain the approximations, we first sort the cash flow data from a given decision sequence by the first-period cash flows. It is important to do the sorting on the entire data set so as to preserve the sample paths. Next, we divide the data into three brackets with respect to the first-period cash flow: up until the 33rd percentile of the first-period cash flow constitutes the low bracket; between 33rd and 67th percentiles of the first-period cash flow constitutes the medium bracket; and the remainder constitutes the high bracket. Then, we take the median of the first-period cash flow in each of these brackets. For the example problem, these values are calculated as $2.984 million, $3.239 million, and $3.562 million. Thus, the continuous distribution of the first-period cash flow is approximated with three discrete points ($2.984 million, $3.239 million, and $3.562 million), which are equally likely to occur. These values will be plugged into the decision

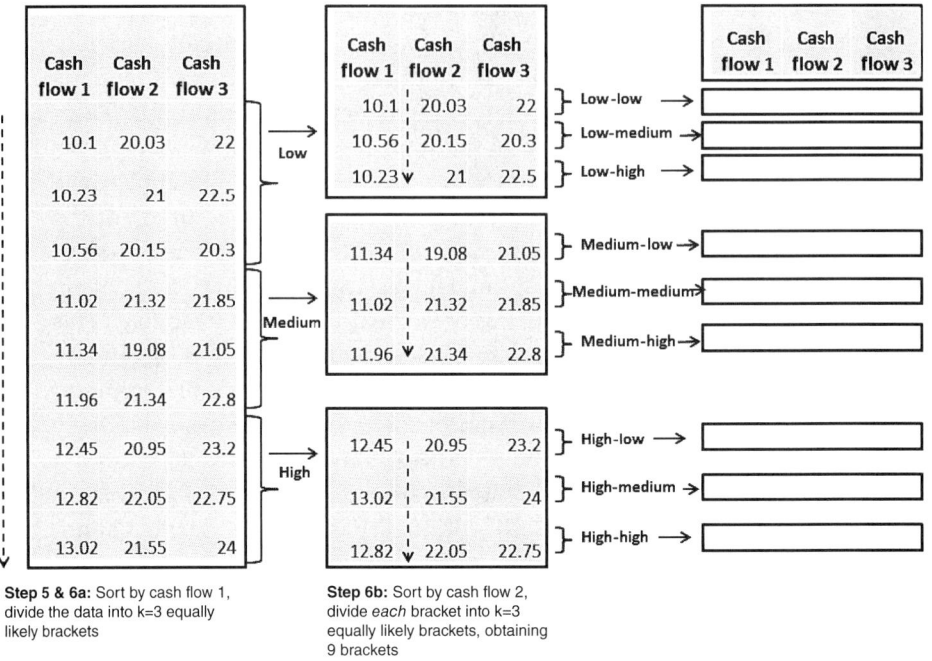

Step 5 & 6a: Sort by cash flow 1,
divide the data into k=3 equally
likely brackets

Step 6b: Sort by cash flow 2,
divide *each* bracket into k=3
equally likely brackets, obtaining
9 brackets

Figure 10.3
Illustration of the conditional distribution computations for a given decision sequence.

tree in step 7 as the chance node for the first-period cash flow. Note that, in principle, one can use a larger number of discrete points to approximate the distribution instead of the three-point discrete approximations used in this example.

In step 6, we discretize the cash flow distributions for the remaining time periods. This step provides the input for the rest of the chance nodes of the decision tree. Note that the cash flow distributions for period t are conditional on the cash flow distributions of period $t-1$, as well as on what decision sequence is chosen. For example, the high level at period t given that the cash flow was low in period $t-1$ is, in general, different from the high level given that the cash flow in period $t-1$ was high. The discrete conditional probability distributions for the remaining cash flows are computed using the following procedure:

Step 6a: Divide the initial data of cash flows into k tables based on the first-period cash flows (e.g., high, medium, low), as explained above. Make sure to preserve the sample path structure: If a realization of $\tilde{C}_1$ falls into the bracket high (as a result of step 5), then carry that entire sample path $\tilde{C}_1, \tilde{C}_2, ..., \tilde{C}_T$ over the table for high. As a result of this step, k tables with (approximately) N/k rows and T columns are obtained, where N is the size of the Monte Carlo simulation (in this case, 1,000 simulations) and T is the number of periods (in this case, 20 years).

Step 6b: Apply the bracket median approximation to each table obtained in step 6a (e.g., high, medium, low) in order to discretize the second-period cash flows. That is, sort within the tables with respect to the second-period cash flow $\tilde{C}_2$. Then, further divide *each* table into k equally likely brackets with respect to the second-period cash flows. At the end of this step, the simulation data is divided into k^2 tables (e.g., high-high, high-medium, high-low, medium-high, medium-low, etc.) with (approximately) N/k^2 rows and T columns. The process is illustrated in figure 10.3. The median of the second-period cash flow for each bracket provides the approximated value for that bracket, which will be plugged into the decision tree in step 7.

Step 6c: Repeat this procedure until the last period. If there are no decision nodes after a certain period t, one can apply the procedure to the present value of the cumulative cash flows after t instead of carrying the procedure until the last period. This would result in losing the volatility information after t but would not change the optimum decision sequence and the expected net present value. For the example problem, there are no decisions after the fourth period. So, for simplicity, we can discretize the present value of the cash flows from the third period to the last period instead of repeating the procedure for all the remaining periods.

Note that these operations for obtaining the conditional probabilities must be carried out *for each decision sequence*. Fortunately, this can be accomplished efficiently with the help of a simple Visual Basic for Applications macro.

The last step of the algorithm is translating the valuation problem into a decision tree and solving the decision tree using backward induction. In practice, this step is handled easily by using decision analysis software such as DPL or Precision Tree.

The example problem is solved using DPL with a discount rate equal to the weighted average cost of capital of the firm, which is assumed to be 10%. The optimal policy is highlighted in bold in figure 10.4. The expected present value (PV) is \$135.58 million and the expected net present value (NPV) of the project is \$21.58 million. The expected NPV of the project without options (which is the expected NPV of the strategy "invest-suspend-suspend") is –\$11.95 million; hence, the combined value of the options is approximately \$33.53 million. The optimal policy suggests that the firm should invest immediately and expand capacity in both the third and fourth periods, irrespective of the cash flow outcomes. Note that in general the optimal course of action tends to be contingent upon how the uncertainty is resolved, an example of which can be seen in the modified model (table 10.5) in the discussion.

The value of a specific option associated with each decision node may be determined simply by rebuilding the tree without that option

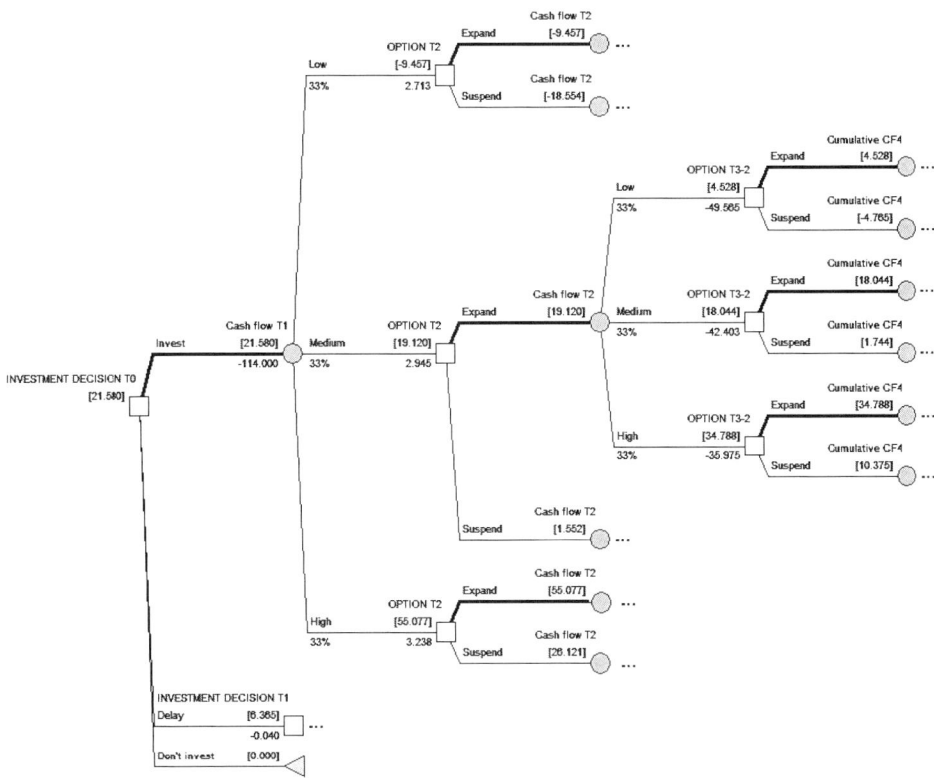

Figure 10.4
Optimal policy under the SD-based decision tree approach.

and solving that simplified tree. An estimation of the expected value of that option is the difference between the value of the project with the option and without the option. If the option has a cost, the firm should obtain the option only if this estimated value is higher than the cost of the option.

The SD-based decision tree algorithm provides a rigorous way to improve decision making in a variety of domains from corporate investment projects to public policy interventions. However, the method has a major limitation when used in financial valuation: it uses the risk-adjusted discount rate for the project without options as the discount rate for the entire decision tree. The problem here is with the poor treatment of changing risk characteristics from a financial perspective. Essentially, to value the project accurately, a financial analyst would usually have to choose different discount rates according to whether the project is being evaluated with options or without options, a choice that depends on each alternative's particular risk levels (Teisberg 1995). The SD-based decision tree algorithm ignores this complexity. This limitation is less significant when evaluating managerial options based on nonfinancial metrics, because in those settings the discount rate typically does not play a critical role; however, it may significantly affect the results in a financial valuation. Note that the fidelity of the model structure (particularly with respect to nonlinearities) is degraded somewhat by the available methods that can accurately capture the effect of changing risk profiles on discount rates. Thus, we suggest that for valuations that are based on nonfinancial metrics, it may be better to use the SD-based decision tree approach. However, for the valuation of corporate investment projects in which financial accuracy is essential, it may be better to use another valuation method, such as the diffusion approximation approach.

The diffusion approximation approach (Tan 2010; Tan et al. 2012) is an SD-based valuation method that overcomes the poor treatment of discount rates by adapting a method in the real options literature (Copeland and Antikarov 2001; Brandao, Dyer, and Hahn 2005) to the SD environment. Since the diffusion approximation approach heavily relies on finance theory, we omit the details here and refer the interested reader to the appendix at the end of this chapter. In what follows, we compare the results from the two methods and discuss their limitations.

Table 10.4
Expected net present value (NPV) of the project and the optimal policy under the two methods

	Expected NPV	*Optimal Policy*
SD-based decision tree approach	$21.58 million	*Period 1: Invest *Period 3: Expand *Period 4: Expand
Diffusion approximation approach	$19.94 million	*Period 1: Invest *Period 3: Expand *Period 4: Suspend if Period 1 cash flow is low and investment cost stays high. Expand otherwise.

Comparison of the Results

We observe that, for the example project, the diffusion approximation approach and the SD-based decision tree approach provide similar estimates for the project value (table 10.4). Furthermore, the two methods suggest similar optimal policies.

It is fairly common to see a similarity in project value estimates. Nonetheless, the optimal policies will not always be in agreement. To illustrate how recommendations from the two methods may diverge, we slightly modify the project assumptions. The base model assumes that the solar power plant operates in a market with sufficient demand for its generation, meaning that electricity generated can always be sold. We modify this assumption such that if the gas price is relatively low, demand for the firm's generation drops after a delay. This additional feedback increases the overall risk faced by the firm: on top of price and cost uncertainty, the firm also suffers from uncertainty in demand. In response to this increased volatility in project returns, the diffusion approximation approach produces a very different prescription from the SD-based decision tree approach.[1]

For the modified project, the diffusion approximation method calculates the expected PV of the investment as $113.22 million, which translates to an expected NPV of –$780,000 (figure 10.5). Thus, the investment should *not* be undertaken. In contrast, for the same project the SD-based decision tree approach calculates an expected PV of $118.761 or, equivalently, an expected NPV of $4.761 million (figure 10.6). To achieve that, the optimal policy suggests investing and then expanding capacity unless the first-period cash flow is low (table 10.5).

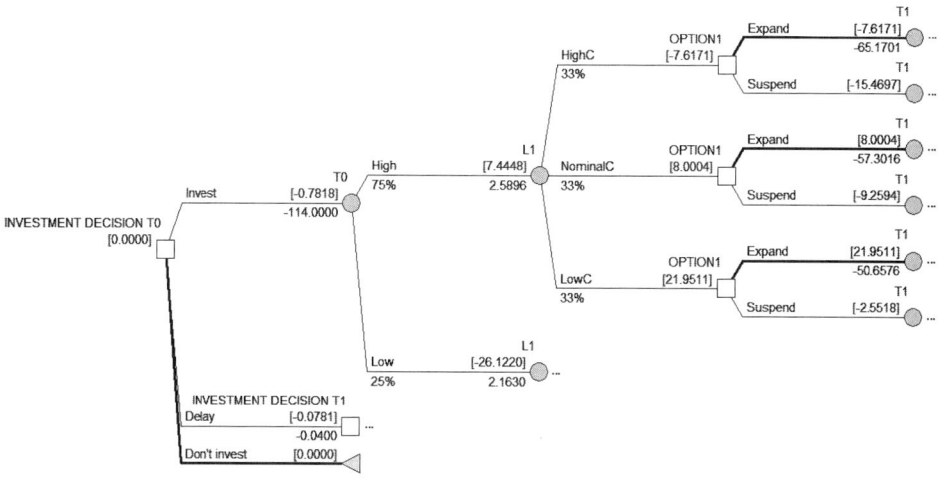

Figure 10.5
Modified model-optimal policy under the diffusion approximation approach.

Table 10.5
Modified project—expected net present value (NPV) and the optimal policy under the
two methods

	Expected NPV	Optimal Policy
SD-based decision tree approach	$4.761 million	*Period 1: Invest *Period 3: Suspend if Period 1 cash flow is low; Expand otherwise *Period 4: Suspend if Period 1 cash flow is low and Period 2 cash flow is low or medium; Expand otherwise
Diffusion approximation approach	0	*Don't invest

The PVs estimated by the two methods are relatively close. However, the strategies prescribed are drastically different. So, which one should the firm follow?

Given that both methods have strengths and weaknesses (covered in the discussion), there is no correct answer, per se; but we can argue that the diffusion approximation approach is more reliable for this problem because it incorporates risk more accurately. The additional risk caused by demand uncertainty is compounded when the expansion options are exercised. Because the SD-based decision tree approach

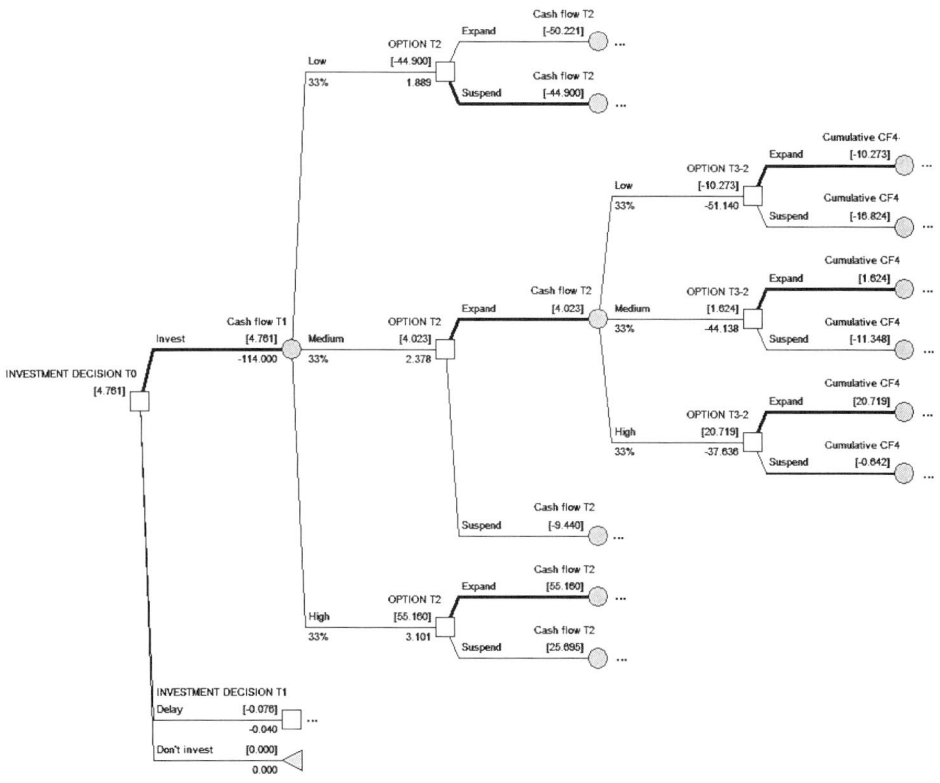

Figure 10.6
Modified model-optimal policy under the SD-based decision tree approach.

cannot properly take into account these changing risk characteristics,
it overestimates the value of the project compared to the diffusion
approximation approach. For the current example, this discrepancy
implies that the firm actually faces more risk than the SD-based decision tree approach can accurately measure and that, by following the
recommendations of that approach, the firm may end up investing in
a project that it should not. Under different circumstances, that same
flaw in incorporating risk could also underestimate the upside risk
associated with an option, which could cause the firm to forgo a worthwhile investment opportunity.

Discussion

Under the base model assumptions, we observe that the diffusion
approximation algorithm and the SD-based decision tree approach

Table 10.6
A guideline describing when each method is more appropriate to use

	Diffusion approximation Approach	SD-Based Decision Tree Approach
Projects with long time horizons	☑	
Projects whose risk-profiles change significantly when managerial options are exercised	☑	
Projects whose PVs do NOT follow a lognormal distribution		☑
Projects for which the cash flows with the option are NOT proportionate to the cash flows without the option		☑

provide similar results. When this is the case, the analyst can feel fairly confident in the policy prescriptions generated by the model. However, if the prescriptions from the two methods diverge significantly, as illustrated with the modified example, the analyst needs to make a careful assessment (table 10.6).

The diffusion approximation approach overcomes the major limitation of the SD-based decision tree approach in the domain of corporate investment valuation, which is the use of the same risk-adjusted discount rate for the project with and without options regardless of the changing risk character. In general, the errors caused by using the wrong discount rate are magnified for projects with long lives and for projects whose risk profiles change significantly when options are exercised (Smith and McCardle 1998; Teisberg 1995). In those cases, the differences between the two methods become significant, and the diffusion approximation approach may become preferable.

However, the diffusion approximation approach has limitations of its own. The algorithm makes two strong assumptions. One is that the value of the project without options can be represented by a geometric Brownian motion. This implies that the method will yield a reasonably accurate valuation only if the project value at each period (calculated by equation [1] in the appendix at the end of this chapter) has an approximately lognormal distribution. Fortunately, this often holds because of the various influences upon the individual cash flows that can be represented in an SD model. Still, a goodness-of-fit test such as chi-square, Kolmogorov–Smirnov, or Anderson-Darling should be applied to check the validity of this assumption (e.g., DeGroot and Schervish 2002).[2]

A second restriction is that the diffusion approximation approach requires the cash flow of the project with an option to be proportionate to the cash flow of the project without the option. For example, if the firm exercises an expansion option to increase its capacity by x%, the revenues should roughly change by a linear function of x. The analyst should confirm this relationship with a sensitivity analysis. If the sensitivity analysis reveals that the option changes the cash flows in a complex, nonlinear fashion, then the diffusion approximation approach may not be the most appropriate modeling approach. In such cases, the SD-based decision tree approach may be a better alternative, because it provides a greater fidelity to the details of the project. Yet when using the SD-based decision tree approach, especially in the valuation of corporate investment projects, the analyst needs to make sure that the results are not highly sensitive to the choice of the discount rate. This can easily be done by evaluating the decision tree under plausible discount rates and verifying that the optimal policy remains the same.

We note that the choice of discount rate typically plays a less critical role for decision-making problems that are based on nonfinancial metrics (e.g., Hovmand and Ford's (2009) analysis of mandatory arrest policies in domestic violence); thus, in those settings the diffusion approximation approach does not offer a significant advantage over the SD-based decision tree approach. Further, diffusion approximation reduces the level of fidelity to model details, as highlighted above. Thus, we suggest that for decision-making problems where financial accuracy is not essential, it may be better to use the SD-based decision tree approach.

Conclusion

Traditional methods to value projects that involve managerial options typically use analytically tractable stochastic processes to model the underlying uncertainty; yet these processes may not always capture the complex behavior of uncertainties that result from nonlinear feedback structures, such as learning curves and rework. Modeling the underlying uncertainty within the SD framework may remedy this situation, thanks to the unique capabilities of SD in handling nonlinearity and path-dependence. However, the SD modeling environment is not particularly suited to optimizing a sequential decision process, which is essential in evaluating a project with multiple managerial options. Fortunately, this problem can be resolved by using decision tree analysis, which provides an intuitive approach to model sequential decision processes.

In this chapter, we have shown how to use formal algorithms that take advantage of the complementary strengths of SD models and decision trees in evaluating projects that involve managerial options. The methods we have described are based upon formulating an SD model of the project, running Monte Carlo simulations of the SD model to generate cash flow distributions (or distributions of other variables of interest), and then transforming the cash flow data into a decision tree.

Even though we have drawn our example problem from renewable energy investment projects, application of these methods is not necessarily limited to the domain of corporate project management. The algorithms can be used to improve decision making in any domain (e.g., public policy, environmental policy, etc.), as long as the problem at hand involves managerial options and significant uncertainties. When applied outside the domain of corporate project management, the analyst needs to replace "cash flow" with an appropriate metric of interest and construct the decision tree based on the distribution of that metric in each period.

Appendix: Diffusion Approximation Approach

In its simplest form, the diffusion approximation approach assumes that the changes in the project's value over time approximately follow a geometric Brownian motion (GBM) diffusion process, which is the same process used in the Black-Scholes (Black and Scholes 1973) option pricing model. The method relies on the marketed asset disclaimer assumption, which argues that the value of the project without options is the best unbiased estimator of the market value of the project. Hence, the present value of the project without options is taken as the market price of the project as if it were traded (Copeland and Antikarov 2001). Then, the value of the project without options is assumed to change over time according to a GBM process. In the decision tree representation, the project values are discounted with the risk-free rate, as opposed to risk-adjusted discount rates, since risk-neutral probabilities are used in the GBM approximation.[3] In other words, using risk-neutral probabilities eliminates the need to estimate different risk-adjusted discount rates as options are added to the project, which overcomes the major

Table 10.7
Steps of the diffusion approximation algorithm (Tan 2010; Tan et al. 2012)

Steps of the Diffusion Approximation Algorithm	
Step 1	Identify the managerial options
Step 2	Build the deterministic SD model that captures the project dynamics
Step 3	Model the uncertainty by specifying the random variables and their distributions
Step 4	Run Monte Carlo simulations of the SD model for the project without options and calculate the present value
Step 5	Obtain the cash flow payout rates for each period
Step 6	Estimate the volatility of the project returns
Step 7	Calculate the parameters of the binomial approximation to the GBM process (u, d, and p)
Step 8	Build the binomial tree
Step 9	Add the options and solve the decision tree by using the risk-free rate

limitation of the SD-based decision tree method in the valuation of corporate investment projects. Note that the assumptions behind the GBM model may not hold for all projects, in which case other models of stochastic processes (e.g., mean reverting) may be used (Brandao and Dyer 2005; Hahn and Dyer 2008; Wang and Dyer 2009).

The first three steps of the diffusion approximation approach (table 10.7) are similar to the SD-based decision tree approach: First, identify the managerial options. Second, build the deterministic SD model. Third, specify the distributions of the uncertain variables. The remaining steps differ, however, because they involve calculating the parameters of a GBM approximation of the project uncertainty. The critical parameters required to model this approximation are the PV of the project without options, the cash flow payout rate in each period t, δ_t (to be defined shortly), the volatility of the project returns σ, and the risk-free rate r.

Step 4 of the diffusion approximation approach is to use a discounted cash flow analysis to calculate the expected PV of the project without options. To do that, we run Monte Carlo simulations of the project without options. For the example project, this corresponds to assuming the firm invests immediately and does not exercise any expansion options. Then, we calculate the PV for each sample path using a risk-adjusted discount rate μ, such as the weighted average cost of capital of the firm. The average of these PVs is used as an estimate of the PV of the project without options. In this case, the expected PV is found to be \$102.05 million.

In step 5, we obtain the cash flow payout rate δ_t in each period. The cash flow payout rate is the ratio of the cash flow in a given period t to the value of the project in that period. The project value at time t is the present value of the *remaining* project cash flows. Note that each iteration (sample) j of the Monte Carlo simulation provides a realization of the random cash flow obtained in period t, denoted by $C_{t,j}$. Let $\tilde{V}_t$ and $\tilde{C}_t$ be random variables representing the remaining project value and the cash flow in period t, and $\bar{V}_t$ and $\bar{C}_t$ be their corresponding means. For each iteration j of the Monte Carlo simulation, the project value $V_{t,j}$ at time t is given by

$$V_{t,j} = \sum_{i=t}^{T} \frac{C_{i,j}}{(1+\mu)^{i-t}},$$ (1)

where μ is the risk-adjusted discount rate and T is the number of periods. Accordingly, the cash flow payout rate in period t is defined as:

$$\delta_t = \frac{\bar{C}_t}{\bar{V}_t}.$$ (2)

For the example project, cash flow payout rates (δ_t) for each time period are calculated as shown in table 10.8.

The cash flow payout rates are used to calculate the cash flows that are paid out at the end of each time period as a function of the project value (similar to the dividend distribution rate of financial assets). In line with previous work, the method assumes that the cash flows vary over time, reflecting the uncertainty in the project value, but that in each time period they are a constant fraction (δ_t) of the remaining value

Table 10.8
Cash flow payout rates for the example project

Period		Period		Period		Period	
1	0.029	6	0.127	11	0.162	16	0.244
2	0.103	7	0.138	12	0.161	17	0.299
3	0.121	8	0.142	13	0.176	18	0.365
4	0.122	9	0.141	14	0.192	19	0.525
5	0.133	10	0.152	15	0.217	20	1.000

of the project (Copeland and Antikarov 2001; Brandao, Dyer, and Hahn 2005).

The sixth step of the algorithm is to estimate the volatility (σ) of the project returns. Note that the GBM model (equation [3]) is based on the assumption of constant volatility. Smith (2005) suggests an approach to estimating this volatility that can be adapted to the SD simulation environment. First, we model the GBM approximation of the project value, setting $V_0 = PV$ and assuming that the change in project value follows

$$dV_t = \mu V dt + \sigma V dz , \tag{3}$$

where $dz = \varepsilon\sqrt{dt}$ and $\varepsilon \sim N(0,1)$. Note that, at the end of each period, the project value is diminished by the cash flow accrued to the firm, C_t, which can easily be modeled by using the assumption $C_t = V_t \delta_t$. Converting these equations into a discrete-time framework yields:

$$\begin{aligned} V_t &= (V_{t-1}(1+\mu)\Delta t + \sigma V_{t-1}\varepsilon\sqrt{\Delta t})(1-\delta_t) \\ C_t &= (V_{t-1}(1+\mu)\Delta t + \sigma V_{t-1}\varepsilon\sqrt{\Delta t})\delta_t \end{aligned} . \tag{4}$$

One can use Microsoft Excel's built-in functions or a statistical simulation software package such as @Risk to simulate this GBM process for a given volatility, σ. Then, the next step is searching for the volatility that best mimics the uncertainty in the original SD model; that is, the volatility that minimizes the difference between the cash flow distributions generated by the original SD simulation and the cash flow distributions obtained by the GBM approximation.

Following Smith (2005), the method suggests comparing the 10th, 50th, and 90th percentiles of the cash flow distributions obtained through the GBM approximation *for a given volatility* to the corresponding values given by the original SD model. Note that this process results in $3 \times T$ pairs to be compared, where T is the number of periods. The sum of squared errors between these pairs provides a measure of fit for the GBM approximation under the given volatility. One can repeat this procedure for a set of plausible values for volatility and then choose the value that minimizes the sum of squared errors.

For the example project, we determined a set of candidate values for volatility. For each candidate value, a GBM approximation is modeled using equation (4) along with the estimates of PV (step 4) and cash flow payout rates (table 10.8). We used @Risk to simulate the GBM processes and to estimate the cash flow percentiles. We also calculated the same percentiles for the Monte Carlo simulation of the original SD model.

Table 10.9
Sum of squared errors of the GBM approximation for different values of volatility

Volatility	Sum of Squared Errors
4%	33.7132
5%	8.6913
6%	**6.3220**
7%	22.1460
8%	43.5354
9%	74.6664
10%	135.8993
11%	221.1119

Then, for each candidate value, we calculated the sum of squared errors between the original SD model and the corresponding GBM approximation (table 10.9) and determined the volatility that minimizes the sum.

We find that when $\sigma = 0.06$ (bolded), the GBM approximation quite closely mimics the cash flow distributions given by the original SD model (figure 10.7). Note that when more precision is required, the search can be widened.

Once the volatility of the project returns is determined, we have all the parameters required to model the GBM that characterizes the project uncertainty. However, in order to represent this continuous stochastic process with a decision tree, a discrete approximation needs to be used. The binomial approximation to the GBM process serves this purpose by representing the GBM process with a binomial lattice. A binomial lattice is a probability tree with binary chance branches that go up (u) or down (d), with the unique feature that the outcome resulting from moving up and then down is the same as the outcome from moving down and then up (figure 10.8). In particular, the binomial lattice model assumes that with probability p the value of the project V will go up to Vu, and with probability $1-p$ it will go down to Vd at the end of one period. The parameter u is greater than 1 (reflecting a proportional increase), whereas $d = 1/u$ (reflecting a proportional decrease). A binomial lattice may also be "unfolded" and represented as an equivalent binomial tree. Although unfolding increases the number of endpoints in the model and thus decreases computational

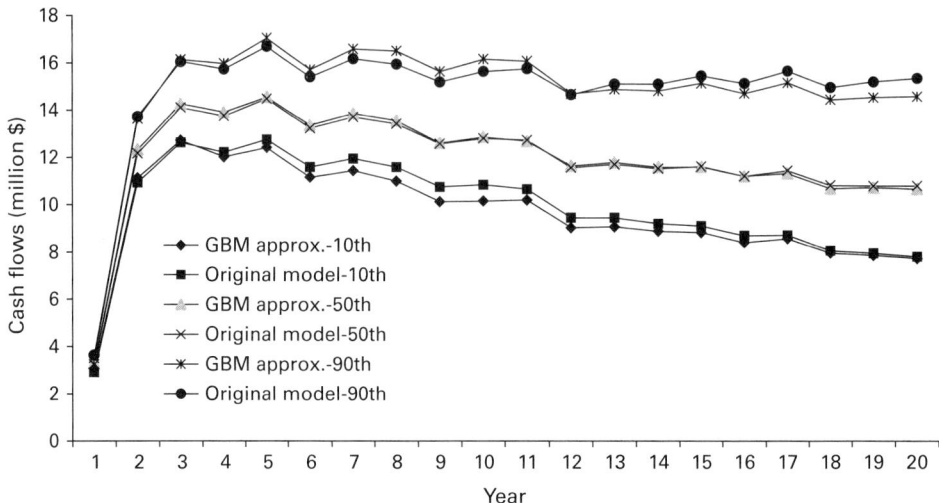

Figure 10.7
GBM approximation vs. the original model cash flows.

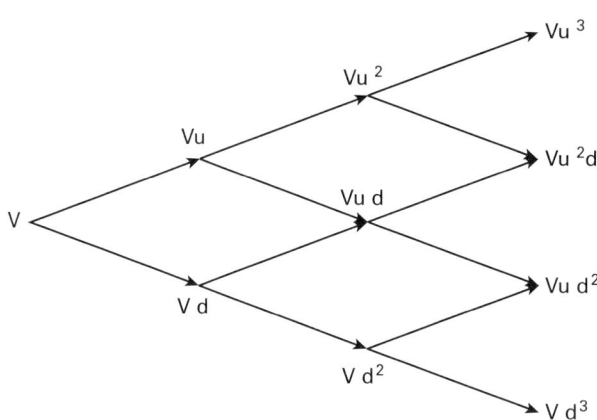

Figure 10.8
A binomial lattice.

efficiency, it allows these problems to be solved using decision tree software with a simple and intuitive visual representation.

Step 7 of the algorithm calculates the parameters of the binomial lattice. The initial value V of the project is approximated by the PV of the project without options obtained in step 4. To calculate the remaining three parameters u, d, and p, it is sufficient to know the volatility σ of the GBM process, which is estimated in step 6, and the risk-free discount rate r:

$$
\begin{aligned}
u &= e^{\sigma\sqrt{\Delta t}} \\
d &= 1/u \\
p &= \frac{1 + r\Delta t - d}{u - d}
\end{aligned}
\qquad (5)
$$

where Δt is the time period used in the binomial lattice. The probabilities p and $1-p$ are the probabilities that a risk-neutral investor would assign to the two outcomes; therefore, they are often called *risk-neutral probabilities*. Assigning risk-neutral probabilities enables the use of the risk-free rate to discount the cash flows; hence, we avoid estimating different risk-adjusted discount rates as options are added to the project.

For the example project, we set $\Delta t = 1$ year and assume a 5% risk-free rate to obtain $u = 1.062$, $d = 0.942$, and $p = 0.901$. Finally, the initial value V of the project is approximated by the PV of the project without options, $102.05 million.

In step 8, we build the binomial tree based on the parameters obtained in step 7. This requires calculating the project value as well as the cash flow at each period, at each node (in general there is more than one chance node at each period, as can be seen in figure 10.6), and for each state up (u) and down (d). First, note that at each period a fraction δ_t of the project value is paid out as cash flow, diminishing the value of the project to $V_t(1-\delta_t)$. Accordingly, the project value in the subsequent period is calculated as follows:

$$
\begin{aligned}
V_{t+1}^{u} &= V_t(1-\delta_t)u \\
V_{t+1}^{d} &= V_t(1-\delta_t)d
\end{aligned}
\qquad (6)
$$

where u and d are given by equation (5). The cash flow $C_{t,k} = V_t^k \delta_t$ at state k and period t can be discounted at the risk-free rate r because risk-neutral probabilities are used. Hence, the *discounted* cash flow that is paid out at each time period t and at each state k is given by

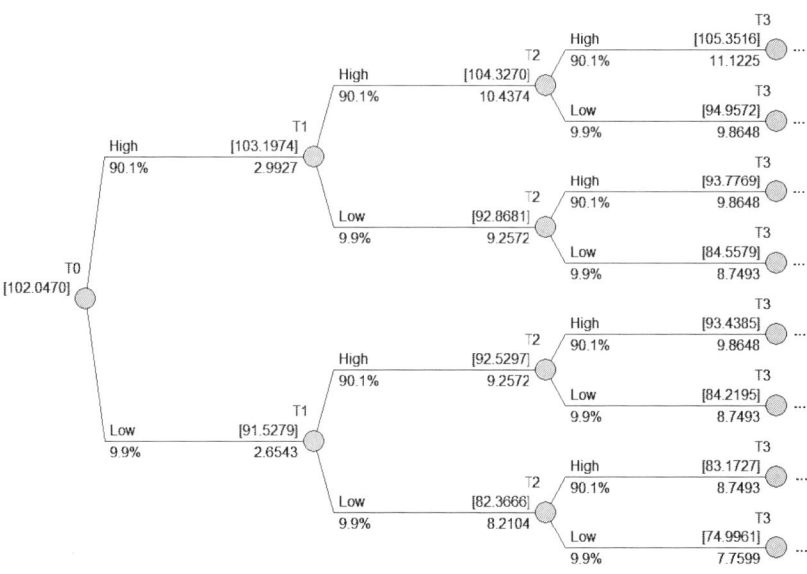

Figure 10.9
Binomial tree model of the project without options.

$$C_{t,k} = \frac{V_t^k \delta_t}{(1+r)^t} \, . \tag{7}$$

Using equations (6) and (7), and the parameters u, d, p, PV, and δ_t calculated above, one can build the binomial tree. Figure 10.9 depicts the first three periods of the binomial tree for the example project.

Finally, once the project without options is modeled with a binomial decision tree, options can be added in the form of decision nodes. For example, an "abandon" option can be modeled by adding a decision node without any subsequent chance nodes (i.e., no further cash flows), whereas simple expansion and contraction options can be modeled as percentage changes in the cash flows (for details, see Brandao, Dyer, and Hahn 2005). For the example project, the operating cash flow (i.e., cash flow exclusive of capital costs) changes proportionally with the changes in the capacity. For example, the option to expand by 25 MW ("expand low") increases the capacity by 83%. The revenues from electricity generation are roughly proportional to the capacity; hence, the increase in revenue once the option is exercised can be modeled by simply increasing the cash flows (equation [7]) by 83%.

Yet this scheme does not directly allow for incorporating the uncertainty in the capacity cost (the learning curve uncertainty) because the cost of capacity is a one-time payment at the exercise of the option, which does not affect the cash flow stream afterward. To handle this wrinkle, we modeled the learning curve uncertainty as a private (project-specific) risk. In many capital investment projects, certain project-specific risks, such as technological risks, cannot be hedged by trading securities. In view of that restriction, real options studies distinguish between public (market-priced) risks and private (project-specific) risks (Smith and Nau 1995). Fortunately, treating different types of risks in the valuation of corporate investment projects is straightforward once a decision tree is created. Public and private risks may be represented with separate chance nodes. Risk-neutral probabilities are used for the former, and subjective probabilities are used for the latter. Note that such a distinction is not relevant outside the domain of corporate investment projects.

For the example project, the cost of capacity is discretized so that at any period t, it is high, nominal, or low. The values for each of these branches were obtained by discretizing the Monte Carlo simulation data for the cost of capacity following the bracket median method described in the section "SD-based Decision Tree Approach." A three-point bracket median method is used while preserving the path-dependence of the learning curve uncertainty by carefully computing the conditional probabilities of the branches. For example, chance node branches HighC, NominalC, and LowC that come out of the node L1 in figure 10.10 discretize the learning curve uncertainty for the first period.

For the example project, the decision tree was built and solved using DPL (figure 10.10). The expected NPV is found to be $19.94 million. The optimal policy suggests investing immediately and then expanding capacity in the following period. If the cash flow in the first period is low and investment costs stay high, the firm should not expand any further. Otherwise, it is optimum to undertake one more stage of capacity expansion.

Challenge: In the example problem, after the solar plant comes online, the firm has the option to expand by adding 25 MWs of capacity with the option to build another 25 MWs in the successive year. What if the firm has an additional option to undertake the expansion all at once; that is, add 50 MWs instead of 25 MWs after the solar plant comes online? Is it optimal to exercise this new option to "expand high"?

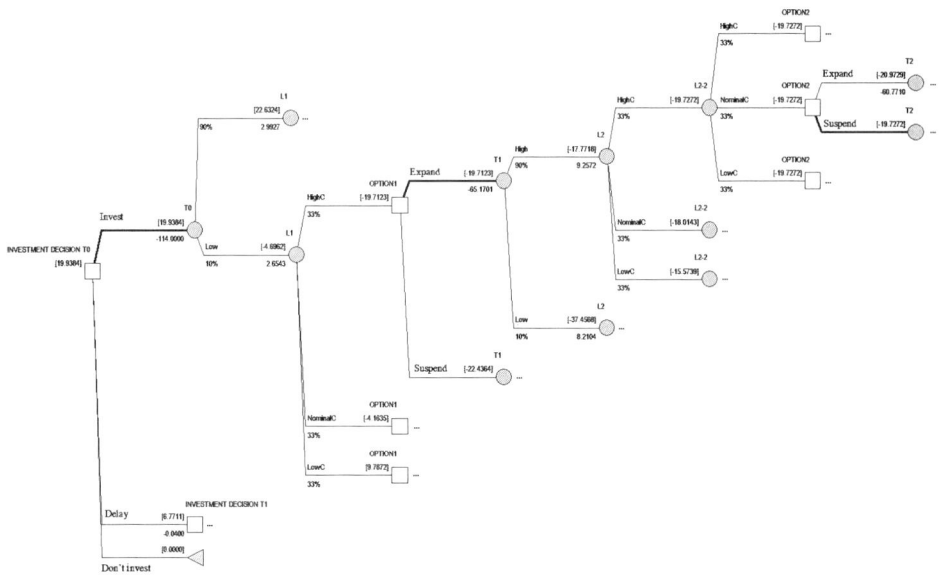

Figure 10.10
Optimal policy under the diffusion approximation approach.

Notes

1. Details of the analysis are available from the first author upon request.

2. Note that if the variable x follows a lognormal distribution, then $ln(x)$ follows a normal distribution. Hence, to check whether the project value V_t is lognormal, one can test the normality of $ln(V_t)$ by using standard tests for normality that comes with any statistical software package.

3. Risk-neutral valuation is an important principle in option pricing (Hull 2006), which is based on using a risk-neutral measure in valuing the asset. A risk-neutral probability measure is in general different from the "physical" probabilities. For example, if the NPV of a project is V when using a risk-adjusted discount rate, a set of risk-neutral probabilities would give the same project value V when discounting the cash flows at the risk-free rate of return. If one can find a set of risk-neutral probabilities for the project without options, then the problem of estimating the correct risk-adjusted discount rates for the project with options is eliminated, because the project NPV with options can then be estimated by using the risk-free rate of return (which can simply be observed in the market).

References

Argote, L. 1999. *Organizational learning: Creating, retaining, and transferring knowledge.* Boston, MA: Kluwer Academic.

Bertsekas, D. 2005. *Dynamic programming and optimal control, Vol.I.* Belmont, MA: Athena Scientific.

Black, F., and M. Scholes. 1973. The pricing of options and corporate liabilities. *Journal of Political Economy* 81 (3): 637–654.

Brandao, L. E., and J. S. Dyer. 2005. Decision analysis and real options: A discrete time approach to real option valuation. *Annals of Operations Research* 135 (1): 21–39.

Brandao, L. E., J. S. Dyer, and W. J. Hahn. 2005. Using binomial decision trees to solve real-option valuation problems. *Decision Analysis* 2 (2): 69–88.

Capozza, D., and Y. Li. 1994. The intensity and timing of investment: The case of land. *American Economic Review* 84 (4): 889–904.

Clemen, R. T. 1997. *Making hard decisions: An introduction to decision analysis*. Belmont, CA: Duxbury Press.

Copeland, T., and V. Antikarov. 2001. *Real options: A practitioner's guide*. New York: Texere LLC.

DeGroot, M. H., and M. J. Schervish. 2002. *Probability and statistics*. Boston, MA: Addison-Wesley.

Dixit, A. K., and R. S. Pindyck. 1994. *Investment under uncertainty*. Princeton, NJ: Princeton University Press.

Ford, D. N., and S. Bhargav. 2006. Project management quality and the value of flexible strategies. *Engineering, Construction, and Architectural Management* 13 (3): 275–289.

Ford, D., and S. Sobek. 2005. Adapting real options to new product development by modeling the second Toyota paradox. *IEEE Transactions on Engineering Management* 52 (2): 175–185.

Forrester, J. W. 1961. *Industrial dynamics*. Cambridge, MA: MIT Press.

Hahn, W. J., and J. S. Dyer. 2008. Discrete time modeling of mean-reverting stochastic processes for real option valuation. *European Journal of Operational Research* 184 (2): 534–548.

Hearps, P., and D. McConnell. 2011. Renewable energy technology cost review. Melbourne Energy Institute technical paper series. Melbourne, Australia.

Hovmand, P., and D. N. Ford. 2009. Computer simulation of innovation implementation strategies. In *Proceedings of the 2009 Winter Simulation Conference*, Austin, TX.

Hull, J. 2006. *Options, futures and other derivatives*. Upper Saddle River, New Jersey: Prentice Hall.

Johnson, S. T., T. Taylor, and D. N. Ford. 2006. Using system dynamics to extend real options use: Insights from the oil and gas industry. In *Proceedings of the System Dynamics Conference*, Nijmegen, The Netherlands.

Kogut, B., and N. Kulatilaka. 1994. Operating flexibility, global manufacturing, and the option value of a multinational network. *Management Science* 40 (1): 123–139.

Martin, D. 2010. A steeper learning curve for PV. Available at http://www.exposolar.org/2012/eng/center/contents.asp?idx=94&page=5&search=&searchstring=&news_type. Accessed June 8, 2012.

McDonald, R., and D. Siegel. 1986. The value of waiting to invest. *Quarterly Journal of Economics* 101: 707–727.

Myers, S. C. 1987. Finance theory and financial strategy. *Midland Corporate Finance Journal* 5: 6–13.

National Renewable Energy Laboratory. 2006. A review of PV inverter technology cost and performance projections. Subcontract report: NREL/SR-620–38771.

Osgood, N. 2005. Combining system dynamics and decision analysis for rapid strategy selection. In *Proceedings of the 23rd International Conference of System Dynamics Society*, Boston.

Osgood, N., and G. Kaufman. 2003. A hybrid model architecture for strategic renewable resource planning. In *Proceedings of the 21st International Conference of System Dynamics Society*, New York.

Paddock, J. L., D. R. Siegel, and J. L. Smith. 1988. Option valuation of claims on real assets: The case of offshore petroleum leases. *Quarterly Journal of Economics* 103 (3): 479–508.

Smith, J. 2005. Alternative approaches for solving real-options problems. *Decision Analysis* 2 (2): 89–102.

Smith, J., and K. McCardle. 1998. Valuing oil properties: Integrating option pricing and decision analysis approaches. *Operations Research* 46 (2): 198–217.

Smith, J., and R. Nau. 1995. Valuing risky projects: Option pricing theory and decision analysis. *Management Science* 14 (5): 795–816.

Sterman, J. D. 2000. *Business dynamics: Systems thinking and modeling for a complex world.* New York: McGraw-Hill/Irwin.

Tan, B. 2010. Risk mitigation strategies for project management, platform development, and supply chain design. PhD thesis, University of Texas at Austin.

Tan, B., E. G. Anderson, J. S. Dyer, and G. G. Parker. 2009. Using binomial decision trees and real options theory to evaluate system dynamics models of risky projects. In *Proceedings of the 27th International Conference of System Dynamics Society*, Albuquerque, NM.

Tan, B., E. G. Anderson, J. S. Dyer, and G. G. Parker. 2010. Evaluating system dynamics models of risky projects using decision trees: Alternative energy projects as an illustrative example. *System Dynamics Review* 26 (1): 1–17.

Tan, B., E. G. Anderson, J. S. Dyer, and G. G. Parker. 2012. Using binomial decision trees and real options theory to evaluate system dynamics models of risky projects. Working paper.

Teisberg, E. O. 1995. Methods for evaluating capital investment decisions under uncertainty. Real options. In *Capital investment: New contributions*, ed. L. Trigeorgis. Westport, CT: Praeger Publishing.

Trigeorgis, L. 1988. A conceptual options framework for capital budgeting. *Advances in Futures and Options Research* 3 (3): 145–167.

Trigeorgis, L., and S. P. Mason. 1987. Valuing managerial flexibility. *Midland Corporate Finance Journal* 5: 14–21.

Wang, T., and J. Dyer. 2009. Valuing multifactor real options using a binomial tree. *Decision Analysis* 7 (2): 185–195.

11 Optimal Control for Complex Systems

Edward G. Anderson Jr. and Nitin R. Joglekar

System dynamics (SD) models typically examine the observed behavior of human actors (Forrester 1961; Sterman 2000). However, there are a number of cases when it is useful to know not what human agents actually do in a given situation, but rather what would be the best thing for them to do. That is, we would like to create a prescriptive rather than a descriptive model. There are three major reasons for this. One is the usefulness of creating a benchmark to assess the performance of human actors' decisions and policies (Sterman 1989). A second is to derive an optimal policy that can form the basis for a managerial policy to be used in practice. A third reason is that it is occasionally worthwhile to examine a closed-form analytic simplification of an SD model. In that case, concentrating solely on the optimal control policy, rather than analyzing the space of all plausible policies, can sometimes simplify the analysis of behavior modes.

In this chapter, we will use examples drawn from the literature, particularly the staffing and project management literature, to describe how to determine optimal control policies for deterministic systems. We will then also show how to determine optimal control policies when systems are driven by random inputs, when there are multiple state variables involved, and when the values of the stock variables are not known with certainty. Many of these solutions assume linearity, normally distributed stochastic disturbances, and other conditions. So we will discuss how to adapt these techniques to situations when these conditions do not hold, as well as explore recently developed approaches (such as machine learning) to cope with these issues. Finally, we will also highlight some of the important works in the SD literature that use optimal control theory.

Theoretical Background

There is a rich literature on control theory in electrical and mechanical engineering (see e.g., Stengel 1994). It is crucial to realize, as evidenced by these books, that most work in the engineering realm is concerned with achieving good performance with respect to some criterion while avoiding instability or other undesirable performance outcomes. In other words, its goal is not necessarily to achieve optimal performance, but rather to achieve good performance in a wide range of situations. Even when engineering work does develop optimal formulae, the assumption is that the modeled system is only an approximation of the real system, and hence such "optimal" control is, in practice, only approximately optimal.

This is the reason why optimal control in the engineering realm almost always concentrates on modeling problems as a linear system with quadratic weights for penalties on control and state variables in the objective function. As a short aside, state variables describe those aspects of the system that can potentially have some inertia. Intuitively, they have a "memory" and can be used to forecast the future of the system. For a rocket, this would include its spatial coordinates and altitude as well as how much fuel it has. In system dynamics parlance, these variables are "stocks." On the other hand, control variables are those decision variables, such as rocket thrust, that can be used to change the state of the system. Typically, it is assumed that they can be varied instantaneously. They are typically represented by "flows" in system dynamics models.

Engineers do not assume that real-world models are necessarily linear or that penalties are precisely quadratic. Rather, they work on the assumption that such a model can approximate reality in the neighborhood in which the system is operating. Then a control policy derived from these assumptions will be reasonably good, or, as Stengel (1994) puts it, the policy will be in the "neighborhood of optimality."

A final advantage of linear-quadratic (LQ) assumptions is that the policy derived is usually a linear combination of state variables. This sort of policy is common in SD literature; although in practice most of these SD policies are typically *not* formally derived by deploying an explicit performance criterion. These practices are discussed in Özveren and Sterman (1989) and Mohapatra and Sharma (1985).

One alternative to derivation and formal analysis, often used by managers, is to put linear weights on states or control variables in their objective functions. If this is done for state variables, the optimal control policy is a "bang-bang" solution in which you merely push the control variable as far as possible in the appropriate direction until you achieve the results you want. While this is in itself not necessarily a bad thing, often there are limits to how hard you can push a system. If these limits are not captured, the resulting policy will not be realistic. Furthermore, in our experience, given the delays in feedback loops present in system dynamics models, bang-bang policies often result in boom-and-bust behavior from over-controlling the system. This behavior is akin to how people accidentally burn and then freeze themselves when they adjust the temperature of a shower while standing in said shower. Linear weights on state variables (actually, the weights on deviations of those state variables from their targets) create their own issues, as they generally result in one variable being optimized at the cost of all the other variables in the system. Hence, quadratic weights make sense more often than one might initially expect for optimizing SD models.

Economics and operations research (see e.g., Kamien and Schwartz 1981; Sethi and Thompson 2000) also studies optimal control, though more for insight than to develop actionable control policies, because their models are typically highly stylized. They also assume more general functional forms, although in practice the forms that they do choose (such as square roots, exponentials, etc.) are somewhat restricted (Sethi and Thompson 2000). Thus, in many cases they offer little help to those wanting to go beyond the traditional neighboring-optimal LQ assumptions just described. On the other hand, more flexible forms of optimal or quasi-optimal control such as reinforcement learning, approximate dynamic programming, neuro-dynamic learning, and so on, have recently been developed, which show much promise (Bertsekas and Tsitsiklis 1996; Bertsekas 2007; Jiang and Jiang 2012).

In the field of system dynamics, there are a number of articles and other pieces of research that employ optimal control methods. This work arguably begins with Coyle's (1985) study of the optimization of a standard system dynamics production model. We have already mentioned Özveren and Sterman (1989) and Mohapatra and Sharma (1985). However, there is a whole related branch that leverages eigenvalues (or "pole" analysis in some engineering texts), beginning with Forrester (1982) and continuing with Kampmann and Oliva (2006) and Gonçalves (2009), among others, which is described in chapter 7 of this book. Kivijärvi and Tuominen (1986) and Macedo (1989) anticipate some of the features of approximate dynamic programming in their work. More recently, Joglekar and Ford (2005) explore optimal control in the framework of project management. Anderson and Fine (1999), Anderson (2001a, 2001b), and Joglekar, Anderson, and Shankaranaray-anan (2013) study optimal control for staffing (formerly "manpower") planning. Anderson, Morrice, and Lundeen (2005, 2006) use optimal control to study service supply chains, and Gonçalves, Hines, and Sterman (2005) and Gonçalves (2006) examine the interaction of phantom orders with the bullwhip effect. Very recently, Rahmandad (2008) and Rahmandad, Repenning, and Sterman (2009) have begun to apply reinforcement learning (a form of approximate optimal control) concepts to system dynamics models.

A Simple Personnel Model Under Demand Growth

We next preset a simple system dynamics model of a firm, drawn from Anderson (2001a), which seeks to manage a rookie-pro personnel struc-ture in the face of exponential growth. We will use optimal control theory to develop an optimal policy for this model to show how to use optimal control principles in practice. It is important to note that, while we concentrate in this section and the rest of the chapter on examples drawn from staffing planning, the techniques can be used in all sorts of other domains, such as inventory, capacity, intellectual property, economics, or public policy management (Kamien and Schwartz 1981; Gaimon 1997; Sethi and Thompson 2000).

It is important to note that this particular system is nonlinear, because the heart of the model is exponential demand growth. The reason we use this example is to show that even traditional optimal control methods can be used in many nonlinear situations.

Employees

Highly skilled workers in many industries (e.g., engineers, computer programmers and analysts, skilled trades) require periods of both formal and on-the-job training after hiring before they become fully productive. Thus we have two classes of employees within firms. One class consists of fully productive employees who have completed their classroom and on-the-job training, whom we will term "experienced." They are assumed to be permanently engaged by the company. The other category will consist of "apprentice" employees who are not yet fully productive and are relatively easier to hire or terminate. In many system dynamics models, the former class is called "pros" and the latter class "rookies."

Let $x(t)$ = the number of experienced employees at time t,
$n(t)$ = the number of apprentice employees at time t,
$c(t)$ = the capacity of employees at time t expressed in full-time experienced-employee equivalents,
δ = the productivity of apprentices relative to experienced employees, and
λ = the training completion rate of apprentices.

The productivity of new employees $n(t)$ who are still undergoing training will be discounted to reflect their lesser productivity:

$$c(t) = x(t) + \delta n(t). \tag{1}$$

Here, δ is assumed to be less than unity. Employee "capacity" $c(t)$ can be thought of as the rate at which the workforce can complete individual skilled tasks. Apprentices are allowed to have negative productivity, because the capacity gained by utilizing an apprentice is in many instances more than offset by the mentoring and error-correction they require from experienced employees. If there were no apprentices to train, these experienced employees could use their "mentoring" time performing immediately productive work. Note that since this is a model that assumes demand growth (see below), $n(t)$ is always nonnegative.

Following the "fractional-flow" assumption common to most staffing and system dynamics papers, a constant fraction of apprentices will complete training each period and move into the experienced category. This rate, λ, we will term the training completion rate.

$$\frac{dx(t)}{dt} = \lambda n(t) \tag{2}$$

For expositional simplicity, we also assume no employee attrition and no losses or gains of experienced employees to or from competitors. Note that the number of experienced employees $x(t)$ in this model is a stock variable (or state variable in control theory parlance) because it cannot instantaneously vary either up or down (the "snapshot" test from Sterman 2000). On the other hand, it is assumed that apprentice employees $n(t)$ can be hired or fired instantaneously (at least relative to the other delays in the model). Here $n(t)$ is referred to as a control variable, because it makes changes to the experienced employee stock instantaneously and is a decision under managerial control (see figure 11.1).

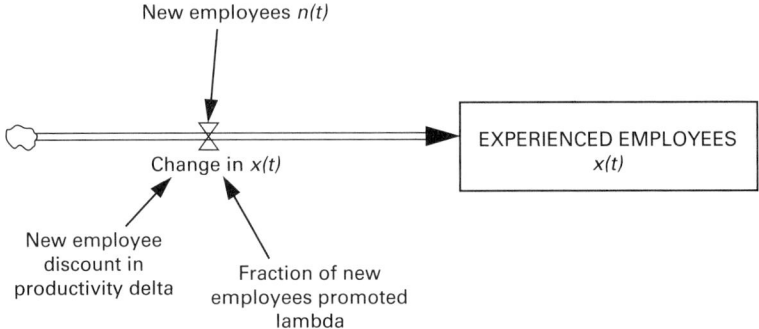

Figure 11.1
Stock and flow of simple personnel model under growth.

With respect to demand for employees,

let $r(t)$ = the demand at time t expressed in experienced employee equivalents,
r_0 = the initial demand at time $t=0$, and
κ = the instantaneous demand growth rate per period.

Without loss of generality, each experienced employee is assumed to have a productivity of one task per unit time. Thus, demand $r(t)$ can be in experienced employee equivalents. Demand is assumed to grow exponentially:

$$r(t) = r_0 \exp(\kappa t),\tag{3}$$

where κ and r_0 are assumed nonnegative.

Objective Function
All optimization models must have some objective function (often called a penalty function) to optimize. The objective function for the current model is below.

Let ζ = the penalty cost of mismatches between demand and capacity,
β = the cost of holding employees, and
ρ = the fractional penalty discount rate per unit time.

$$Z = \min \int_0^\infty \exp(-\rho t)\left\{\zeta[r(t)-c(t)]^2 + \beta[(1-\delta)n(t)]^2\right\}dt\tag{4}$$

subject to equations (1)–(3),

where ρ, ζ, and β are assumed nonnegative. The first factor of the integrand discounts the objective function by a constant fraction ρ each year. Note that we limit $\rho \leq 2\kappa$ in order to make the integral finite. However, this condition is reasonable given the business situation modeled (Anderson 2001a). The second factor balances two terms. The first term is how much capacity $c(t)$ deviates from demand $r(t)$. Note that $c(t) = x(t) + \delta n(t)$, because apprentice employees are not as cost-effective as experienced employees, hence they are discounted by δ, where $0 \leq \delta < 1$. The second term captures the amount of labor lost to low apprentice productivity. Note that under the demand assumption in equation (3), the number of new workers cannot be negative. Finally, ζ is assumed without loss of generality to be unity for the remainder of this section.

There are typically a number of "judgment calls" in setting up any objective function. For example, in equation (4), we need to capture the fact that having too much capacity or too little are both problematic. Yet one could argue that the cost of one too many experienced workers should be different from one too few. Also, we need to capture that there is a penalty associated with equation (1) employing apprentice workers, because they are inefficient. There are good reasons for using the quadratic function in equation (4) as discussed in Anderson (2001a), but other formulations are possible as well. Given that this is a tutorial, we will not go into them here, but note only that (1) the "correct" objective function is not often immediately obvious; and (2) not all objective functions are equally tractable. In this model, the quadratic productivity penalty leads to an objective function that effectively strikes a *consistent, stationary* balance between service quality and productivity. However, if one is not careful, one can easily end up with an objective function that is tractable but not useful for management purposes.

An Optimal Control Policy
Given the constraints in equations (1)–(3), we can minimize the objective in equation (4) by using optimal control theory (Kamien and Schwartz 1981).

Let $\mathcal{H}_c$ = the Hamiltonian, and
$m(t)$ = the adjoint variable associated with equation (2) at time t.

The Hamiltonian is much like a dynamic version of a Lagrangian because it combines the objective function with a weighted combination of the dynamic constraints for each state variable. In general, each state variable (i.e., a stock in SD parlance) has one dynamic constraint, which represents all the in- and out-flows to this state variable. In this model, the sole state variable (stock) is the number of experienced workers $x(t)$, and its dynamic constraint is $\dfrac{dx(t)}{dt} = \lambda n(t)$ from equation (2). Each dynamic constraint in the Hamiltonian is weighted by an adjoint (or co-state) variable, which is similar to a shadow price in linear programming. For this model, then, the Hamiltonian is

$$\mathcal{H}_c = \exp(-\rho t)\left\{\left[r_0 \exp(\kappa t) - x(t) - \delta n(t)\right]^2 + \beta\left[(1-\delta)n(t)\right]^2\right\} + \lambda m(t)n(t) \quad (5)$$

for $t \geq 0$.

The interpretation of the adjoint variable $m(t)$ is that it represents the cost of deviating one unit of its stock variable from the optimal path per unit time. For example, if the value of x(t) at time 4 should be 6 experienced employees and is instead 4.5 experienced employees, and the shadow price m(4) is $2.0/experienced employee/hour, then the objective function Z will be approximately (1.5 experienced employees) * $2.0/experienced employee/hour = $3.0/hour.

Any solution to this problem must satisfy the following (Kamien and Schwartz 1981) conditions:

1. **Maximize the Hamiltonian with respect to each control variable**. In this case, the control variable is the number of apprentice employees $n(t)$ employed by the firm. Hence, for $t \geq 0$:

$$\frac{\partial \mathcal{H}_C}{\partial n(t)} = 0 \Rightarrow -2\exp(-\rho t)\delta\left[r_0 \exp(\kappa t) - x(t) - \delta n(t)\right] \quad (6)$$

$$+ 2\exp(-\rho t)\beta(1-\delta)^2 n(t) + \lambda m(t) = 0$$

2. **For each state variable, satisfy its corresponding adjoint (or co-state) equation**. Hence, for $t \geq 0$:

$$\frac{dm(t)}{dt} = -\frac{\partial \mathcal{H}_c}{\partial x(t)} = 2\exp(-\rho t)\left[r_0 \exp(\kappa t) - x(t) - \delta n(t)\right]. \quad (7)$$

3. **Satisfy each dynamic constraint**. Formally, this can be recovered by setting the time derivative of each state variable equal to the deriva-

tive of the Hamiltonian with respect to its corresponding adjoining variable. Hence, for $t \geq 0$:

$$\frac{dx(t)}{dt} = \frac{\partial \mathcal{H}_c}{\partial m(t)} = \lambda n(t) \cdot \tag{8}$$

4. **Each adjoint variable must satisfy a terminal (often referred to as the transversality) condition**. In this case, as in many well-behaved models, eventually the state variable must reach an equilibrium that minimizes the static value of the penalty function Z. In other words,

$$\lim_{t \to \infty} m(t) = 0 . \tag{9}$$

Note that other boundary conditions can apply, such as the state variables having a terminal value. See Kamien and Schwartz (1981) for an extended discussion.

5. **Guarantee optimality**. Satisfying the conditions in equations (6)–(9) only guarantees that a solution is an extremal, which is a dynamic analogue of an extremum. This extremal can in fact be either a maximum, a minimum, or the dynamic equivalent of a saddle point. To ensure optimality, some other condition must be invoked. There are a number of conditions that will satisfy this (see Kamien and Schwartz 1981 for an extended discussion). The one most commonly used in practice, however, is this: *if a solution satisfies conditions [1] through [4] listed above and the integrand and state constraint(s) are all jointly convex functions of the state and control variables, then that solution will be optimal*. This criterion is universal in that it is a sufficiency condition that will work for any optimal control problem. Fortunately, the objective function's integrand and state constraint are indeed jointly convex functions of $x(t)$ and $n(t)$, so any solution that satisfies equations (6)–(9) will be optimal.

Solving for an equation that satisfies equations (6)–(9) yields the following optimal policy for this particular problem:

$$x(t) = Ar(t) + (x_0 - Ar_0)\exp(-Bt) \text{ for } t \geq 0 \tag{10}$$

$$n(t) = \kappa\lambda^{-1}Ar(t) - B\lambda^{-1}(x_0 - Ar_0)\exp(-Bt) \text{ for } t \geq 0, \tag{11}$$

where x_0 and r_0 are, respectively, the initial values of experienced employees and capacity requirements, and where

$$A = \frac{1 + \lambda^{-1}\delta(\rho - \kappa)}{\lambda^{-2}\kappa(\rho - \kappa)\left[\delta^2 + \beta(1 - \delta)^2\right] + \lambda^{-1}\delta\rho + 1} \tag{12}$$

and

$$B = -\frac{\rho}{2} + \frac{\lambda}{2}\left[\rho^2\lambda^{-2} + 4\frac{\lambda^{-1}\rho\delta + 1}{\delta^2 + \beta(1 - \delta)^2}\right]^{\frac{1}{2}}. \tag{13}$$

This is the optimal policy for the model described in equations (1)–(3). Any deviation from this policy will invoke a cost greater than the minimum necessary, as specified by the objective function in equation (4).

Note that this policy is linear in the state variables. This is, in fact, always true for penalty functions that are quadratic in the state and control variables, which explains in part why quadratic penalty functions are so often used in the literature.

Discussion

It is worthwhile to note a couple of properties of this solution that are common among many optimal solutions for dynamic systems. One is that, while this model is relatively simple, the policy at first glance appears complex. However, this appearance is deceptive. The policy breaks into two components: a transient component, which involves B and dies out relatively quickly, and a permanent component, which does not involve B and persists so long as $t > 0$. This is a common feature of many optimal control policies. Secondly, the persistent component of this policy essentially involves setting experienced employees $x(t)$ and apprentice employees $n(t)$ at a constant fraction of demand. That fraction is constant and can be easily analyzed for insights. For example, it is straightforward to demonstrate using comparative statics that the fraction of all employees who are experienced increases in the training completion rate λ, decreases in the demand growth rate κ, and is unaffected by the discount rate ρ and the relative productivity of apprentice employees δ. These findings can thus provide a guide to design a policy to manage new employees under a growth scenario. In general, such findings that supply the building blocks of a superior policy to control a dynamic system are often the most important fruits of an application of optimal control to a simplified system's model.

Of course, most real-life systems incorporate nonlinearities and other complexities beyond what is shown in figure 11.1. Still if the system even approximates that in figure 11.1 to a material degree, these insights can be used to design a policy that is approximately optimal. In fact, this is quite common in the realm of control theory, both in engineering and its applications to system dynamics. In particular, as we shall show in an example below, the most common method for dealing with nonlinearities is to create a linear approximation (Özveren and Sterman 1989; Stengel 1994). The guiding principle is that, if one can create a dynamic model that is tractable and materially approximates the real system even for a short time, one can often determine an optimal control policy that will provide insights into how best to manage the real system. Of course, such a policy must be tested using standard system dynamics policy sensitivity tests (Sterman 2000) in order to guarantee the robustness of any such policy derived from optimal control analysis.

Having now demonstrated the use of optimal control in a simple deterministic model, we now turn to one in which stochastic variables play a part.

Optimal Control with Stochastic Disturbances

Employees
The model below is adapted from Anderson (2001b). It is quite similar to the model just described. However, there are some differences, particularly with respect to the demand formulation and the objective function. A stock-and-flow diagram is in figure 11.2. Note that equation (15) guarantees that layoffs will never occur when there are apprentice employees, and vice versa.

Let $x(t)$ = the number of experienced employees at time t,
$n(t)$ = the number of apprentices at time t,
$l(t)$ = the rate of experienced employee layoffs per unit time at time t,
ε = the fractional turnover rate of experienced employees per unit time, and
λ = the fractional rate at which apprentices complete their apprenticeship and become experienced employees. We will call this the "training completion" rate.

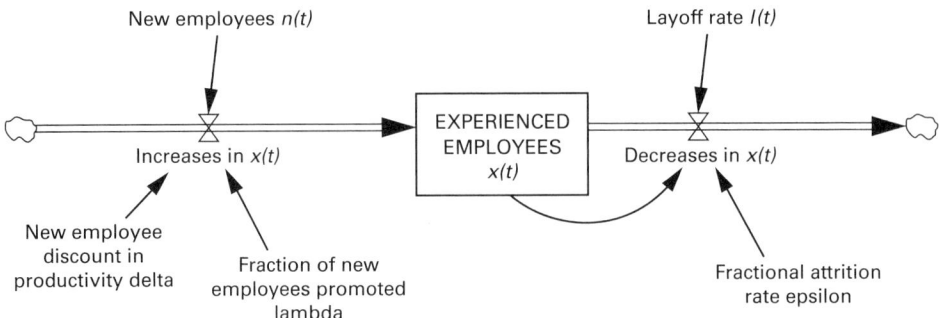

Figure 11.2
Stochastic optimization model.

$$\frac{dx(t)}{dt} = \lambda^{-1} n(t) - l(t) - \varepsilon x(t) \tag{14}$$

$$n(t)l(t) = 0 \tag{15}$$

$$x(t), n(t), l(t) \geq 0 \tag{16}$$

Demand

Demand is now a stochastic process. In particular, it is modeled by a Brownian motion with a drift process. This process mimics observed business cycles over the short term (Anderson 2001b).

Let t = time,
$r(t)$ = demand at time t,
$b(t)$ = a Brownian motion process with drift underlying demand,
γ = the drift rate of the underlying Brownian motion process,
β = a parameter controlling the degree of serial autocorrelation in demand (in essence, the "pink noise" coefficient),
$v(t)$ = independent, identically distributed zero-mean Gaussian process driving $b(t)$ (i.e., a "white-noise" process),[1] and
σ_v = the standard deviation of $v(t)$.

Formally, the equations for the demand process underlying figure 11.3, which is a Brownian motion with drift process, are

$$\frac{db(t)}{dt} = \gamma + v(t) \tag{17}$$

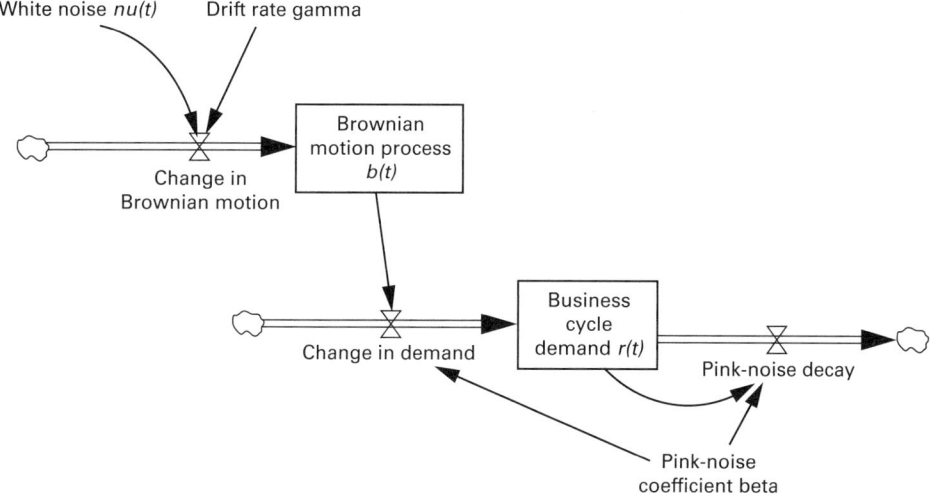

Figure 11.3
Model of business cycle demand.

$$\frac{dr(t)}{dt} = \frac{b(t) - r(t)}{\beta}. \tag{18}$$

The Objective Function

The objective function is similar to that of the previous problem. It is a weighted combination of the gap between demand and actual employee capacity and the rate of change in employee capacity. However, there are always compromises that must be made between fidelity of the model and its tractability when using analysis as opposed to simulation. For example, we now have to not only hire but also fire employees, because demand may decrease over time. The issue is that the number of new employees was a control variable in the previous problem. We could approximate firing in the new problem by allowing new employees $n(t)$ to be negative, which is a bit bizarre on its face and would imply the rate of firing has a negative effect on total employee productivity. Or we could make new employees a state variable, although this makes finding an analytic solution more difficult. One way to resolve this problem is to simplify it by assuming that new employees $n(t)$ do not contribute to employee capacity because of their inefficiency and need for mentoring by experienced employees, which has been justified by many interviews carried out by the authors with industry experts. The parameter α puts a weight on how important the gap in experienced employees is relative to the cost of hiring or firing them. There is no real need to square α; we only do this because it reduces the number of radicands in the solution, thus making it more readable. Finally, the objective function is also discounted by the rate ρ. Note that here "E" is the expectation operator.

Thus, the objective function is

$$Z = \min E \int_0^\infty \exp(-\rho t) \left\{ \alpha^2 \left[r(t) - x(t) \right]^2 + \left[\lambda \frac{dx(t)^2}{dt} \right]^2 \right\} dt \ . \tag{19}$$

Forecasting and Separability

We now face a problem because, generally speaking, the problem of forecasting the random processes that drive inputs to the system in the future cannot be solved separately from the control problem. In other words, an intuitive way to solve control problems with stochastic inputs is to (1) figure out the forecast of the state variables without concern for the control policy and (2) then develop an "optimal" solution using that forecast, assuming that the resulting control policy does not interact with the forecast. Using a forecast in this way is often referred to as treating it in a certainty-equivalence manner. Using certainty-equivalent control is common practice, and it often works quite well. But it is not, in general, optimal because the forecast and control policy often mutually interact (Stengel 1994).

However, there is an important exception that can make the certainty-equivalence approach optimal. If (1) the objective function for the control problem is a linear combination of terms that are quadratic functions of the control and state variables; (2) the changes in state variables are linear in the control variables and the random inputs; and (3) the random inputs are Gaussian processes, then the forecast and control problems can be solved separately, and optimal control will be achieved (Stengel 1994). This is often referred to as the separability principle. This was part of the reason we simplified the model earlier to ensure these conditions would be met without sacrificing too much fidelity to reality.

A Gaussian process is, formally speaking, a process in which any collection of realizations drawn from that process are (1) separately each normally distributed and (2) together are characterized by a multivariate normal distribution. In practice, this means that a Gaussian process must result from summing white-noise processes (i.e., each realization is normally distributed, and each realization is statistically independent of the others) or linear functions such as integration, differentiation, and so on of them. Such a problem is so common in the literature that it is often referred to as the linear-quadratic-Gaussian (or LQG) problem. Importantly, even when the system to be controlled deviates from these three assumptions, an approximation (most often by linearizing the system) is used to achieve near-optimal control in many cases (Stengel 1994).

Note that not any forecast can be used. In particular, the forecast must be a minimum, mean square error (or MMSE) forecast (sometimes known as Wiener predictor). Assuming some standard regularity conditions (such as $r(t)$ being continuously differentiable), we now find the MMSE forecast for demand for the current problem.

Let $\hat{r}(t,u)$ = the expected demand for time $t+u$ forecasted at time t. To obtain an MMSE prediction of demand at time $t+u$ forecast at time t for a linear system, whose parameters are constant, the error in the forecast must be uncorrelated with past demand and unbiased (Shanmugan and Breipohl 1988). That is, given $r(t)$:

$$\text{cov}[r(t+u)-\hat{r}(t,u), r(t-y)] = 0 \ \text{ for t,u,y} \geq 0. \tag{20}$$

To find such a forecast, we transform equation (18) into an integral equation,

$$r(t+u) = r(t) + \int_0^u \frac{b(t+w)-r(t+w)}{\beta} dw \ \text{ for t,u,w} \geq 0. \tag{21}$$

The expectation of $r(t+u)$ forecast at time t is (using previously developed notation):

$$\hat{r}(t,u) = E\left[r(t) + \int_0^u \frac{b(t+w)-r(t+w)}{\beta} dw \right] \ \text{ for t,u,w} \geq 0. \tag{22}$$

Because $\hat{r}(t,u)$ is the expectation of $r(t+u)$, it will be unbiased. Furthermore, both $b(t)$ and $r(t)$ are continuous. Also, because $b(t)$ is Markovian, its expectation is

$$E[b(t+w) \,|\, b(t)] = b(t) + \gamma w \ \text{ for t,w} \geq 0. \tag{23}$$

Thus, one can move the expectation operator inside the integral in equation (22):

$$\hat{r}(t,u) = r(t) + \int_0^u \frac{[b(t)+\gamma w] - \hat{r}(t,w)}{\beta} dw \ \text{ for t,u,w} \geq 0. \tag{24}$$

Solving this integral equation for $\hat{r}(t,u)$,

$$\hat{r}(t,u) = r(t) + \gamma u + [b(t) - r(t) - \gamma\beta]\left[1 - \exp\left(-\frac{u}{\beta}\right)\right] \quad \text{for } t,u \geq 0. \tag{25}$$

Substituting in from equation (18) for $b(t)$, and canceling terms:

$$\hat{r}(t,u) = r(t) + \gamma u + \beta\left[\frac{dr(t)}{dt} - \gamma\right]\left[1 - \exp\left(-\frac{u}{\beta}\right)\right] \quad \text{for } t,u \geq 0. \tag{26}$$

This yields the final required expression:

$$\hat{r}(t,u) = r(t) + \gamma u + \beta\left[\frac{dr(t)}{dt} - \gamma\right]\left[1 - \exp\left(-\frac{u}{\beta}\right)\right] \quad \text{for } t,u \geq 0. \tag{27}$$

Some notes on this forecast are in order, because they are typical of MMSE forecasts in general. The forecast is initially determined at time t by current demand. Then there is a linear component, which is a function of the mean stochastic trend γ. Finally, there is also a transient term (the third term on the right-hand side of equation [27]). This dies out after a period of time (approximately 2β years for this problem) as the effect of any recent autocorrelation decays (see Oppenheim, Willsky, and Young 1983 for details). This three-component structure is typical of many MMSE forecasts in practice.

Control Policy
Equipped with the MMSE forecast, we can now proceed to determine the optimal control policy for this problem.

Let $x(t,u)$ = number of experienced employees (or capacity) at time $t+u$ given a policy determined at time t,
$n(t,u)$ = number of apprentices at time $t+u$ given a policy determined at time t, and
$l(t,u)$ = rate of experienced employee layoffs at time $t+u$ given a policy determined at time t.

The $x(t,u)$, $n(t,u)$, and $l(t,u)$ notations above indicate that these are the optimal policies at time t for u time units ahead. Conceptually, this would be similar to a u-step look-ahead policy in a discrete model.

Then, given the initial number of experienced employees, x(t,0), we can rewrite the certainty equivalent problem as follows:

$$\min \int_0^\infty \exp[-\rho(t+u)]\left\{\alpha^2[\hat{r}(t,u) - x(t,u)]^2 + \left[\lambda^{-1}\frac{dx(t,u)}{du}\right]^2\right\}du, \tag{28}$$

subject to $\dfrac{dx(t,u)}{du} = \lambda^{-1}n(t,u) - l(t,u) - \varepsilon x(t,u)$ for $u \geq 0$,   (29)

$n(t,u)l(t,u) = 0$ for $u \geq 0$, and   (30)

$x(t,u), n(t,u), l(t,u) \geq 0$ for $u \geq 0$.   (31)

Solving the problem using the same methodology shown for the earlier deterministic model in equations (1)–(4) yields the following optimal control policy.

$$\dfrac{dx^*(t,u)}{du}\bigg|_{u=0} = K_0[r(t) - x(t,0)] + K_G\gamma + K_A\beta\left[\dfrac{dr(t)}{dt} - \gamma\right] \text{ for } t \geq 0 \qquad (32)$$

$$K_0 = \left[\left(\dfrac{\rho}{2}\right)^2 + \alpha^2\lambda^2\right]^{\frac{1}{2}} - \dfrac{\rho}{2} \qquad (33)$$

$$K_G = 1 + \rho\dfrac{\rho\lambda^{-1} - \sqrt{\rho^2\lambda^{-2} + 4\alpha^2}}{2\alpha^2\lambda} \qquad (34)$$

$$K_A = \dfrac{1}{2} \bullet \dfrac{\lambda^{-1}\sqrt{\rho^2\lambda^{-2} + 4\alpha^2} + \dfrac{2\alpha^2\beta}{1+\rho\beta} - \rho\lambda^{-2}}{\lambda^{-2} - \dfrac{\alpha^2\beta^2}{1+\rho\beta}} \qquad (35)$$

Setting u to zero gives us the optimal control policy, when used in conjunction with the MMSE forecast, as is true for all LQG problems, as discussed in the prior section (Stengel 1994).

$$\dfrac{dx^*(t)}{dt} = \dfrac{dx^*(t,u)}{du}\bigg|_{u=0} = K_0[r(t) - x(t,0)] + K_G\gamma + K_A\beta\left[\dfrac{dr(t)}{dt} - \gamma\right] \qquad (32b)$$

This implies that the optimal policy for control variables, which are the number of apprentice workers and the layoff rate, at time t are

$$n^*(t) = \lambda^{-1}\max\left[\dfrac{dx^*(t)}{dt} + \varepsilon x^*(t), 0\right] \qquad (36)$$

$$l^*(t) = \min\left[\dfrac{dx^*(t)}{dt} + \varepsilon x^*(t), 0\right]. \qquad (37)$$

Discussion

A couple of notes arising from this policy are appropriate. First, we are implicitly linearizing the model by assuming that the number of experienced workers never approaches zero. If such a situation were to happen, this policy would no longer be appropriate. Fortunately, empirical studies show that this condition is relatively rare with experienced workers in ongoing technical businesses, for which this model was specified (Morrice, Anderson, and Bharadwaj 2004). Nonetheless, a robustness test of how well this policy performs under extreme conditions is called for. More generally, the structure of the control variables is relatively simple. For both new employees and layoffs, the control variable is a linear combination of state variables and linear operations (such as integration or differentiation) on the MMSE forecast. This is the general form for optimal policies for LQG problems (Sethi and Thompson 2000; Stengel 1994). Only the derivation of the parameters is complex. Again, performing an analysis of how these constants change as a function of their parameters can yield interesting insights. Finally, because the optimally controlled state variables for LQG problems are invariably Gaussian processes themselves, predictions and descriptive statistics of their future behavior are straightforward (Anderson 2001b).

Multiple Variables and Measurement Error

Many problems have multiple state variables, which makes the methods developed for one state variable more complex. We illustrate in this next example how to handle multiple variables using the state-space method to represent systems and its Riccati equation. The Riccati equation is a formulation that simplifies the process for finding the optimal solution for linear systems with quadratic penalties. We illustrate this method using a project management example drawn from Joglekar, Anderson, and Shankaranarayanan (2013). (Note that the portions adapted in this paper come primarily from an earlier working paper, Anderson and Joglekar 2010.)

Another problem we face is that even the measurements of the state variables we seek to control are only approximate (Stengel 1994), as is discussed in chapter 4. For example, in Joglekar, Anderson, and Shankaranarayanan (2013), the status of the number of projects remaining to be completed is only approximately known. The reason for this is the presence of many projects that are, for example, 40% complete, in which case 0.6 projects remain to be completed. This number tends to be only approximate because of mistakes resulting from aggregating numerous progress status reports, the difficulty estimating the number of hours needed to complete any given project *ex ante,* and so on. The point is that one might think that, for example, 40% of a project is completed, when in fact only 30% of it actually has been completed. This creates a problem in which not only are future forecasts uncertain, but so too are the current values of the state variables that must be managed. In this example, we also show how to leverage the Kalman filtering from chapter 4 to overcome this problem.

The Problem

The base model which we seek to optimally control is shown in figure 11.4. Standing at the center of this model is a software engineering firm that has a number of project tasks that need to be completed to finish its current projects. We refer to this firm as the principal. The principal has a constant capacity (number of employees). The demand for new work feeds the principal's work in progress (WIP). For convenience, but without loss of generality, we now measure the projects, as adjusted by their progress, by the estimated work hours necessary to complete them (i.e., two projects of 5,000 hours each, each of which are 30% completed, would represent a WIP of 3,000 hours) in keeping with the practice of the firm modeled in Joglekar, Anderson, and Shankarana-rayanan (2013). The principal keeps and executes work internally ("insourcing") and/or outsources to a partner. The partner has flexible capacity—it can change the number of its employees each period so that it can accept all work given and complete each after a constant flow time—consistent with current practices in software outsourcing. It is important to note that, without loss of generality, this partner could represent the aggregate characteristics of a number of multiple partners. Both the principal and the partner can complete work; however, a fraction of the work completed by the partner must pass back to the principal for integration, such as testing, which is performed internally (Ford and Sterman 1998). Each project thus completed induces a reintegration penalty, which is typical of real-world outsourcing of product and process (including software) development (Parker and Anderson 2002).

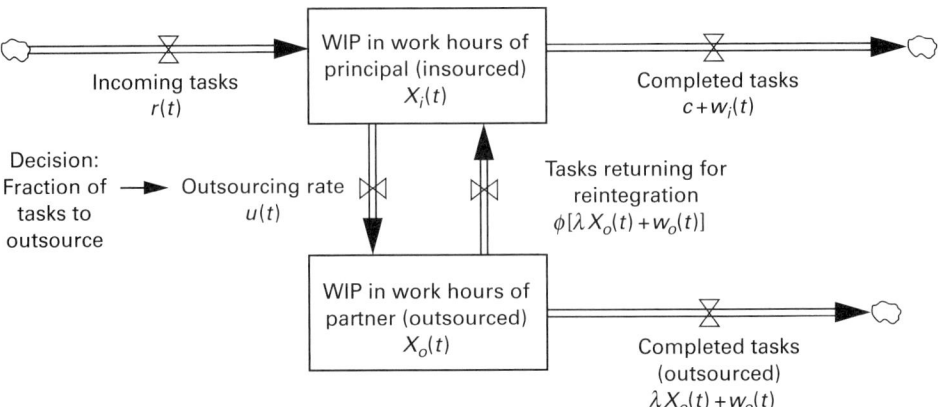

Figure 11.4
Model of outsourcing projects.

Let

$r(t)$ = end-customer demand rate,

c = completion rate at the principal, a constant,

λ = fractional completion rate of WIP per unit time at the partner,

$w_i(t)$ = the net variation in the changes in stock of WIP at the principal,

$w_i(t) \sim N(0, \sigma_i^2)$ and $\text{cov}(w_i(t), w_i(t+\varepsilon)) = 0$, for $\varepsilon \neq 0$,

$w_o(t)$ = the net variation in the changes in stock of WIP at the partner,

$w_o(t) \sim N(0, \sigma_o^2)$ and $\text{cov}(w_o(t), w_o(t+\varepsilon)) = 0$, for $\varepsilon \neq 0$,

$x_i(t)$ = the WIP, measured in number of hours of work, at the principal at time t,

$x_o(t)$ = the WIP, measured in number of hours of work, at the partner at time t,

ϕ = the fraction of projects completed at the partner requiring reintegration at the principal $(0 \leq \phi < 1)$, and

$u(t)$ = the number of projects transferred from the principal to the partner at time t.

Note when $u(t) < 0$, re-insourcing of previously outsourced projects occurs.

The states of the system evolve as follows:

$$\frac{dx_i(t)}{dt} = r(t) - c - w_i(t) - u(t) + \phi[\lambda x_o(t) + w_o(t)] \tag{38}$$

$$\frac{dx_o(t)}{dt} = u(t) - [\lambda x_o(t) + w_o(t)]. \tag{39}$$

Equation (38) tracks the rate of change of the insourced WIP. For ease of understanding, $x_i(t)$ is the work in progress, measured in number of hours of work, at the principal at time t. Referring to the arrows in figure 11.4, $r(t)$ is the incoming tasks, $c(t)$ is the completion rate of tasks, $u(t)$ is the fraction of tasks outsourced to partners, and the last term represents the tasks that return to the principal for reintegration. Further, $w_i(t)$ is the net variability created by rework at the principal site itself. Equation (39), which tracks the rate of change of the out-sourced WIP, is analogous. Equations (38) and (39) are linked by (1) the decision variable $u(t)$ and (2) the amount of reintegration work needed, which is determined by the fraction ϕ.

Optimal Multivariate Control

First, let us assume we have perfect data. The objective function looks forward from the present time $t = 0$ seeking to balance WIP, capacity, and capacity-adjustment costs.

Let

β_i^2 = penalty on insourced backlog (square is merely for analytic convenience), $\beta_i{>}0$,

β_o^2 = penalty on outsourced backlog (square is merely for analytic convenience), $\beta_o{>}0$,

$\tilde{x}_i$ = target insourced WIP, $\tilde{x}_i \geq 0$, and

$\tilde{x}_o$ = target outsource WIP, $\tilde{x}_o \geq 0$.

Note β_i^2 and β_o^2 are squared only for convenience in expressing the control law. Hence, the objective function is still of the quadratic form, and the problem as a whole is LQG, just as in the previous section. The objective for aggregate planning in this case is to trade-off our desire to keep the two WIPs near their nominal "target levels," thus "smoothing" them, against the fact that outsourcing and re-insourcing projects to make these adjustments is also expensive in terms of effort, resources, and so on. The target WIPs are assumed to be exogenous to this calculation, which accords with our experience with many firms as well as the research literature on this topic (Morrice, Anderson, and Bharadwaj 2004).

Thus our objective function is

$$\min_{u(t)} J = \lim_{t_f \to \infty} \frac{1}{t_f} \int_0^{t_f} \left\{ u^2(t) + \beta_i^2 [X_i(t) - \tilde{X}_i]^2 + \beta_o^2 [X_o(t) - \tilde{X}_o]^2 \right\} dt . \tag{40}$$

Note that we are using J instead of Z to represent the objective function for this section to avoid confusion with the noisy measurement variables, which will be introduced later. The first term of the objective function reflects the quadratic cost associated with the control effort, and the second and third terms account for the quadratic costs associated with deviations from the target insourced and outsourced WIP. Fundamentally, the objective function in equation (40) performs an infinite horizon trade-off between smoothing each stage's WIP around a target level and minimizing costs arising from shifting projects between principle and partner. Interestingly, the form of this penalty is *exactly analogous* to that of a pair of classic models in the operations management literature, the HMMS model (Holt et al. 1960) and Sethi and Thompson's (2000) inventory model. In these models, the penalty for undershooting the WIP "target" existed to avoid idling workers through WIP starvation. Somewhat more complex, yet analogous issues apply to WIP in professional services. Because the opportunity cost of lost revenue relative to salary costs is on the order of a factor of 20, generally development firms consider zero-WIP (which implies idling capacity) unacceptable. The cost of switching projects is also quadratic-weighted—as is typical in the capacity planning (e.g., Holt et al. 1960 and Sethi and Thompson 2000), for a number of reasons. Moving projects from one site to another often involves some sort of set-up cost (Anderson and Joglekar 2010). Further, transferring five projects per month from the principal to a partner is logically much simpler than 50. Hence, the coordination problem tends to escalate in a convex manner, suggesting the quadratic form. In summary, the objective function is designed to balance project-switching costs against service capacity utilization at both principal and partner.

The model in figure 11.5 builds on that in figure 11.4 by allowing inaccuracies in project-status data, albeit the figure is now more conceptual for simplicity. The data are filtered through a mechanism to control the outsourcing decision using a control law derived assuming perfect data. The managers keep aggregate statistics (means and variance) on the estimation and measurement errors. These measurement errors are assumed to be independent, white-noise processes, as described in the previous section.

The Kalman (or Kalman-Bucy) filter shown in figure 11.5 is similar to that in chapter 4 of this book. However, it has a different goal, which is to ameliorate these estimation and measurement errors by producing an estimate of the states of the model that is "better" than directly

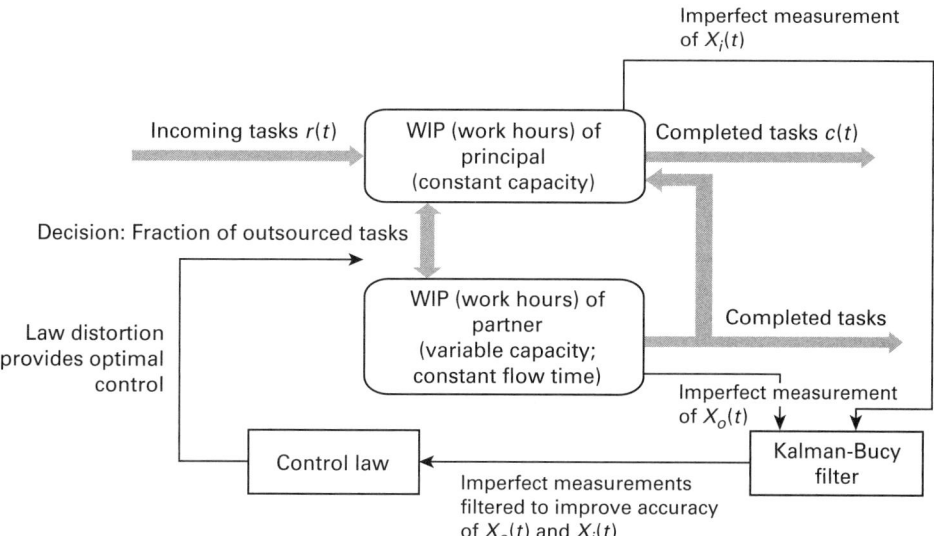

Figure 11.5
Closed loop control with inaccurate data.

looking at the measurements with errors. In fact, using a Kalman-Bucy filter in this manner exploits the separability principle discussed in the prior section. Specifically, we use the filter to obtain optimal MMSE estimates of the state variables. Once the filter provides these estimates, we will then use a previously developed control law for a deterministic system (without stochastic disturbances) on those MMSE estimates to provide an optimal result. We will also present the generalizable Matrix Riccati equations that embody many of the necessary conditions for dynamic optimality discussed in the prior sections.

First, we derive the optimal control law based on the objective function in equation (40). We have to deal with two state variables, noise variables, and so on. For convenience, we write them in matrix form (Stengel 1994). Let

$$\mathbf{x}(t) = \begin{bmatrix} x_i(t) \\ x_o(t) \end{bmatrix}, \mathbf{w}(t) = \begin{bmatrix} w_i(t) \\ w_o(t) \end{bmatrix}, \mathbf{z}(t) = \begin{bmatrix} z_i(t) \\ z_o(t) \end{bmatrix}, \mathbf{n}(t) = \begin{bmatrix} n_i(t) \\ n_o(t) \end{bmatrix}, \tag{41}$$

and

$$\mathbf{F} = \begin{bmatrix} 0 & \phi\lambda \\ 0 & -\lambda \end{bmatrix}, \mathbf{G} = \begin{bmatrix} -1 \\ 1 \end{bmatrix}, \mathbf{Q} = \begin{bmatrix} \beta_i^2 & 0 \\ 0 & \beta_o^2 \end{bmatrix}, \mathbf{R} = [1], \mathbf{W} = \begin{bmatrix} \sigma_i^2 & 0 \\ 0 & \sigma_o^2 \end{bmatrix}, \mathbf{N} = \begin{bmatrix} v_i^2 & 0 \\ 0 & v_o^2 \end{bmatrix}. \tag{42}$$

Recall that the vectors $\mathbf{x}(t)$, $\mathbf{w}(t)$, $\mathbf{z}(t)$, and $\mathbf{n}(t)$ represent WIP; variation in changes to WIP; WIP that is observed and has measurement errors (i.e., is inaccurate); and the amount of measurement noise with respect to WIP because of quality errors at a given time t. Here $\mathbf{F}$, $\mathbf{G}$, $\mathbf{Q}$, $\mathbf{R}$, $\mathbf{W}$, and $\mathbf{N}$ represent the matrices for completion-rate fractions; effect of the control variables on the rate of change of the state variables; the cost penalties for the state variables' deviations from targets; the cost penalty for the control variable; the variance in $\mathbf{w}(t)$ defined earlier; and the variance in $\mathbf{n}(t)$ defined earlier. Accordingly, $n_i(t)$ = the noise in the estimation at the principal firm, $n_i(t) \sim Norm(0, \upsilon_i^2)$ and $\mathbf{corr}(n_i(t),$ $n_i(t+\varepsilon)) = 0$, for $\varepsilon \neq 0$. The variable $n_o(t)$ is the noise in the data at the partner site. It is distributed as $Norm(0, \upsilon_o^2)$ and $\mathrm{corr}(n_o(t), n_o(t+\varepsilon)) = 0$, for $\varepsilon \neq 0$. We write the evolution of the state variable (i.e., number of insourced and outsourced projects) as

$$\begin{bmatrix} x_i'(t) \\ x_o'(t) \end{bmatrix} = \mathbf{F}[\mathbf{x}(t) - \tilde{\mathbf{x}}(t)] + \mathbf{G}u(t) + \mathbf{w}(t) . \tag{43}$$

If the information is imperfect due to measurement error, the observed state variables are $\mathbf{z}(t)$:

$$\mathbf{z}(t) = \mathbf{x}(t) + \mathbf{n}(t). \tag{44}$$

Using the notation developed in equations (41)–(44), we can rewrite the objective function equation (40) in matrix notation as

$$\min_{u(t)} J = \lim_{t_f \to \infty} \frac{1}{t_f} \int_0^{t_f} \left\{ \mathbf{R}u^2(t) + \mathbf{Q}[\mathbf{x}(t) - \tilde{\mathbf{x}}]^T [\mathbf{x}(t) - \tilde{\mathbf{x}}] \right\} dt, \tag{40b}$$

where $\mathbf{R} = [1]$ and $\mathbf{Q} = \begin{bmatrix} \beta_i^2 & 0 \\ 0 & \beta_2^2 \end{bmatrix}$.

The matrix notation developed in equations (41)–(44) is referred to as the state-space form of an optimal control problem and is commonly used by engineers and control theorists (Stengel 1994). (The superscript T indicates that the matrix or vector is transposed.)

Optimal Control with Perfect Data

Given this formulation, there is an equation that guarantees an optimal control law called the Riccati equation. That equation effectively compresses the work resulting from the minimum principle of the prior two sections. In essence, if we can find an $\mathbf{S}_c$ that satisfies the Riccati equation for this system (and any other system that can be written in this manner) and is positive-definite (for any nonzero vector of real numbers $\mathbf{a}$, $\mathbf{a}^T\mathbf{S}_c\mathbf{a} > 0$; or, alternately, all the eigenvalues—see chapter 7—are strictly positive), we can find the optimal control law. (In general, these are a system of linear equations that can be solved either through matrix manipulations or separately in the standard manner.) In this case, $\mathbf{S}_c$ must satisfy

$$\mathbf{F}^T\mathbf{S}_c + \mathbf{S}_c\mathbf{F} + \mathbf{Q} = \mathbf{S}_c\mathbf{G}\mathbf{R}^{-1}\mathbf{G}^T\mathbf{S}_c. \tag{45}$$

Solving for $\mathbf{S}_c$ yields:

$$\mathbf{S}_c = \frac{1}{\lambda(\phi-1)} \begin{bmatrix} \beta_i(\beta_i - \tilde{\beta}) & \beta_i\left[\beta_i + \lambda(1-\phi) - \tilde{\beta}\right] \\ \beta_i\left[\beta_i + \lambda(1-\phi) - \tilde{\beta}\right] & (\beta_i + \lambda)^2 - \lambda\phi(2\beta_i + \lambda) - [\beta_i + \lambda(1-\phi)]\tilde{\beta} \end{bmatrix}, \tag{46}$$

where $\tilde{\beta} = \sqrt{\beta_o^2 + \beta_i^2 + \lambda^2 + 2\lambda\beta_i(1-\phi)}$. Now that we have found $\mathbf{S}_c$, we can then find the optimal control policy by substituting it into the following equation, which is true for any positive-definite matrix that satisfies the Riccati equation.

$$u^*(t) = -\mathbf{R}^{-1}\mathbf{G}^T\mathbf{S}_c\left[\mathbf{x}(t) - \tilde{\mathbf{x}}\right] \tag{47}$$

This results in the following optimal control law for our current problem:

$$u^*(t) = \mathbf{k}_c\left(\mathbf{x}(t) - \tilde{x}\right), \tag{48}$$

where

$$\mathbf{k}_c = \left[\frac{\beta_i}{\beta_i + \lambda - \sqrt{\beta_o^2 + \beta_i^2 + \lambda^2 + 2\lambda\beta_i(1-\phi)}}\right]^T.$$

It is important to realize that, *if this model has no measurement error, we can use this equation directly as our control policy.* It will result in an outsourcing rate that is a weighted proportion of the two state variables: that is, the WIPs at the principal and at the partner.

Optimal Control with Inaccurate Data

Equipped with the optimal control law, we may still need to find an estimate of the state variables if we have inaccurate data with respect to the state variables, which in this example are the WIPs. If we can find the MMSE estimates, moreover, and substitute those estimates in for $x(t)$ in equation (48), then the resulting combined estimation and control policies will be jointly optimal because of the separability principle discussed in the previous example. The Kalman-Bucy filter (from chapter 4) provides the required MMSE necessary to satisfy the separability principle. Using the results of that chapter, we find that $\hat{x}(t)$, which represents the MMSE estimate of $x(t)$, evolves according to the following equation (note that these equations are in general form in equation [49]):

$$\hat{x}'(t) = \mathbf{F}[\hat{x}(t) - \tilde{x}(t)] + \mathbf{G}u(t) + \mathbf{K}_e[z(t) - \hat{x}(t)], \tag{49}$$

where

$$\mathbf{K}_e = \mathbf{P}\mathbf{N}^{-1}.$$

Now that we have the general form, we substitute in the particular parameters from our model. Note that $\mathbf{P}$ must, like $\mathbf{S}_c$, be positive-definite in order to yield the MMSE estimator. (One way to ensure this is to set the fraction of reintegration work as a result of outsourcing $\phi = 0$. Of course, this compromise does limit somewhat the applicability of our results). Then the solution for $\mathbf{K}_e$ is the following:

$$\mathbf{K}_e = \begin{bmatrix} \dfrac{\sigma_i}{v_i} & 0 \\ 0 & -\lambda + \sqrt{\lambda^2 + \dfrac{\sigma_o^2}{v_o^2}} \end{bmatrix}. \tag{50}$$

Now we have the MMSE estimates for the state variables. In order to find the optimal control policy, all we need do is to substitute the MMSE estimates for the state values in the control law derived for the system with perfect estimation. This yields:

$$u^*(t) = \mathbf{k}_c(\hat{x}(t) - \tilde{x}) \text{ where } \mathbf{k}_c = \left[\frac{\beta_i}{\beta_i + \lambda - \sqrt{\beta_o^2 + \beta_i^2 + \lambda^2 + 2\lambda\beta_i(1-\phi)}} \right]^T. \tag{51}$$

Combined with the MMSE estimator from equation (49), we now have a combined optimal control estimation and control policy.

Near-Optimal Control
At this point, we have learned to optimally control LQG models with random inputs and measurement error. Finding these results is, in

general, only manageable because of the separability principle. However, even if the system is not perfectly linear or Gaussian, if we can linearize the model in some manner (see Forrester 1982 or Özveren and Sterman 1989 for excellent examples), we can still obtain a near-optimal policy (Stengel 1994), which can provide the basis for a robust control policy for more complex system dynamics models. These approaches work reasonably well in many cases. When they do not, however, there are alternatives. We now turn to a discussion of these methods.

Modern Nonlinear Control Approaches

Near-optimal control policies can work in situations where the underlying model can be linearized suitably. However, nonlinear behavior must be retained at the core of many other SD formulations, which do not result in tractable control policies using the traditional methods. One way to overcome the limitations associated with linearization is to introduce a learning mechanism that allows the system to adapt to the changes in the state-space while accounting for the underlying nonlinear formulation. Key advances have been made in the field of reinforcement learning (RL) in this regard. A canonical form for RL formulation is shown in figure 11.6 (adapted from Sims 2013; arrows indicate the causal influence such as learning; also see Sutton and Barto 1998).

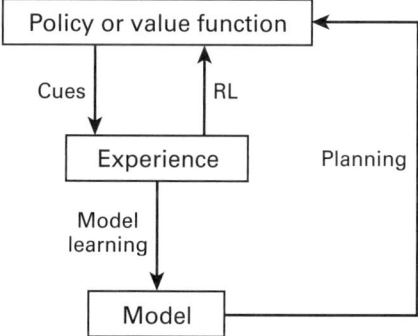

Figure 11.6
Relation between policies and models in RL.

Such learning mechanisms can either involve *model-based* or *model-free* formulations. Model-based learning assumes structural relationships and attempts to impute key parameters through the learning process. The filtering formulation described earlier (in equations [49] through [51]) is an example for the model-based approach. Model-free methods, on the other hand, treat the model as a black box and build the learning process around the inputs and outputs that are associated with such models. We limit the discussion in this section to the setup and implementation of model-free RL algorithms because the RL part of the formulation can be developed in a generic manner in order to improve a wide variety of SD models. That is, these algorithms do not assume that the learning process has access to either the structural details or specific parameters for the model whose behavior has to be controlled.

In order to describe RL logic, we draw upon a framework introduced by Rahmandad and Fallah-Fini (2008) that links the control mechanism (**u**) with a simulation model and a learning algorithm, as shown in figure 11.7. A comparison of this framework with the closed-loop control mechanism (shown previously in figure 11.5) is instructive. The two-stock structure on the top of figure 11.5 is replaced by a system dynamics simulation model (that may feature an arbitrary number of stocks involving either linear or nonlinear formulation). The Kalman-Bucy filter of the bottom of figure 11.5 is replaced by a learning mechanism based on temporal difference (TD). This learning mechanism involves two blocks: (1) state and action estimation process and (2) a Q-learning algorithm. The first block generates a look-up table of available states and the action pairs (**X**,**u**). The variable Q provides an estimate for the value function based on all projected state-action pair alternatives that are available. The system learns these Q values through an algorithm that is implemented in the second block. The intuition behind Q-learning is that the value of any state-action pair (Q(**X**,**u**)) depends on both the immediate reward received (R), and the value (g) of the best state-action pair accessible from the resulting state-space. That is, the model-free RL formulation decomposes the problem of estimating the value of different state-action pairs into two subproblems: an estimation of immediate reward from the SD simulation model, and an iterative estimation of the value associated with the resulting state-action pair.

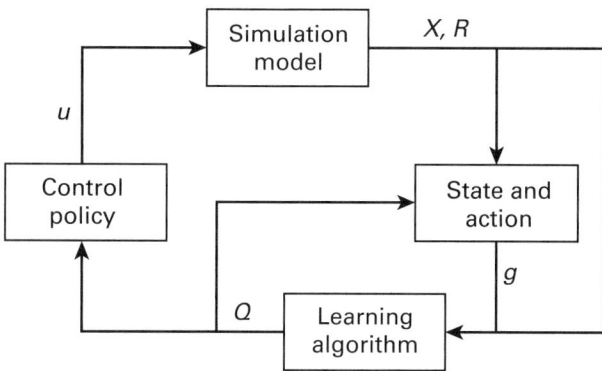

Figure 11.7
Learning and state-action aggregation process.

Much of the literature on Q-learning has been developed for discrete-time systems (Watkins and Dayan 1992). For instance, Rahman-dad and Fallah-Fini (2008) build their algorithm using a discretization formulation and integrate it with an SD model that features a nonlinear relationship between tasks completed and tasks under development. Their learning algorithm is implemented using the Anylogic 6.2 Advanced environment. The Q-learning ideas have also been applied to continuous-time systems (Doya 2000). Such formulations can be implemented within a meta-model, for example, a basic SD structure, and the Q-learning blocks are both incorporated within a Vensim simulation.

We refer readers to Doya (2000) for the details on how to recast the relevant objective function (e.g., equations [5] or [17] or [40]) into a Hamilton-Jacobi-Bellman value function. In order to gain notational clarity, we define $V^{\mu}(\mathbf{X})$ as a value function associated with state $\mathbf{X}$ for an action μ. The goal is to find an optimal policy $u(t) = \mu(\mathbf{X}(t))$ through a learning process. This learning problem can be formulated as the process of bringing the current policy and its value function estimate V closer to the optimal policy, so that it lies within an acceptable range of performance. There are two ways in which Q-learning can be accomplished.

1. Value iteration: estimate and improve the value function based on changing the current policy repeatedly.
2. Policy iteration: improve the policy by making it greedy with respect to the current estimate of the value function V.

While implementation of value and policy iterations are common in RL settings (Russell 2013), they are yet to be explored systematically within the SD literature. Hence, we provide a summary of the alternatives without implementation examples.

Value Iteration

For the learning of the value function in a continuous state-space, it is mandatory to use some form of function approximator. We denote the current estimate of the value function as

$$V^\mu(\mathbf{X}) \cong V(\mathbf{X}(t),\mathbf{w}), \tag{52}$$

where $\mathbf{w}$ is the parameter of the function approximator. The estimate of the value function is updated using a self-consistency condition:

$$d(V^\mu(\mathbf{X}))/dt = V^\mu(\mathbf{X})/\tau - R(t). \tag{53}$$

Here τ is the time constant for discounting the future rewards, and R is the immediate reward, as shown in figure 11.7. The inconsistency in the estimation, a.k.a. TD error, is measured in term of $\delta(t)$:

$$\delta(t) = R(t) + d(V^\mu(\mathbf{X}))/dt - (V(t))/\tau. \tag{54}$$

There are a number of alternative ways in which the reduction in TD error can be implemented. For instance, it has been done by systematically updating the level (V) and its slope: $d(V^\mu(\mathbf{X}))/dt$ (Baird 1993). An alternative method involves changing the V values as a function of time, such that the values further away in the future are exponentially discounted (Sutton and Barto 1998). The choice of the implementation method ought to be consistent with the way in which relevant organizations set up their learning processes. For instance, modelers ought to consider whether managers prefer to discount the inconsistencies that will creep in farther into the future, and then select a suitable rate (τ) for such discounting.

Policy Iteration

One way to improve the policy stochastically is to use the TD error as a reinforcement signal. This might be done by setting up the RL blocks as an associate search element and another adaptive critic element, respectively. The adaptive critic element receives the signal and decides on an improved signal that it sends to the associate search element for next cycle of search. Such a formulation is termed as the actor-critic method (Lendaris et al. 2004). Another approach for implementing policy iteration is based on setting up greedy policies with respect to the current value function. In continuous time, such greedy searches can be computationally expensive. However, for a special set of decision rules (e.g., managers may assume that reward R is convex in **X** and **u**), it is possible to implement efficient algorithms.

We have outlined two major approaches (e.g., value vs. policy iteration) for coming up with approximate, but acceptable, solutions to the control of nonlinear SD formulations. Beyond these approaches, there have also been a host of developments in the field of machine learning. For instance, modelers may take multiple cues from different parts of a model and set up neural networks to improve the estimation of state-action pairs. Within each of these approaches, there are a range of alternatives (e.g., actor-critic versus greedy searches) available to SD modelers. The efficacy of the selected approach depends on the nature of the search landscape as well as the objective for which the SD model is being formulated. As mentioned at the outset of this chapter, a key goal of the optimal control studies is to help managers in improving their behavioral choices associated with the underlying decision-making processes (Sterman 2000). Implementation of RL algorithms and allied studies of behavioral decision making are deemed to be a fruitful research frontier for SD scholars.

Roadmap

To sum up, in solving an analytic model for optimal control:

- One approach that often works is to use an LQG system, as described in the section "Multiple Variables and Measurement Error."
- If you are using an LQG system, you can use the Riccati equation to obtain an optimal control policy and, if necessary, an optimal estimator.
- If you have a nonlinear system, try to determine an LQG system that approximates the behavior of your model in the area of interest.
- If that fails and an LGC approximation will not work, try the methods described in the first section of this chapter ("A Simple Personnel Model Under Demand Growth") for control of nonlinear systems.
- If neither of those methods work, then use the reinforcement methods discussed in the previous section.

Conclusion

Many times, in system dynamics, it is not enough to describe what managers actually do. Instead, we need to characterize what they should do. To aid system dynamicists in this endeavor, we introduced the basic principles of stochastic optimal control. In particular, we first learned how to develop optimal control policies. Then we showed how this could be extended to linear-quadratic models with stochastic inputs (and even endogenous random variables) using the separability principle and an MMSE forecast. Then we gave an example of managing systems in which even the state variables (stocks) cannot be known with precision using the separability principle and the Kalman-Bucy filter. Throughout this chapter, we have discussed how these techniques could be used even with SD models that are not perfectly linear. Finally, we describe some newer techniques to supplement these traditional methods that may be more appropriate in some situations. Armed with these techniques, we hope that system dynamicists can supplement their assessment of managerial decision making observed in practice using a scientific approach that is grounded in control theory formulations.

Exercise

A practice exercise is offered below. You must manage a backlog $b(t)$ with a white-noise input of orders $r(t)$. As in many services, you can only control the backlog by adjusting your capacity upward or downward. You cannot smooth out the workflow with an inventory. Ideally, you want to drive the backlog toward a target level (many service firms do this), which we will call $\tilde{b}$. Below this level, the usage of capacity becomes inefficient (i.e., "too many cooks in the kitchen"). Above that level, backlog becomes ever more costly. In the long run, your capacity will necessarily equal your orders: $E[c(t)]=E[r(t)]$. However, you will need to adjust capacity in the short run to manage your backlog, and changing capacity $c(t)$ is costly. In essence, you will need to balance these two costs. Without loss of generality, we can set the cost of deviating from the target backlog at β and the cost of adjusting capacity at 1. To further simplify the problem—again without loss of generality—we will assume that each unit of capacity can process one unit of backlog per unit time.

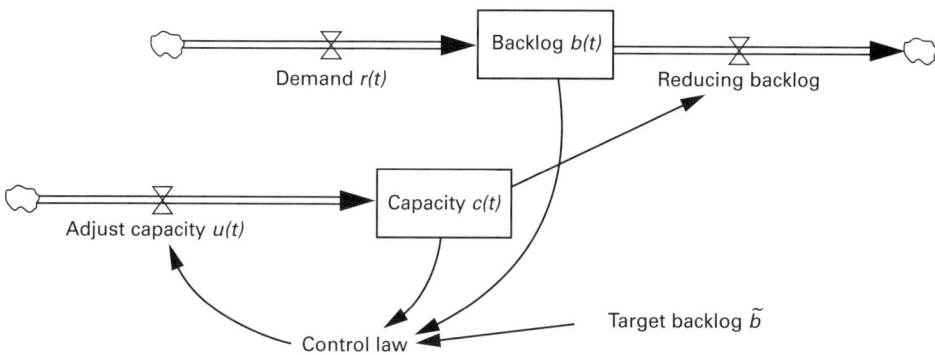

Figure 11.8
Example problem.

To summarize, your problem is

$$\min_{u(t)} J = \lim_{t_f \to \infty} \frac{1}{t_f} \int_0^{t_f} \left\{ u^2(t) + \beta \left[b(t) - \tilde{b} \right]^2 \right\} dt ,$$ (55)

where

$$\frac{db(t)}{dt} = r(t) - c(t) \text{ and } \frac{dc(t)}{dt} = u(t).$$ (56)

Answer: Because you do have perfect estimation, you do not need an estimator. You can solve it by the Hamiltonian described in the section "Optimal Control with Stochastic Disturbances," but the easiest method is to use the Riccati equation described earlier and set the estimate to be equal to the actual value. The resulting control law is

$$u(t) = \beta^{\frac{1}{2}} \left[b(t) - \tilde{b} \right] - \beta^{\frac{1}{4}} \sqrt{2} \left[c(t) - r_{avg} \right],$$ (57)

where r_{avg} is the long-run average of demand. Note that the control law is *not* dependent on the standard deviation of noise, which is typical of LQG controllers, as described earlier.

Note

1. Note that care must be taken when simulating a continuous "white-noise" process with a system dynamics simulation package. In particular, a continuous white-noise process decorrelates every instant, while the simulated white-noise process decorrelates only once every time step. This technical issue can be solved in two ways in the simulation. One is to use a "pink-noise" process (Sterman 2000) with a time constant of correlation much shorter (four times shorter is a good heuristic) than any other time constant in the model. The other is to scale the simulated white-noise standard deviation. This

can be accomplished by defining the standard deviation of the simulated white-noise process σ_{sim} such that $\sigma_{sim} = \sigma_{cont} / \sqrt{dt}$, where σ_{cont} is the standard deviation of the continuous white-noise process to be simulated and dt is the simulation time step.

References

Anderson, E. G. 2001a. Managing the impact of high market growth and learning on knowledge worker productivity and service quality. *European Journal of Operational Research* 134 (3): 508–524.

Anderson, E. G. 2001b. The nonstationary staff-planning problem with business cycle and learning effects. *Management Science* 47 (6): 817–832.

Anderson, E. G., and C. H. Fine. 1999. Business cycles and productivity in capital equipment supply chains. In *Quantitative models for supply chain management*, ed. S. Tayur, M. Magazine, and R. Ganeshan, 381–415. Dordrecht, The Netherlands: Kluwer Press.

Anderson, E. G., and N. R. Joglekar. 2010. Managing accuracy of project data in a distributed project setting. University of Texas at Austin Working Paper.

Anderson, E. G., D. J. Morrice, and G. Lundeen. 2005. The "physics" of capacity and backlog management in service and custom manufacturing supply chains. *System Dynamics Review* 21 (3): 217–247.

Anderson, E. G., D. J. Morrice, and G. Lundeen. 2006. Stochastic optimal control for staffing and backlog policies in a two-stage customized service supply chain. *Production and Operations Management* 15 (2): 262–278.

Baird, L. C., III. 1993. *Advantage updating*. No. WL-TR-93–1146. Wright Lab, Wright-Patterson Air Force Base, OH.

Bertsekas, D. P. 2007. *Dynamic programming and optimal control*. Belmont, MA: Athena Scientific.

Bertsekas, D. P., and J. N. Tsitsiklis. 1996. Neuro-dynamic programming. Optimization and Neural Computation Series, 3. *Athena Scientific* 7: 15–23.

Coyle, R. G. 1985. The use of optimization methods for policy design in a system dynamics model. *System Dynamics Review* 1 (1): 81–91.

Doya, K. 2000. Reinforcement learning in continuous time and space. *Neural Computation* 12 (1): 219–245.

Ford, D. N., and J. D. Sterman. 1998. Dynamic modeling of product development processes. *System Dynamics Review* 14 (1): 31–68.

Forrester, J. W. 1961. *Industrial dynamics*. Cambridge, MA: MIT Press.

Forrester, N. B. 1982. A dynamic synthesis of basic macroeconomic theory: Implications for stabilization policy analysis. PhD thesis, Massachusetts Institute of Technology, Alfred P. Sloan School of Management.

Gaimon, C. 1997. Planning information technology—knowledge worker systems. *Management Science* 43 (9): 1308–1328.

Gonçalves, P. 2006. The impact of customer response on inventory and utilization policies. *Journal of Business Logistics* 27 (2): 103–128.

Gonçalves, P. 2009. Behavior modes, pathways and overall trajectories: Eigenvector and eigenvalue analysis of dynamic systems. *System Dynamics Review* 25 (1): 35–62.

Gonçalves, P., J. Hines, and J. Sterman. 2005. The impact of endogenous demand on push-pull production systems. *System Dynamics Review* 21 (3): 187–216.

Holt, C. C., F. Modigiliani, J. F. Muth, and H. A. Simon 1960. *Production Planning, Inventories, and Workforce.* New York, NY: Prentice-Hall.

Jiang, Y., and Z. P. Jiang. 2012. Robust adaptive dynamic programming for nonlinear control design. In *Decision and Control (CDC), IEEE 51st Annual Conference*, 1896–1901.

Joglekar, N. R., E. G. Anderson, and G. Shankaranarayanan. 2013. Accuracy of data in distributed project settings: Model, analysis and implications. *ACM Journal of Data and Information Quality* 4 (3): 13–34.

Joglekar, N. R., and D. N. Ford. 2005. Product development resource allocation with foresight. *European Journal of Operational Research* 160 (1): 72–87.

Kamien, M. I., and N. L. Schwartz. 1981. *Dynamic optimization.* Amsterdam: Elsevier.

Kampmann, C. E., and R. Oliva. 2006. Loop eigenvalue elasticity analysis: Three case studies. *System Dynamics Review* 22 (2): 141–162.

Kivijärvi, H., and M. Tuominen. 1986. Solving economic optimal control problems with system dynamics. *System Dynamics Review* 2 (2): 138–149.

Lendaris, G. G., J. C. Neidhoefer, J. Si, A. Barto, W. Powell, and D. Wunsch. 2004. Guidance in the use of adaptive critics for control. In *Handbook of learning and approximate dynamic programming*, ed. J. Si, A. Barto, W. Powell, and D. Wunsch, 97–124. Hoboken, NJ: Wiley.

Macedo, J. 1989. A reference approach for policy optimization in system dynamics models. *System Dynamics Review* 5 (2): 148–175.

Mohapatra, P. K., and S. K. Sharma. 1985. Synthetic design of policy decisions in system dynamics models: A modal control theoretical approach. *System Dynamics Review* 1 (1): 63–80.

Morrice, D. J., E. G. Anderson, and S. Bharadwaj. 2004. A simulation study to assess the efficacy of linear control theory models for the coordination of a two-stage customized service supply chain. In *Proceedings of the 2002 Winter Simulation Conference*.

Oppenheim, A., A. Willsky, and I. Young. 1983. *Signals and systems.* Englewood Cliffs, NJ: Prentice-Hall.

Özveren, C. M., and J. D. Sterman. 1989. Control theory heuristics for improving the behavior of economic models. *System Dynamics Review* 5 (2): 130–147.

Parker, G. G., and E. G. Anderson, Jr. 2002. From buyer to integrator: The transformation of the supply chain manager in the vertically disintegrating firm. *Production and Operations Management* 11 (1): 75–91.

Rahmandad, H. 2008. Effect of delays on complexity of organizational learning. *Management Science* 54 (7): 1297–1312.

Rahmandad, H., and S. Fallah-Fini. (2008). Learning control policies in system dynamics models. In *Proceedings of the 26th International System Dynamics Conference*, Athens.

Rahmandad, H., N. Repenning, and J. Sterman. 2009. Effects of feedback delay on learning. *System Dynamics Review* 25 (4): 309–338.

Russell, I. 2013. Value iteration, policy iteration, and q-learning. Available at uhaweb.hartford.edu/compsci/ccli/projects/QLearning.pdf. Accessed November 4.

Sethi, S. P., and G. L. Thompson. 2000. *Optimal control theory: Applications to management science and economics.* Boston, MA: Kluwer Academic.

Shanmugan, K. S., and A. M. Breipohl. 1988. *Random signals: Detection, estimation, and data analysis.* New York: Wiley.

Sims, C. 2013. Reinforcement learning: Model-based. Available at www.bcs.rochester.edu/people/robbie/jacobslab/cheat_sheet/ModelBasedRL.pdf. Accessed November, 4.

Stengel, R. F. 1994. *Optimal Control and Estimation.* Mineoloa, NY: Dover.

Sterman, J. D. 1989. Modeling managerial behavior: Misperceptions of feedback in a dynamic decision making experiment. *Management Science* 35 (3): 321–339.

Sterman, J. 2000. *Business dynamics: Systems thinking and modeling for a complex world.* Boston: McGraw-Hill/Irwin.

Sutton, R. S., and A. G. Barto. 1998. *Reinforcement learning: An introduction.* Cambridge, MA: MIT Press.

Watkins, C. J., and P. Dayan. 1992. Q-learning. *Machine Learning* 8 (3–4): 279–292.

12 Modeling Competing Actors Using Differential Games

Hazhir Rahmandad and Raymond J. Spiteri

From competition among firms to government policy setting in a national economy, and from playing competitive sports to pursuit-evasion problems in missile defense, many real-world problems include stakeholders with competing interests engaged in managing a complex dynamic system. For example, each firm attempts to maximize its profits by continuously changing its pricing and investment decisions, among others. These decisions influence the state of the firm and the market (e.g., customer perceptions and choices), therefore influencing not only the profits of the focal firm but also other players in the market. In such a setting, the reactions of each player may depend on the actions of the other player(s), creating a complex problem to analyze and solve. The theory of noncooperative dynamic games provides one framework to analyze these settings, with the hope of offering internally consistent solutions to the question: how do the dynamics unfold if all players are rational and fully informed about the structure of the system and try to maximize their own payoffs, understanding that the other players are doing the same?

Dynamic modeling practice typically attempts to capture such competitive settings by extracting decision rules for different players and allowing the dynamics to evolve based on those decision rules. For example, using empirical data, a functional form is found for pricing decisions of each firm that relates the firm's price to inventory levels, competitor's prices, and other empirically identified information cues. Once such functions are estimated for different players and market dynamics (e.g., adoption rates by customers given the prices for different firms) are specified, one can simulate the resulting system and project the time trajectories of expected outcomes. In this approach, the resulting trajectories may or may not be consistent with the objectives of the players. For example, by following those decision rules, the firms may end up taking actions that hurt their own profit in the short or the long run.

In contrast, dynamic game theory assumes that players are rational; that is, they only take actions that are consistent with maximizing their payoff. It also assumes that all the players consider other players to be rational as well and attempts to find out the action policies that are consistent with those assumptions. As a result, the dynamic game diverges from dynamic modeling practice not only in making the rationality assumption but also in its attempt to solve optimization problems that follow the rationality assumption and lead to Nash equilibria (i.e., a set of policies that, when adopted by each player, will allow none of the players to improve their payoff by deviating from that policy). Finding Nash equilibria has been theoretically attractive to many because of their internal consistency, avoidance of ad hoc assumptions in specifying decision rules, and their normative power to guide action.

Although much can be debated about the validity of the rationality assumption in different settings, knowing about the basic concepts of dynamic game theory allows dynamic modelers to make informed decisions about the suitable theoretical foundation upon which to build their work for each application. Moreover, much research in economics, strategy, political science, ecology, and evolutionary biology has adopted the dynamic game framework for understanding noncooperative, multistakeholder dynamic settings. Understanding this literature and communicating with scholars engaged in this research require a familiarity with dynamic game theory.

In this chapter, we provide a brief overview of differential games, an important subset of dynamic games in which the actions of players unfold continuously over time and the dynamics of the system can be represented as a system of differential equations. Differential games are theoretically and analytically complex, requiring significant mathematical background, and are only solvable analytically in special formulations. We therefore focus on providing a conceptual overview and discussing simple numerical methods for solving these games, with the understanding that covering the state-of-the-art analytical methods goes beyond the scope of current chapter, and even numerical methods may be lacking for complex games.

The field of differential games combines elements from game theory and optimal control, each a substantial field extensive enough to fill multiple graduate-level courses. Therefore, this short overview is only able to scratch the surface of the topic, and interested readers are referred to more advanced resources introduced in the discussion section. This chapter also relates to this volume's chapter 11, which introduces optimal control theory and its applications for dynamic modelers. We therefore limit our discussion of solution methods for the optimal control problems to simple concepts useful for the solution methods discussed in the current chapter. Nevertheless, this chapter is intended to be standalone, allowing readers to start applying the differential game concepts to simple problems they may face in their research. To this end, we provide step-by-step instructions, implementation notes, implementation code, and an exercise with solution.

Theoretical Background and Formulation

A differential game captures a dynamic system in which Y players[1] interact with each other by taking actions that influence a dynamic system, leading to payoffs for themselves and others. To fully define a game, the time span of the game (typically the value T, i.e., the final time for the game), a set of rules (what actions are feasible given the state of the system and how those influence the dynamics) and an information structure (what is known to each player before making decisions) should also be specified. Formally, we define the state of the system, $x(t)$, as a vector in the state-space of the system ($x(t) \in X, X \subseteq \mathbb{R}^n$), itself a subset of $\mathbb{R}^n$. For most practical applications, $x(t)$ is the vector that holds the values of all the stock variables in a system dynamics model. Therefore, n is the number of stocks (i.e., the size of the system of differential equations) in the system that represents all the players. For example, if we are modeling the competition between two firms with stocks of inventory and labor for each, as well as two stocks for customers loyal to each firm, we have a six-dimensional state-space ($X \subseteq \mathbb{R}^6$), and six numerical values in the vector $x(t)$ fully define the state of the system at time t. At time t, player i selects a control action $u^i(t)$ from his set of feasible actions, U^i. These feasible actions are a function of $x(t)$, the actions taken by other players (which we denote as $u^{-i}(t)$), and time ($U^i(x(t), u^{-i}(t), t)$. For example, the firm's decisions may include what price to set at time t and how many workers to hire/fire, a two-dimensional $u^i(t)$ vector in this example (for each of the two firms, $i = [1, 2]$). Function f specifies the dynamics of the system in the form of a system of differential equations that connects the rate of change in stocks (state vector) to the current state and the actions taken by different players:

$$\dot{x}(t) = f(x(t), u^1(t), \ldots, u^Y(t), t) \cdot \tag{1}$$

In the system dynamics modeling tradition, this function f is the model in a reduced form where all auxiliary variables are collapsed to make the net rate of change in each stock (summarized in vector $\dot{x}(t)$) only a function of other stocks and players' actions. The initial state of the system, $x(0) = x0$, corresponds to the initial value of stocks and is typically assumed to be known.

Each player has an objective function, $J^i(.)$, which depends on a discount rate (r); a continuous payoff function for player i, $F^i(.)$ (e.g., expected profit given the value of stocks, time, and actions taken by different players); and a scrap value, $S^i(x(T))$, that represents the value associated with landing at state $x(T)$ at the final time (e.g., the value of final inventory levels). Formally,

$$J^i\left(u^i(.)\right) = \int_0^T e^{-rt} F^i(x(t), u^1(t), \ldots, u^Y(t), t) dt + e^{-rT} S^i(x(T)) \cdot \tag{2}$$

In writing $J^i(.)$, we emphasize its dependence on $u^i(.)$ because that is the only factor directly controlled by player i. The firm can only directly control its own pricing and hiring decisions; the rest unfolds outside of the direct control of the firm, yet others' actions also influence $J^i(.)$.

Finally, the information structure of the game specifies what information is available to each player at the time she is taking her actions. Specifically, if $u^i(t)$ is expressed as a function $\phi^i(.)$, then player i's information set (I^i) specifies the input domain for this function:

$$\phi^i : I^i \to U^i. \tag{3}$$

The game's information structure (I) is the set of information domains for all players. For example, if each player knows the current state and time (a common assumption in many theoretical applications)— that is, she can fully observe the state of the system, $x(t)$ —then the game is called a *Markovian game* ($I^i = \{x(t), t\}$; $u^i(t) = \phi^i(x(t), t)$). In such a case, our example firms know the values of inventory and labor stocks for themselves and the other player, as well as the stocks of loyal customers for each. However, they do not know anything else; for example, they have no memory of the past actions taken by the other player. It is critical to note that many different information structures are conceivable, and they have a significant effect on the results of a game.

Given the system's dynamics (function f), the feasible actions (U^i), and the information structure, each player attempts to maximize its objective function (J^i) by selecting the best-response policy function ϕ^i. In doing so, the player takes into account the fact that all other players are attempting to maximize their own objectives as well and have access to their own information sets; therefore, the optimization over ϕ^i should account for their moves and countermoves. A set of policy functions $\phi^* = (\phi^{1*}, \phi^{2*}, \ldots, \phi^{Y*})$ is called a *Nash equilibrium* for a differential game if each ϕ^{i*} function maximizes the objection function of player i, given that the other players are following the rest of the policy functions in the set (ϕ^{-i*}). In deterministic differential games— that is, when the function f has no stochastic components—the optimum policy function set ϕ^* often defines a unique time path for the evolution of the system over time ($x^*(.)$).[2] In stochastic differential games, the function ϕ^i is optimized over the expected pathways that could emerge from the system's dynamics. In this overview, we limit our discussion to deterministic games, noting that most of the insights carry over to stochastic games, but with additional complexities for solving those games.

The Nash equilibrium for a differential game is generally not unique. On the one hand, it depends on the information structure assumed. An *open-loop Nash equilibrium* emerges when we assume ϕ^i is only a function of time for a given initial condition ($I^i = [t, x(0)]$); that is, the players, knowing the initial state of the system, specify every action they will take over time at the very beginning of the game and will not change those selections based on any other information. Even though not always realistic, open-loop equilibria are the simplest to determine (because $u^i(t)$ is only a function of time) and therefore popular in the literature. A Nash equilibrium is closed-loop if some information about the current or past state of the system is available to the players. A Markovian game assumes players have full information about the current state of the system $x(t)$. An open-loop Nash equilibrium can be considered as a degenerate Markovian equilibrium, in which the players choose to only use time in their policy functions. Still more comprehensive information structures could be conceived in which players can remember the history of the game (what actions were taken by each player at different points in time, a piece of information not embedded in $x(t)$), allowing them to punish other players for their deviations from mutually beneficial pathways. In addition, mixed information structures, in which different players have different information (e.g., some players play open-loop and others are Markovian) could be conceived. Therefore, the choice of information structure is a critical modeling assumption that should be made based on characteristics of the case, such as the availability of information, the rationality of players, the costs of adjustment to the policy in the real world, and the analytical complexity of the problem. In essence, solving a differential game for two different information structures is better seen as solving two different games with different key assumptions, rather than finding two different solutions to the same game.

Even after fixing the information structure, there are often many Nash equilibria that are mathematically valid. For example, in a deterministic system, if two policies take identical actions on the optimal path for the system's dynamics (x^*(.)) but vary *off* the optimal path, they could both be acceptable Nash equilibria because the players can foresee that off-equilibrium path actions are irrelevant to what happens when the game is played. Yet not all mathematically valid equilibria are informative for a given problem. Therefore, additional criteria have been established to differentiate more realistic equilibria. An equilibrium is designated as *time-consistent* if the optimal actions do not change should the game be restarted from any point on the optimal path, $x^*(s)$, re-solving the game for time s to T. It can be shown that all Markovian Nash equilibria are time-consistent, but if an equilibrium depends on information about the history of the game, beyond what is embedded in the state vector $x(s)$, it likely will not be time-consistent because the restarting of the game will erase that memory and thus change the optimal policy function for the remainder of the game. An equilibrium is called *subgame perfect* if restarting the game from any point in the state-space, X, and continuing until time T does not change the optimal policy function set ϕ^*. By definition, a subgame-perfect equilibrium is also time-consistent. Subgame-perfect equilibria are often sought after because they are robust to initial conditions and random shocks that may take the system off the predicted path.

Solving Differential Games

In essence, solving a differential game requires finding the policy function set (ϕ^*) consistent with the information structure of the game that satisfies the Nash equilibrium condition. To solve such a problem, one needs to overcome two distinct challenges. First, finding an optimal policy function for any player, ϕ^{i^*}, requires a mapping from the available information set (I^i; e.g., time and state-space X in the case of Markovian games) to feasible actions U^i. As a concrete example, let us assume that an example firm only consisted of two stock (state) variables, labor and inventory, and we could simplify these continuous stock variables into three discrete states each (low, medium, high). Assume we could also simplify the price and hiring decisions (typically continuous) to only two discrete values (increase or decrease). Let us also assume that only two time periods are considered. We would then have 9 (=3*3) distinct states and 4 (=2*2) distinct actions feasible from each state. A Markovian policy mapping ϕ^i needs to specify which of the four actions would be taken from each state (i.e., the space of information known to player) at times 1 and 2, separately—a choice from over 68 billion (4^{2*9}) distinct policies. To reduce the complexity, one can come up with functional forms for mapping state-time to actions, yet it is hard to pre-specify the restriction on such a mapping, and in principle any functional form can fit the bill. This simple example shows how the complexity of finding optimal policies grows exponentially with the size of the problem, making it an analytically intractable problem in its general form. In fact, this component of solving a differential game is equivalent to an optimal control problem; thus, the close relationship between differential games and optimal control.

A second challenge in solving differential games arises from the fact that optimum policy functions should be found simultaneously for all the players involved. In essence, we are required to solve Y coupled optimization problems because the policy of each player influences the optimization problem of the others. This challenge may be simplified if the game has a special structure; for example, it is symmetric for two players, so that the loss of one player is the gain of the other. As an example, pursuit-and-evasion games (e.g., aircraft vs. missile) can be formulated as symmetric games, leading to a single minimax optimization problem (Isaacs 1999). But more general formulations lack generic solutions and require iterative methods to find the Nash equilibrium.

Two general approaches can be pursued for solving the differential games. The analytical approach builds on the Hamilton-Jacobi-Bellman optimality conditions for an optimal control problem and provides solution structures for some special classes of differential games. The most famous class of problems for which analytical solutions are available (under certain regularity conditions) is linear-quadratic games. In linear-quadratic games, the state transition (f) is linear with respect to state (x) and control (u) variables, and payoff (F) functions are quadratic with respect to state and control variables. The details of these analytical solutions go beyond the scope of the current paper and can be found in most texts discussing differential games (e.g., Dockner 2000). Moreover, a lot of research goes into finding solutions to other subclasses of problems or transforming new problems into an analytically solvable form. In fact, specific applied mathematics journals, such as *Dynamic Games and Applications* and *Journal of Dynamics and Games*, are solely dedicated to mathematical analysis of dynamic games, of which differential games are an important subset.

Numerical solutions to differential games seek to solve problems that are not analytically tractable. These methods often build on advances in numerical solutions to the optimal control problems. Many successful methods for numerically solving optimal control problems combine the core idea of dynamic programming (Bellman 1957; i.e., to learn an optimal value function that updates the value of each state based on the value of states that results from following a policy on that state[3]) with an iterative improvement of the policy based on the value of states to which it leads. Cycling through policy and value updates, the algorithm estimates approximate functional forms (e.g., using neural networks) for value functions and can extract the optimum policies from the optimum values. These methods are described in detail under the topic headings of approximate dynamic programming (Powell 2011), neurodynamic programming (Bertsekas and Tsitsiklis 1996), and reinforcement learning (Sutton and Barto 1998).

Another set of solution methods focuses on approximations in the policy space (Baldwin and Sims-Williams 1968). Avoiding the use of value function, these methods directly estimate the parameters of a (flexibly designed) policy function form to maximize the expected discounted objective function. A nonlinear optimization engine is used to conduct the optimization in this setting. The strength of this method is in the direct estimation of the optimum policy function; this may be more efficient computationally, especially if the available information (I^i) is rather limited (e.g., in an open-loop control problem). The downside is that without significant insights into the structure of the policy function, the algorithm may converge to a policy that is far from the global optimum, and it would be hard to recognize this occurrence. This risk is especially significant in light of the fact that optimum policy functions routinely include discontinuities and trigger points, even if the underlying dynamic system and payoff functions are smooth and differentiable.

Numerical methods also exist for overcoming the second challenge, the interdependency of the optimal control problems different players are facing. One typical way to deal with this challenge is to iterate across different players, consecutively solving their optimal control problems based on the best policies found for the other players through the previous steps. If these iterations converge—that is, all individual optimal policies remain stable in consecutive iterations—the resulting policy set is a Nash equilibrium by definition. In practice, it is often more productive to find the optimal policy over multiple parallel games, only differing in initial states, simultaneously. This idea can overcome the challenge that in iterating between only two players, the resulting policies may be sensitive to small changes from the other player's actions, and therefore the iteration may not converge. By optimizing the policy over multiple instances of the game (starting from different initial conditions on the state-space), the optimal policies obtained become more robust because now the estimated policy can be expected to work under many different time trajectories of competition. Iterations of these robust policies are more likely to converge. A secondary benefit of policies found through this method is that they are more likely to be subgame perfect, that is, not sensitive to the initial conditions of the game.

Different techniques for solving the optimal control problem are found under the general headings of optimal control, approximate dynamic programming, neurodynamic programming, and reinforcement learning, among others, and a vast literature in machine learning and operations research discusses the various alternatives (Powell 2011; Bertsekas and Tsitsiklis 1996; Sutton and Barto 1998; Bertsekas 2007). Advanced methods for numerically solving games are less widely available. A notable exception is the level set method toolbox developed for MATLAB and available with requisite documentation (Mitchell and Templeton 2005). Yet even the most advanced numerical algorithms must compromise between guaranteeing the (global) optimality of solution and the size of problems they can address. In practice, most methods that guarantee an optimal solution cannot handle more than a handful of stock variables. In this paper, we use approximations in policy space for solving the optimal control for each player. This choice is largely informed by the conceptual simplicity and computational efficiency of this technique, which makes it feasible for larger problems typically faced by dynamic modelers. However, the methods provided here do not guarantee the optimality of the resulting solution. We will also use iterations among players to converge to the Nash equilibrium. We limit the analysis here to two players.

Previous Applications in System Dynamics

Although a significant literature on differential games exists, the dynamic modeling literature in general and the system dynamics literature in particular have had rather limited exposure to this topic. In fact, a review of the system dynamics literature on game-theoretic concepts finds only a few applications. The few applications that make game theoretic considerations (Wang, Cai, and Zeng 2011; Mukherjee 1995; Kim and Kim 1997) focus on static games, where each player takes a single action. Notable exceptions exist in applications to supply chains and software development (Gonçalves and Sterman 2005; Rahmandad and Sibdari 2012). Gonçalves and Sterman consider static and dynamic games in a supply chain and show a prisoner-dilemma type of equilibrium could exist that forces all players to order using an aggressive policy, hurting everybody involved. In the second study (Rahmandad and Sibdari 2012), a system dynamics model of software product development, open-source contributors to the development process, and network and complementary product effects are developed. The model is then used to analyze the competition between two identical firms, with each making pricing decisions over time to maximize their net present value, facing the competition's best response. The firms also determine their level of software openness at the beginning of the game. The study finds that network effects and the existence of complementary products encourage more open product platforms and heavier discounts. Firms in such markets are forced to start the competition with significant discounts or offering their software products for free, only raising their price later when they have established a strong installed base that can sustain their products' utility in tight competition. This chapter only analyzes the open-loop Nash equilibrium for this differential game and does not discuss more complex equilibrium concepts, where firm policies are a function of both time and the state of the system.

A Simple Example

In this section, we introduce a simple differential game to help build intuition regarding the basic solution concepts. For simplicity, we normalize all the variables so that we use the fewest number of parameters without limiting the model's degrees of freedom. Consider two firms, each represented by a single state variable, which we call *quality* (shown in figure 12.1 as "*Quality Q_i*" for firm i[4]). Firms are competing in a limited market, and their market share is proportional to their quality divided by the sum of the qualities for both firms. Qualities increase when firms invest an amount ("*Investment u i*") between 0 and 1 into them and depreciate with a time constant "*Quality life f i.*" Investments also have a normalized unit cost (c). Firms' profits are calculated based on their market share and their costs. We assume that firms are engaged in an infinite-horizon competition (no finite final time) with continuously discounted profits. Figure 12.1 provides the causal diagram of this simple model, and the full equations are available in table 12.1.

We note that if u_i is constant, this system reaches an equilibrium in which $Q_i(t) = u_i.f_i$ due to the first-order dynamics. We further simplify the problem by assuming that information domain for each firm is empty; that is, the firms have no knowledge about what is happening in the game beyond the relations in table 12.1; in particular, they do not know the initial state of the system.[5] This assumption significantly simplifies the game. Combined with the infinite-horizon nature of the game, it forces the firms to pick a constant value for u_i because there

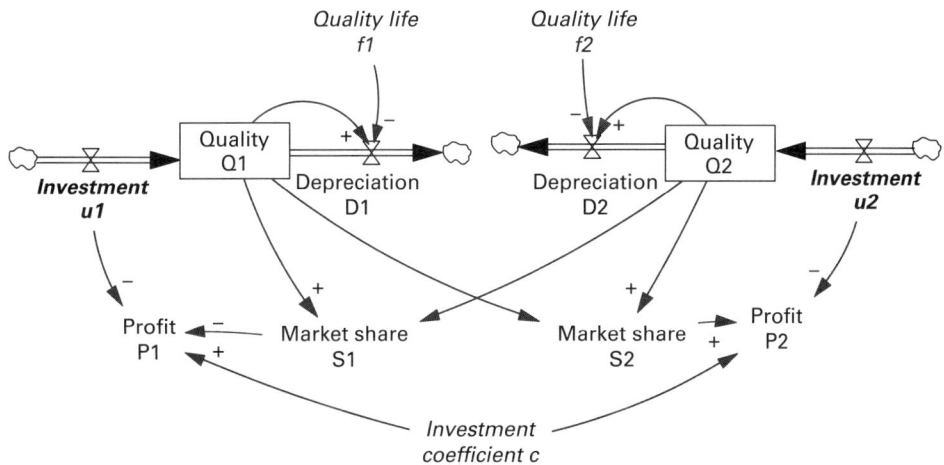

Figure 12.1
Causal diagram for the simple example model with a single state variable for each firm.

Table 12.1
Equations for the simple competition with single-stock firms

Equation	Units
$Q_i(t) = \int\limits_0^t (u_i(s) - D_i(s)) ds + Q_i(0)$	Dimensionless
$D_i(t) = Q_i(t) / f_i$	1/Year
$S_i(t) = \dfrac{Q_i(t)}{Q_1(t) + Q_2(t)};$	Dimensionless
$S_i(t) = 0.5 \; if \; Q_1(t) = Q_2(t) = 0$	
$P_i(t) = S_i(t) - u_i(t).c$	Dimensionless
$f_1 = 1 \; ; \; f_2 = 2$	Year
$c = 0.5$	Year

is no information upon which to adjust the value of u_i. Therefore, the problem simplifies into finding the values of u_i that lead to a Nash equilibrium.

To solve this problem, we first find firm profit as a function of fixed u_i values, assuming that the system has reached equilibrium:

$$P_i(t) = u_i f_i / (u_1 f_1 + u_2 f_2) - u_i c. \tag{4}$$

Now the problem for each firm is to find the value u_i that maximizes its profit given what the other firm is expected to do. Figure 12.2a shows the contours for values of P_1 as a function of different values of u_1 and u_2 (using the parameter values specified in table 12.1). One can expect that for each value of u_2, firm 1 will pick the value of u_1 that maximizes P_1. This is equivalent to solving the problem to maximize P_1 given some value of u_2. We repeat this optimization procedure many times to graph the optimum response of the first firm as a function of second firm's investment rate; that is, u_2 values (see the thick dotted line).

The same procedure can be repeated to find the optimum response function for the second firm, u_2, this time as a function of the first firm's investment rate, u_1. These two best-response curves can be then superimposed on the policy space to find the point where they cross, $[u_1^*, u_2^*]$. This point is by definition a Nash equilibrium because if either firm follows the investment level given by this policy, the other firm has no incentive to change its investment policy. Therefore, we have solved this simple differential game, and the solution is for both firms to follow a policy of $u_i = 0.45$.

The intuition behind a few concepts can now be clarified using this simple example. First, note that our assumptions about the information structure are critical. We assume the two firms have no information about the state of the system and therefore could simplify their resulting policies into a single number, u_i. More often, such an assumption is not empirically warranted, and the optimum policy for each firm takes a more complex shape; that is, a potentially different value of u_i for each point in the available information set for player i (e.g., each point in time in case of an open-loop equilibrium).

Second, here we can review the two basic steps in the solution method for solving differential games. Specifically, the first step, finding the optimum policy for each player, is the same as finding one optimum u_i value given a fixed value for the other player's action. In figure 12.2b, we represent the iteration steps for this optimization problem with the dashed arrows. When these arrows move horizontally, they represent solving the optimization problem for finding u_1, given a fixed u_2, and vice versa when moving vertically. The iterations are thus represented by alternating between horizontal and vertical moves. In the graphs shown, we start (arbitrarily) with $u_2 = 0.67$ and find the optimum $u_1 = 0.31$. In the second step, fixing u_1 at 0.31, we find $u_2 = 0.39$. Next, we optimize for firm 1 to find $u_1 = 0.48$ given $u_2 = 0.39$, and so on, until we converge to an equilibrium, which will be the Nash equilibrium. Note that, in solving realistic problems, the optimization step in each iteration is generally highly complex, even though the basic ideas remain the same. In practice, the optimization step is solved through a dynamic program (or approximation thereof) or a nonlinear policy optimization, in which a policy mapping ϕ^i is estimated that produces a value for u_i given any point in available information set I^i.

Algorithm for Solving Differential Games

In this section, we provide a detailed description of an algorithm for solving differential games using a set of simple heuristics and numerical approaches that can be applied by typical users of dynamic modeling tools. We provide the steps along with a concrete example to better demonstrate the techniques previously described. We break down the process of defining and solving the differential game into the steps of

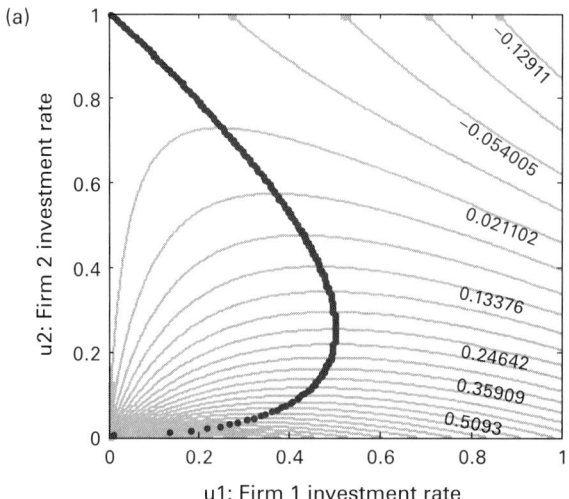

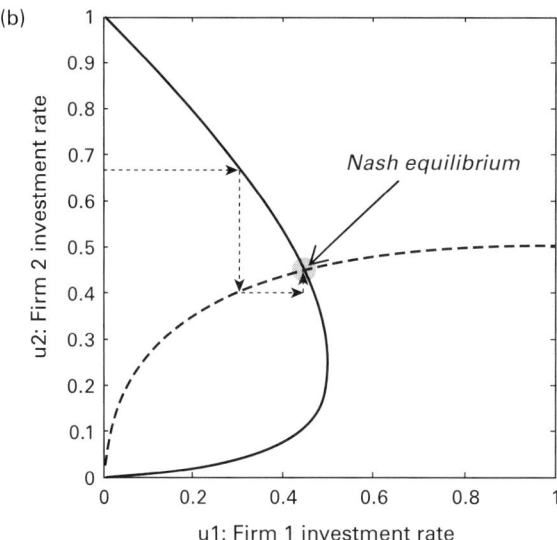

Figure 12.2
Equilibrium analysis for simple two-firm model. (a) Firm 1's payoff contours as a function of strategies of the two firms, as well as the optimal response trajectory (dotted line) for firm 1. (b) The Nash equilibrium and the path of the iterative algorithm to find it.

system dynamics definition and iterative policy optimization solution. At each step, we first define that step generally and then demonstrate it explicitly for an example problem.

Step 1: Building a Model for the System's Dynamics

This step entails the traditional dynamic model building process upon which we build in this paper. For details on this process, we refer to one of the texts that focuses on the dynamic modeling process (e.g., Sterman 2000). Here we assume the model is represented using a set of deterministic ordinary differential equations (ODEs); the steps are similar for working with stochastic ODEs and agent-based and discrete-event models. The key issues to keep in mind in the modeling process are to have a clear representation of the two players (competitors, stakeholders, etc.), their distinct payoff functions, and how they are coupled through some equations in the system. We should also explicitly identify each player's feasible action space and the information available to them. Below, we define the dynamics for an example problem representing competition between two firms in a limited market and identify the action, payoff, and information set.

Step 1 in Action: Example Model Definition

We apply the solution process to a competition between two hypothetical firms. We use simple firms, each with two stock (state) variables that represent their active capability (C) and the capability under development (D). The capability under development becomes active capability with a first-order delay of d, and capability expires after a delay of f.

$$C_i(t) = \int_{s=0}^{s=t} \left(\frac{D_i(t)}{d} - \frac{C_i(t)}{f} \right) ds + C_i(0) \text{ (dimensionless)} \tag{5}$$

$$D_i(t) = \int_{s=0}^{s=t} \left(I_i(t) - \frac{D_i(t)}{d} \right) ds + D_i(0) \text{ (dimensionless)} \tag{6}$$

The key decision for the management of the firm (U) is to allocate a fixed stream of resources (r) between investing in capability development (I) and production. The production function for the firm is an increasing return to scale ($\alpha + \beta > 1$) Cobb–Douglas function of production investment fraction ($1 - U_i$) and current capability (C), leading to potential production (P) for each firm.

$$I_i = r_i U_i \text{ (1/month)} \tag{7}$$

$$P_i = kC_i^{\alpha}(1-U_i)^{\beta} \ (\$/\text{month}) \tag{8}$$

Finally, the two firms are coupled through a market-sharing mechanism, where a fixed market demand (m) is allocated between the two firms in proportion to their potential production. If market demand exceeds total production of the two firms, they both sell all their potential production, leading to the realized revenue (R) for each according to equation (9):

$$R_i = Min\left(\frac{P_i}{P_1 + P_2}m, P_i\right) \ (\$/\text{Month}). \tag{9}$$

Figure 12.3 provides a graphical overview of this model for one firm.

Payoff. Each firm's total payoff (O) is the integral of its revenue over the whole time horizon of the competition (T):

$$O_i = \int_{t=0}^{t=T} R_i dt \ (\$) \tag{10}$$

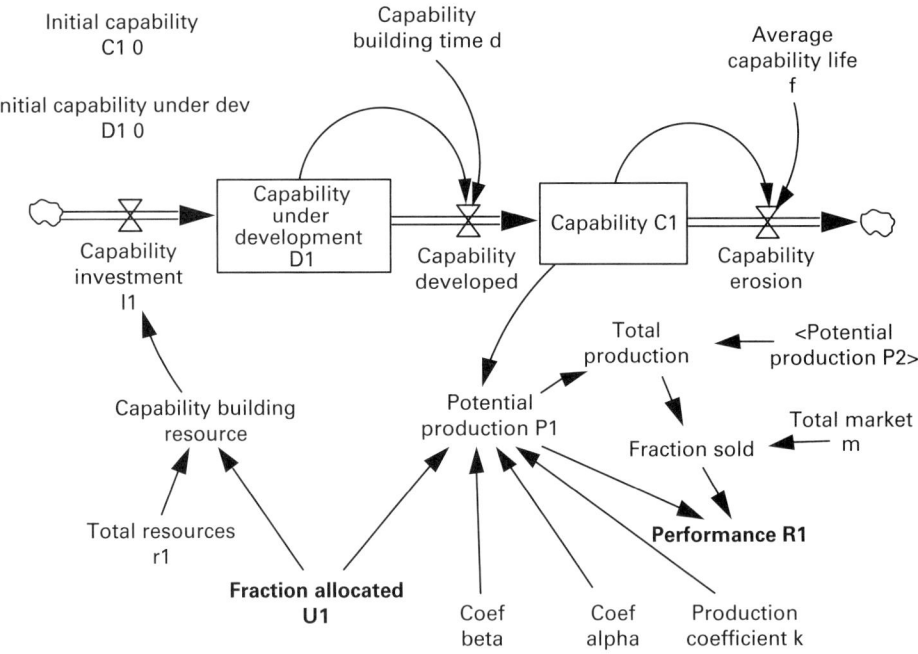

Figure 12.3
Causal diagram for a single firm in the complete competition analyzed in the example.

Feasible action set. At any point in time, each firm is able to pick a $U_i(t)$ value between 0 and 1.

Available information set. Both firms are assumed to know their own state as well as the state of the other firm at any point in time (four state variables plus the time), but they do not keep the history of the actions taken; that is, this is a Markovian game. Formally, $I^i = [C_1, D_1, C_2, D_2, t]$.

To solve this differential game, we assume each firm maximizes its own payoff by finding a best-response policy (ϕ^i) that specifies its decision parameter over time ($U_i(t)$), given the best-response policy expected from the other firm (ϕ^{-i^*}). A Nash equilibrium is found when ϕ^{1^*} and ϕ^{2^*} have converged in successive iterations of the following optimization problem:

$$\max_{U_i(t)=\phi^i(I^i)} O_i \tag{11}$$

Subject to $0 \le U_i(t) \le 1$, given ϕ^{-i^*}.

The two firms start the competition from different initial states ($C(0)$ and $D(0)$). Table 12.2 provides the initial values and parameter values used in the analysis that follows. We have implemented the solution in both the Vensim and MATLAB environments. Full model implementations are available in the e-companion for this chapter.

Table 12.2
Parameters used in the analysis of the full competition

Parameter	Value	Units
f	6	Month
d	3	Month
$r_1 = r_2$	100	1/Month
k	277	$/Month
α	2	Dimensionless
β	2	Dimensionless
m	4E7	$/Month
T	30	Month
$D_1(0)$	10	Dimensionless
$C_1(0)$	0.5	Dimensionless
$D_2(0)$	0	Dimensionless
$C_2(0)$	450	Dimensionless

As a result of the increasing return to scale structure of the production function, the optimum policy requires the management to change the Di periodically, switching between investment in capability development and production. This problem structure makes finding the differential game solution rather complex because the two players should time their investments to both benefit from the oscillating nature of optimum investment policies as well as to avoid saturating the market and therefore losing their potential production in competition.

Step 2: Iteratively Solve the Policy Optimization Problem to Find the Game's Equilibrium

Multiple methods are available for finding the optimum policy, ϕ^{i*}, for each player. In this tutorial, we focus on using direct policy optimization rather than methods based on finding the value function for states of the system and finding the optimum policy based on those value functions. This choice is largely informed by conceptual simplicity and scalability of the direct policy optimization method, even though theoretical support for this approach is more limited. Accordingly, the steps for solving the policy-optimization problem are identified below.

A. Specify a set of features, F^i, for each player that includes the most relevant pieces of information to be used in each player's policy function. For example, features could represent what the real players normally consider in making their decisions. By definition, they are a function of the available information set, I^i: players cannot use information to which they have no access. Although it is possible to use the full information set as input into the policy function, use of features can be an effective way to capture the modeler's qualitative and substantive knowledge about what factors better capture the nonlinearities and threshold effects in decisions. Therefore, we use a mapping, Θ^i, to specify features as a function of each player's information set: $F^i = \Theta^i(I^i)$.

Application to example: In our example application, we normalize the elements of the available information set, I^i, so that the operations in the next steps are more robust and parameter estimates in step E below are easier to compare. These normalized values constitute our features, $F^i = \left[\overline{C_1}, \overline{D_1}, \overline{C_2}, \overline{D_2}, \overline{t} \right]$. The following normalization operations define the normalized features: $\overline{D_i} = \dfrac{D_i}{r_i d} ; \overline{C_i} = \dfrac{C_i}{r_i f}, \overline{t} = \dfrac{t}{T}.$

Other features, such as the ratio between capabilities, could also be useful to include; however, for simplicity we limit our analysis to the above features.

B. Define a functional form for the policy function for each player, ϕ^i, with the input domain of F^i, so that $U^i(t) = \phi^i(F^i(t))$. A few functional forms that could be used depending on the problem structure include linear ($\phi^i(X) = b_0 + \sum\limits_{i=1}^{i=k} b_i X_i$), logistic ($\phi^i(X) = 1/(1 + e^{\varphi(X_i)})$), and two or more layered neural networks (Bertsekas and Tsitsiklis 1996). Here X is the vector of features going into the approximating function and b is the vector of (unknown) parameters that needs to be estimated to find the best-response function.

Application to example: In our example, we use the logistic function, with the input to the logistic function including all the direct interactions among the features (e.g., $\overline{D_1 C_2}, \overline{D_2}^2$). Specifically, defining the augmented feature vector of $F'_i = \left[\overline{C_1}, \overline{D_1}, \overline{C_2}, \overline{D_2}, \overline{t}, 1\right]$, we use the following policy approximation function $\phi^i(.) = 1/(1 + e^{F'_i \cdot B_i \cdot F'^T_i})$. Here, we need to specify the 6*6 parameter matrix B_i to fully specify this function.[6] For example, if all the elements of B were set to zero, the resulting policy for player i would be $U^i(t) = \dfrac{1}{1 + e^0} = 0.5$; that is, distributing resources equality between capability development and production at all times.

C. Vectorize the implementation of the system model (from step 1 above) so that N parallel competitions can be simulated in one simulation run. For these parallel competitions (between $2N$ players), the model parameters are identical except for the initial point in the state-space from which the players start. As a reasonable starting assumption, the initial points could be distributed randomly on the feasible range of the state-space. Optimizing over this vectorized set of competitions makes the resulting policy robust to initial conditions and applicable in any state of the system; that is, similar to a subgame-perfect solution. It also allows for better convergence results and more robust optimization.

Application to example: We start the initial states ($D_i(0)$ and $C_i(0)$) according to the following distributions: $C_i(0) \sim \text{Uniform}[0–500]$, $D_i(0) \sim \text{Uniform}[0–300]$. For each of the N parallel competitions, we pick the initial state values from these two distributions for the two players (four independent random draws).

D. Specify algorithmic parameters, including the number of parallel
   competitions (N), the starting policy function parameters for the
   second player, the optimization parameters (specifically the number of new restarts), and the threshold for stopping the iterative
   algorithm (δ).
   Application to example: We experimented with different parameters and found N values in the range 5 to 20 provide a good balance
between simulation time, fast convergence, and robust results. However,
to also assess how well the algorithm converges to similar policies for
both firms, we used $N=100$ in the results reported below.[7] We found
that, with this value of N, most optimization restarts converge to the
same peak, adding more confidence to the robustness of results. We
used five restarts in each optimization in obtaining the reported results.
Given multiple restarts in policy space, we did not need to worry about
the initial policy adopted by the second firm. In the Vensim implementation, the convergence was assessed by manual inspection, yet a
threshold δ as high as 15% reliably signaled the convergence of the
algorithm.

E. Iterate through the following steps, calling the current iteration m:
   - Maximize F_1 by changing firm 1's functional approximation pa-
     rameters (B_1), while keeping (B_2) (the other firm's policy) at the
     value found at step m-1 (B_2^{m-1*}). Use any optimization algorithm
     deemed fit and any number of start points to balance computa-
     tional costs and the chances that the global optimum is found.
     Call the solution B_1^{m*}.
   - Maximize F_2 by changing B_2, while keeping B_1 at B_1^{m*}.
   - If the maximum of the fractional difference in elements of B_i^{m*}
     and B_i^{m-1*} exceeds the threshold δ, then continue to iteration
     m+1. Otherwise the algorithm terminates successfully; report the
     Nash equilibrium optimum policy functions (B_1^{m*} and B_2^{m*}).

   Application to example: We conducted the optimizations above
using Vensim DSS's stochastic optimization option. This allowed us to
conduct the optimization of the policy function over $N=100$ parallel
competitions, only varying in their initial point in policy space. Each
optimization was restarted five times to increase our confidence in the
generality of the result. In the majority of cases, all five restarts found
the same optimal policy. We ensured that the first iteration started
from the policy found in the previous iteration. After $m=9$ iterations,
we reached convergence using the algorithm described above. Table
12.3 provides the optimum values we found for B_1^{9*} and B_2^{9*}, both

Table 12.3
Parameter values defining the optimum policy for the two firms. See endnote 7 regarding the use of upper-triangular matrices.

B_1^{g*}

	$\bar{C}_1$	$\bar{D}_1$	$\bar{C}_2$	$\bar{D}_2$	$\bar{t}$	1
$\bar{C}_1$	126.7	11.9	-19.2	67.1	-83.3	-19.7
$\bar{D}_1$	0.0	-11.0	25.0	9.8	54.4	-47.9
$\bar{C}_2$	0.0	0.0	-75.7	29.0	-92.9	6.7
$\bar{D}_2$	0.0	0.0	0.0	6.3	11.5	8.1
$\bar{t}$	0.0	0.0	0.0	0.0	125.6	-11.0
1	0.0	0.0	0.0	0.0	0.0	-25.3

B_2^{g*}

	$\bar{C}_1$	$\bar{D}_1$	$\bar{C}_2$	$\bar{D}_2$	$\bar{t}$	1
$\bar{C}_1$	-188.2	-18.7	332.6	-66.3	224.5	-21.4
$\bar{D}_1$	0.0	-2.1	195.3	-30.0	-71.2	-69.0
$\bar{C}_2$	0.0	0.0	143.7	-44.8	-109.5	21.5
$\bar{D}_2$	0.0	0.0	0.0	-11.1	-49.2	9.4
$\bar{t}$	0.0	0.0	0.0	0.0	111.3	-83.5
1	0.0	0.0	0.0	0.0	0.0	-54.7

upper-triangular matrices. For easier reading, the corresponding features for each row and column of the table are also identified. For example, the element in matrix B_2^{9*} that corresponds to coefficient of multiplication of $\overline{D_2}$ and $\overline{C_2}$ is –44.8. Similarly, the coefficient for $\overline{C_2}$ (i.e., corresponding to its multiplication by 1), is 21.5. Using these parameter values, one can reconstruct the optimal policy as $\phi^i(.) = 1/(1 + e^{F'_i \cdot B_i \cdot F'^T_i})$.

Overall, the two policies are similar but not identical.[8] To compare, we also ran the iterative process with $N=1$ and with 20 restarts for each optimization. In this case, it took 25 iterations to converge to the solution. Although this solution is applicable for the problem defined, it is not likely to be close to a subgame-perfect equilibrium; thus, this solution cannot be applied to a competition with different initial points of stocks.

F. Run the simulation model from step 1, with a single competition (i.e., $N=1$) given the initial states reported in the model definition and with final vectors of optimum policy parameters (B_i^{m*}). Report the results of this simulation as the trajectory of the differential game over time and the expected payoffs for each player.

Application to example: Figure 12.4 reports the performance of the two firms in the resulting differential game solution. The optimal policy for both firms dictates periods of full investment in capability development, followed by periods of benefiting from the accumulated capability by investing in performance generation. This strategy leads to oscillations in the capability stocks as well as performance gains. The oscillations are more pronounced in the stock of capability under development than in the stock of capability because the former is directly impacted by investment. Also, performance oscillates significantly because it is directly dependent on the allocation fraction. Such an oscillatory strategy significantly outperforms a fixed strategy. Figure 12.5 compares the revenue trajectories (R_i) for both firms if they follow the best fixed policy of $U^i(t) = 0.5$. The static policy is inferior to the policy that dynamically changes in response to the state of the system and time.

Toward the end of the game, the firms learn to focus only on payoff generation because further investments in the capabilities are too late to improve performance before the end of the game and thus are counterproductive. The resulting optimum policy (specified by the B_i^{m*} matrices) is also applicable to any different starting points in the state-

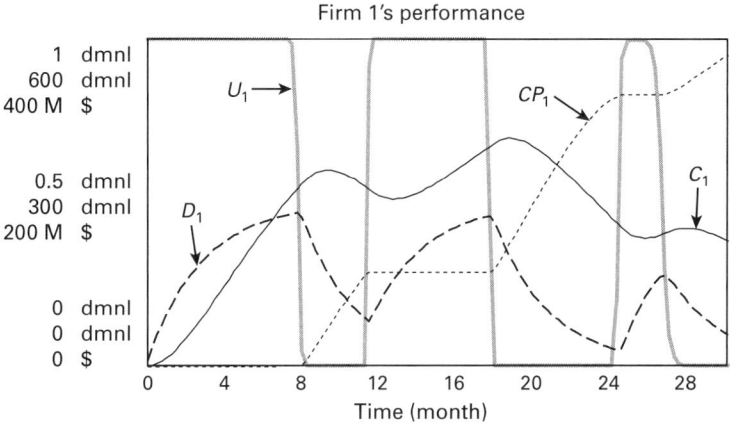

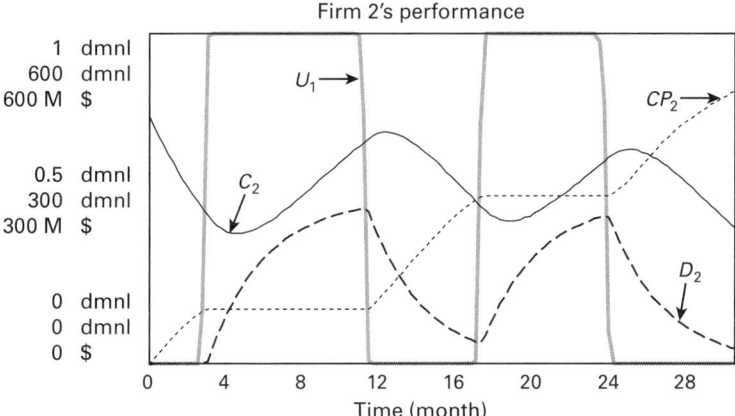

Figure 12.4
Performance of the two firms following policies found in the solution to the differential game. Four metrics are reported for each firm, including the fraction of resources allocated to capability development (U; top scale on y axis); capability and capability under development (C and D; middle scale on y axis); and cumulative payoff (CP; bottom scale on y axis).

space, making it a robust and dependable policy to follow regardless of the current state of competition.

The overall performance, O_i, for the initial conditions defined earlier comes down to \$378 million and \$496 million for firms 1 and 2, respectively. We also solved the problem with $N=1$, converging after $m=25$ iterations, using 20 optimization restarts in each optimization. The resulting optimum policies result in rather similar, but not identical,

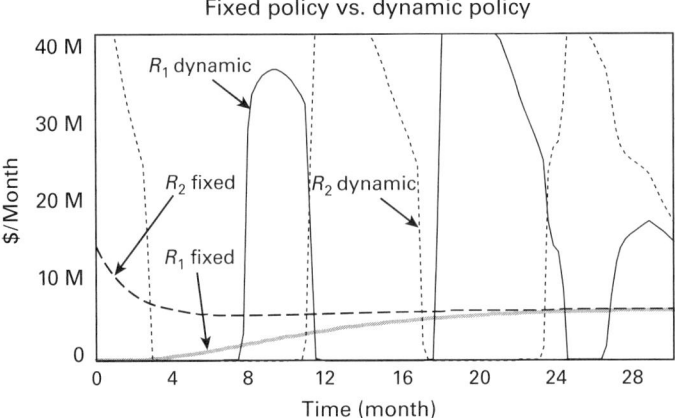

Figure 12.5
Comparison of the performance (R) of best fixed policy (allocating 50% of resources to capabilities) vs. the optimal dynamic policy for the two firms.

performance to the ones reported for $N=100$. The final payoff in this case is $404 million and $458 million for firms 1 and 2, respectively.

Implementation Notes

A few notes are in order regarding implementing the second step of the algorithm above. Identifying good features that capture important nonlinearities can make the optimization problem more smooth and allow for better solutions. Normalizing the features can increase the robustness of numerical optimization method for step E. Normalization can also reduce the number of free parameters in the model, allowing for more efficient estimation of the optimum policy function (Osgood 2009).

Selection of an appropriate functional form is critical. This choice should strike a good balance among two tradeoffs. On the one hand, the more flexible a functional form is, the more likely it is to get close to the real optimal policy, and it is thus more desirable. In the extreme, if we discretized the domain for this function (the feasible range of F^i) and for each point on that set we estimated a distinct parameter in the optimization in step 2.E, the resulting (table) function could potentially match any complex policy structure and thus in theory capture the real optimal policy. On the other hand, by increasing the flexibility of the approximating functional form, we increase the complexity of the optimization problem in step 2.E. Given that we typically do not know much about the structure of the optimization problem, increasing the size of the parameter space to be searched over makes it more likely that the optimization will not find a solution close to the global peak. Therefore, simpler approximation functional forms are valuable in helping solve the optimization problem. In balance, functional forms that provide much flexibility with few parameters are most beneficial.

Computational costs are generally a concern in solving differential games because: (1) each policy optimization may include dozens of parameters to specify the approximation function. Therefore, many restarts of optimization in the parameter space may be required for building confidence in the quality of the results. This can increase the computational cost of the optimization problem significantly. (2) We run the optimization over N parallel competitions to find robust solutions. This increases the simulation time N-fold. However, by aggregating payoffs over multiple competitions, it often reduces the computational cost of each optimization problem and reduces the number of restarts needed to find an optimal solution that is likely to be the global solution. (3) The iterative optimization steps are not guaranteed to find an equilibrium. Therefore, one may need to start the iterations from different initial policies for the second player in step 2.D, leading to additional computational challenges. Overcoming these challenges requires experimenting with algorithmic parameters (number of parallel competitions (N), number of optimization restarts) and approximation functional forms to find a good balance. Typically, increasing N reduces the number of iterations required, but the returns are diminishing for larger N values, so experimenting with N values ~5 to 20 is a good starting point. We used a larger value, $N=100$, to assess the convergence of the optimal policy for the two players in light of the symmetric structure of this game.

Ultimately, the solution to large problems may require resorting to more powerful computational resources than desktop computers, such as computer clusters or cloud computing. In practice, we have found it feasible to solve games with a handful of state variables using the method above on a regular desktop computer. For example, the results obtained above with $N=100$ are generated over approximately eight hours of CPU time running Vensim DSS with the compilation of models.

The algorithmic steps can be implemented in many different software environments. Whereas traditional system dynamics software packages, such as iThink, Vensim, and PowerSim, provide user-friendly environments to facilitate the first step of the algorithm, they are rather inflexible for implementing the second step, where iterative optimizations should be conducted. More general computational platforms such as MATLAB and Anylogic can provide the environment for both steps, but require some additional model translation effort (e.g., from the original simulation modeling environment to the target platform). In the appendix, we provide the files required for implementing the algorithm for the example problem in both Vensim and MATLAB. Note that the Vensim implementation requires stopping the iterations in step 2.E of the algorithm manually. This step could be automated using DLL connections to different software, but to maximize transparency, we stayed within a single environment in both implementations. We hope these files can be useful for those interested in using differential games in their dynamic modeling research. Overall, the learning costs for implementing the algorithms above are rather low; however, learning and implementing more advanced algorithms based on dynamic programming or level-set methods may require a substantial time investment.

Discussion

In this chapter, we provided an overview of the differential game concepts and solution methods relevant to dynamic modelers. Although dynamic games in general and differential games in particular have been around for several decades with much analytical and theoretical research available, the dynamic modeling community has had limited interaction with the associated tools and methods. From an understanding of different equilibrium types in differential games to relatively sophisticated solution methods, this toolbox provides useful concepts and methods that can empower dynamic modelers to tackle new problems where multistakeholder competition is critical. Such problem contexts are common in management research, where competition is a major factor that influences firms' strategies in marketing, pricing, product development, market entry, acquisition, and many other issues. Other topics where multiple stakeholders are involved are also common outside management research. For example, consider in the health care domain the gaming between tobacco companies versus the Federal Food and Drug Administration or the food industry versus healthy eating proponents. Simulation modelers need to incorporate competition and gaming more frequently in their analysis to provide more reliable recommendations. A better understanding of dynamic games also enables dynamic modelers to understand and contribute to research in economics, strategy, finance, ecology, evolutionary biology, and other areas that use the concepts and tools of differential games.

Readers interested in delving more into this research area have many resources at their disposal. Isaacs' original contribution (also in newer reprinted editions) provides a classical reference (Isaacs 1955). Other notable texts are also available for those who are interested in a more in-depth analytical treatment (Dockner 2000; Friedman 2006) or applications to specific domains such as marketing (Jørgensen and Zaccour 2004) and pursuit and evasion (Hájek 2008). The literature on stochastic differential games is also growing, and various references analyze how optimum policies can be designed when random factors outside of the control of the players can influence the dynamics of the game (Bardi, Raghavan, and Parthasarathy 1999). Readers interested in more detailed treatment of optimal control problems can look into both the literature on dynamic programming and optimal control (Bertsekas 2007; Tsitsiklis and VanRoy 1996) as well as approximate dynamic programming (Bertsekas and Tsitsiklis 1996; Powell 2011) and methods derived from reinforcement learning literature (Sutton and Barto 1998; Barto and Mahadevan 2003; Harmon, Baird, and Klopf 1995).

Although this paper focused on an introduction of differential games, evolutionary game theory (Friedman 1991; Samuelson 1997) provides a related domain that may also be of interest to dynamic modelers. Whereas classical games assume rational players with full information about the game structure who devise their optimum policies with foresight, in evolutionary games the players come and go with fixed strategies that are not necessarily optimal. However, new players enter the game with strategies that are more likely to resemble the more successful strategies of their predecessors. They also may try novel components in their strategies. Through this process of random variation and selection among players, new equilibria with stable strategies can emerge in which none of the players benefits from changing their strategy. This approach to games has been successfully applied to many biological problems where genes can be seen as strategies (Maynard-Smith and Price 1973). Applications in other domains, such as economics (Friedman 1998), sociology (Kameda, Takezawa, and Hastie 2003), and philosophy (Gintis 2004) have been on the rise. Besides the potentially more realistic behavioral assumptions that go into formulating evolutionary games, their solution methods are usually less analytically complex and more aligned with the traditional simulation approaches, providing opportunities for dynamic modelers to contribute to this field.

Given the breadth of research on dynamic games, the limited space, and the requisite analytical background, we limited our analysis and methods to basic concepts and simple numerical solution methods. Nevertheless, the solution methods provided here offer several advantages. First, they can be used to find equilibria that are approximately subgame perfect. This means the optimal policies remain valid even if we start the game from a different point on the state-space. Such policies are much desired in nondeterministic real-world settings, where random environmental shocks can take the system to different states that are not predictable a priori. Second, the algorithms scale reasonably well. The computational cost of the optimization step may theoretically grow exponentially with the size of policy space to be explored, but for many practical problems the growth may be much slower. Therefore, the overall computational feasibility of the algorithm can be manageable for medium-sized problems (e.g., dozens of state variables). Third, the algorithm is analytically simple and versatile, requiring limited customization for a new problem. The critical *ad hoc* steps are the selection of appropriate features and approximating functional form, both of which are aided by some insight into the problem structure. All the other steps are easy to codify and automate for any new problem. Finally, the policy-approximation method used here is aligned with the tradition in dynamic modeling of using decision rules to formulate how decision makers act. This not only simplifies formulation of differential game solutions, but it also allows modelers to benefit from their insights into behavioral decision rules and information sources potentially accessible to real-world decision makers, thus leading to behaviorally more realistic differential game solutions.

These advantages come at some cost. Although many users of the differential game framework do not challenge the rationality assumption underlying these solution concepts, management and behavioral decision-making scholars challenge these assumptions based on empirical evidence. Moreover, the solution method we use here implicitly assumes that the decision makers have full information about game structure; that is, they fully know the underlying model that specifies the dynamics, including its parameter values. This assumption is embedded in the optimization step of the process, where each player implicitly observes the results that can be expected from thousands of simulation runs and modifies their policy based on them. As a result, the process through which the players learn about the structure of the game itself is not captured in our method. Online[9] learning methods such as reinforcement learning can be used to capture these learning effects (Givigi, Schwartz, and Lu 2010; Sheppard 1998). Finally, we generally cannot guarantee the global optimality of the solutions found. Not only are multiple equilibria possible, but the policy structure and the features selected are also generally only at best a good approximation for the true optimum policy.

An Exercise

To practice solving differential games, you can find an approximate subgame-perfect Nash equilibrium for the game introduced under the section titled "A Simple Example." Specifically, first limit the game's time horizon to 100 periods, long enough to approximate the infinite time horizon game and remove discounting. Then solve the game for $N = 20$ parallel competitions, starting from randomly drawn initial qualities according to $Q_i(0) = \text{uniform}(0, f_i)$ and using the iterative procedure discussed above. You need to decide on the feature set to be used and the functional form of the policy function. Then simulate the resulting competition using initial quality values of $Q_1(0) = 0.1$ and $Q_2(0) = 1.8$. In the online appendix, we provide a sample solution to this exercise.

Notes

1. The term *player* comes from the game theory literature and generally captures a stakeholder or actor with a distinct preference for certain outcomes and whose actions influence the outcomes of all players.

2. In special cases, the optimum policy may itself be random, e.g., taking one of two actions with 50% probability, leading to a nondeterministic system evolution. For example, consider the rock-paper-scissors game, with deterministic dynamics but random optimum policy.

3. The value function (V) for state x at time t and following policy ϕ^i specifies the accumulated discounted objective (similar to function J) that is obtained from starting at state x at time t and following policy ϕ^i until the final time (T). The Hamilton-Jacobi-Bellman equation specifies a partial differential equation defining optimality conditions for $V(x, t)$:

$$rV(x,t) - V_t(x,t) = \max_u \{ F(x, u, t) + V_x(x, t) f(x, u, t) \mid u \in U(x, t) \} ,$$

where $V_x(x,t)$ and $V_t(x,t)$ denote the partial derivatives of V with respect to x and t.

4. Full variable names are used in the diagrams, with shorthand letter names used in equations.

5. Note that if we provided the players with information on the initial state of the system, given the deterministic nature of the system, they could construct the full future trajectory of the dynamics for any policy, and thus the equilibrium would be open-loop.

6. Note that symmetric elements of the matrix B are added together, and thus only one element is needed for pairs of symmetric elements. We therefore assume **B** to be an upper-triangular matrix and solve to find the 21 independent parameters.

7. Given the symmetric structure of the game, one can expect the optimum policies to be the same for both firms. However, variations in random initial conditions are likely to create some differences in the two optimum policies that emerge.

8. Because feature $\overline{C_1}$ for player 1 is like $\overline{C_2}$ for player 2, the element [1,1] of $\textbf{\textit{B}}_1^{9*}$ (=26.7) corresponds to the element [3,3] of $\textbf{\textit{B}}_2^{9*}$ (=143.7), and so on. Even though different numerically, the two policies produce fairly similar behaviors and thus are sensitive to the variations in initial conditions across the firms in the $N=100$ competitions.

9. The term online is used in the context of learning algorithms, which could be offline (using full data) or online (using data as they come in).

References

Baldwin, J. F., and J. H. Sims-Williams. 1968. An on-line control scheme using a successive approximation in policy space approach. *Journal of Mathematical Analysis and Applications* 22 (3): 523–536.

Bardi, M., T. E. S. Raghavan, and T. Parthasarathy. 1999. *Stochastic and differential games: Theory and numerical methods.* Annals of the International Society of Dynamic Games. Boston: Birkhuser.

Barto, A. G., and S. Mahadevan. 2003. Recent advances in hierarchical reinforcement learning. *Discrete Event Dynamic Systems-Theory and Applications* 13 (4): 343–379.

Bellman, R. 1957. *Dynamic programming.* Princeton, NJ: Princeton University Press.

Bertsekas, D. P. 2007. *Dynamic programming and optimal control*, 3rd ed. 2 vols. Optimization and computation series 1. Belmont, MA: Athena Scientific.

Bertsekas, D. P., and J. N. Tsitsiklis. 1996. *Neuro-dynamic programming.* Optimization and neural computation series 3. Belmont, MA: Athena Scientific.

Dockner, E. 2000. *Differential games in economics and management science.* Cambridge, UK: Cambridge University Press.

Friedman, D. 1991. Evolutionary games in economics. *Econometrica* 59 (3): 637–666.

Friedman, D. 1998. Evolutionary economics goes mainstream: A review of the theory of learning in games. *Journal of Evolutionary Economics* 8 (4): 423–432.

Friedman, A. 2006. *Differential games.* Mineola, NY: Dover Publications.

Gintis, H. 2004. *Moral sentiments and material interests: The foundations of cooperation in economic life, economic learning and social evolution*. Cambridge, MA: MIT Press.

Givigi, S. N., H. M. Schwartz, and X. S. Lu. 2010. A reinforcement learning adaptive fuzzy controller for differential games. *Journal of Intelligent & Robotic Systems* 59 (1): 3–30.

Gonçalves, P., and J. Sterman. 2005. Overordering games in supply chains. In *Proceedings of the 23rd International Conference of the System Dynamics Society*, Boston.

Hájek, O. 2008. *Pu rsuit games: An introduction to the theory and applications of differential games of pursuit and evasion*. Mineola, NY: Dover Publications.

Harmon, M. E., L. C. Baird, and A. H. Klopf. 1995. Reinforcement learning applied to a differential came. *Adaptive Behavior* 4 (1): 3–28.

Isaacs, R. 1955. *Differential games*. 4 vols. Santa Monica, CA: Rand Corp.

Isaacs, R. 1999. *Differential games: A mathematical theory with applications to warfare and pursuit, control and optimization*. Mineola, NY: Dover Publications.

Jørgensen, S., and G. Zaccour. 2004. *Differential games in marketing*. International series in quantitative marketing. Boston: Kluwer Academic Publishers.

Kameda, T., M. Takezawa, and R. Hastie. 2003. The logic of social sharing: An evolutionary game analysis of adaptive norm development. *Personality and Social Psychology Review* 7 (1): 2–19.

Kim, D. H., and D. H. Kim. 1997. A system dynamics model for a mixed-strategy game between police and driver. *System Dynamics Review* 13 (1): 33–52.

Maynard-Smith, J., and G. R. Price. 1973. The logic of animal conflict. *Nature* 246 (5427): 15–18.

Mitchell, I. A., and J. A. Templeton. 2005. A toolbox of Hamilton-Jacobi solvers for analysis of nondeterministic continuous and hybrid systems. *Hybrid Systems: Computation and Control* 3414: 480–494.

Mukherjee, Jaideep. 1995. Coercion, cooperation or indifference? A study of resource use and capital growth policies in the north and the south using differential game theory. *SciTech Journal* 5 (11): 22–25.

Osgood, N. 2009. Lightening the performance burden of individual-based models through dimensional analysis and scale modeling. *System Dynamics Review* 25 (2): 101–134. doi:10.1002/Sdr.417.

Powell, W. B. 2011. *Approximate dynamic programming: Solving the curses of dimensionality*. 2nd ed. Wiley series in probability and statistics. Hoboken, NJ: Wiley.

Rahmandad, H., and S. Sibdari. 2012. Joint pricing and openness decisions in software markets with reinforcing loops. *System Dynamics Review* 28 (3): 206–229.

Samuelson, L. 1997. *Evolutionary games and equilibrium selection*. MIT Press series on economic learning and social evolution. Cambridge, MA: MIT Press.

Sheppard, J. W. 1998. Colearning in differential games. *Machine Learning* 33 (2–3): 201–233.

Sterman, J. 2000. *Business dynamics: Systems thinking and modeling for a complex world*. Boston: McGraw-Hill/Irwin.

Sutton, R. S., and A. G. Barto. 1998. *Reinforcement learning: An introduction*. Cambridge, MA: MIT Press.

Tsitsiklis, J. N., and B. VanRoy. 1996. Feature-based methods for large scale dynamic programming. *Machine Learning* 22 (1–3): 59–94.

Wang, H. W., L. R. Cai, and W. Zeng. 2011. Research on the evolutionary game of environmental pollution in system dynamics model. *Journal of Experimental & Theoretical Artificial Intelligence* 23 (1): 39–50.

Contributors

Wenyi An, University of Calgary

Edward G. Anderson Jr., McCombs School of Business, University of Texas

Yaman Barlas, Bogazici University

Nishesh Chalise, Washington University in St. Louis

Robert Eberlein, isee systems

Hamed Ghoddusi, Stevens Institute of Technology

Winfried Grassmann, University of Saskatchewan

Peter S. Hovmand, Washington University in St. Louis

Mohammad S. Jalali, Virginia Tech

Nitin R. Joglekar, Boston University School of Management

David Keith, MIT Sloan School of Management

Juxin Liu, University of Saskatchewan

Erling Moxnes, University of Bergen

Rogelio Oliva, Mays Business School, Texas A&M University

Nathaniel D. Osgood, University of Saskatchewan

Hazhir Rahmandad, MIT Sloan School of Management

Raymond J. Spiteri, University of Saskatchewan

John Sterman, MIT Sloan School of Management

Jeroen Struben, McGill University

Burcu Tan, A.B. Freeman School of Business, Tulane University

Karen Yee, University of Saskatchewan

Gönenç Yücel, Boğaziçi University

Index

Note: italicized page numbers denote figures and tables.

Acceptance rate, 144–145
Action space, 387
Adaptive decisions, 277
  computational challenge of evaluating
    rules for, 282–283
AMOS, 74, 75
Analysis of covariance (ANCOVA), 71
  in SEM, 72–74
Anylogic, 398
ARMA model, 180
Asymptotic (AS) efficiency, 9, 10–11, 23
Atomic behavior modes, 184–187
Autocorrelation, 5, 11, 25, 28, 146

Backward induction, 291–292
Balance, 246, 255, 281
Base model, loop eigenvalue elasticity
  analysis (LEEA), 221–227
Batch Monte Carlo, 142–143
Bayesian framework for Markov chain
  Monte Carlo (MCMC), 128–129
Behavior analysis
  eigenvalue elasticity method, 208
  pattern recognition, 175, 198–202
Behavior modes, 175, 183–184, *185*
  eigenvalue elasticity analysis of,
    208–209
  SiS evaluation, 191, *192*, 196
Behavior patterns, dynamic, 181–188
  model calibration, 192–198
  model testing, 190–192
Behavior pattern testing (BTS II)
  approach, 42–43
Benchmarks in laboratory experiments,
  269

Benefits and costs tradeoff, 246–247, 271
Bicycle as inverted pendulum, 249
Binomial model, method of moments
  (MM), 45
Bivariate confidence interval, 29–30
Bootstrap percentile (BCa), 12
Bootstrapping, 5, 28–29
  confidence intervals, 11–12, 40
  execution time for, 13
Boundary violations and
  reparameterization in Markov chain
  Monte Carlo (MCMC), 156–160
Bracket median approximation, 315, 317,
  333
Broyden-Fletcher-Goldfarb-Shanno
  (BFGS) method, 20

Calibration. *See also* Markov chain Monte
  Carlo (MCMC); Parameter estimation
  deterministic, 99–100
  divergence and, 117
  with Kalman filtering, 107
  parameter, 126–127
  state resetting, 101–102
Candidate proposal stage, 160–161
Certainty-equivalence, 350
Chance nodes, decision tree, 311
Characteristic polynomial, 215, 238n5
Class identification, 174
  for dynamic behavior patterns, 181–183
Classification in pattern recognition, 177,
  181–183
Climate change, 267
Clustering in pattern recognition, 178,
  198–202